Creep and High Temperature Deformation of Metals and Alloys

Creep and High Temperature Deformation of Metals and Alloys

Special Issue Editors

Stefano Spigarelli
Elisabetta Gariboldi

MDPI • Basel • Beijing • Wuhan • Barcelona • Belgrade

MDPI

Special Issue Editors
Stefano Spigarelli   Elisabetta Gariboldi
Marche Polytechnic University   Politecnico di Milano
Italy   Italy

Editorial Office
MDPI
St. Alban-Anlage 66
4052 Basel, Switzerland

This is a reprint of articles from the Special Issue published online in the open access journal *Metals* (ISSN 2075-4701) from 2018 to 2019 (available at: https://www.mdpi.com/journal/metals/special_issues/creep_deformation_metals).

For citation purposes, cite each article independently as indicated on the article page online and as indicated below:

LastName, A.A.; LastName, B.B.; LastName, C.C. Article Title. *Journal Name* **Year**, *Article Number*, Page Range.

ISBN 978-3-03921-878-3 (Pbk)
ISBN 978-3-03921-879-0 (PDF)

Cover image courtesy of Elisabetta Gariboldi.

Contents

About the Special Issue Editors

Stefano Spigarelli is Full Professor of Metallurgy. His main research activities in recent years have concerned the mechanical properties, microstructure, and heat treatment response of materials such as steels, light alloys, composites, and intermetallics. Prof. Spigarelli's activities includd studies on the characterization of nanostructured coatings. In particular, the creep response and hot workability of several metallic materials have been investigated in depth. The results of these studies have been published in ca. 150 papers in various indexed journals. Prof. Spigarelli has collaborated extensively with researchers from Canada, USA, Japan, Czech Republic, Korea, Norway, Sweden, and Israel.

Elisabetta Gariboldi graduated in Mechanical Engineering in 1990, where she was awarded her Ph.D. in Metallurgical Engineering in 1994. Since 1998, she has been Associate Professor in Metallurgy at Politecnico di Milano, where she teaches "Metallurgy" and "Materials for Energy". The high-temperature mechanical behavior of metals and alloys has been a research field of interest for several years. Investigations have been carried out on the effects of microstructural modification/damage, environmental interactions, and the presence of cracks on the high-temperature behavior of alloys. Several research studies were also carried out to investigate the correlation between the component design, process parameters for fabrication processes, and the resulting microstructure and mechanical properties. Casting of duplex stainless steels as well as high-pressure die-casting of magnesium alloys and plasma-cutting of titanium alloys, coatings, phase change materials, forging of Al alloys in view of the optimization of high-temperature behavior are some of the investigated processes, with the aim to define and optimize the metallurgical and mechanical characteristics of components

metals · MDPI

Editorial

Creep and High-Temperature Deformation of Metals and Alloys

Elisabetta Gariboldi [1,*] and **Stefano Spigarelli** [2,*]

1   Politecnico di Milano, Dipartimento di Meccanica, Via La Masa 1, 20156 Milano, Italy
2   Department of Industrial Engineering and Mathematical Sciences, Marche Polytechnic University, Via Brecce Bianche, 60131 Ancona, Italy
*   Correspondence: elisabetta.gariboldi@polimi.it (E.G.); s.spigarelli@staff.univpm.it (S.S.); Tel.: +39-02-2399-8224 (E.G.); +39-071-220-4746 (S.S.)

Received: 29 September 2019; Accepted: 5 October 2019; Published: 10 October 2019

1. Introduction and Scope

The occurrence of time-dependent deformation of metals and alloys under constant loads or stresses, a phenomenon termed "creep", has been documented for at least two centuries. Yet, its real significance was appreciated only by the late 1940s, when some peculiar features of creep, such as the occurrence of plastic deformation under stresses well below yielding, were investigated in detail.

The continuous development of dislocation theories later enlightened some specific features of creep deformation and gave the basis for correlating the macroscopic creep properties to the time-dependent processes taking place within the metals and alloys. Similarly, the same dislocation theories were used to provide a physical background to the study of metals' and alloys' responses to hot working processes. Stress relaxation effects were also explained and modelled on similar bases.

While progressively more defined experimental and theoretical studies of the creep and hot working process mechanism were carried out, new creep-resistant materials have been developed and/or explained based on the abovementioned microstructure–mechanical behavior correlations. Similarly, new hot-working techniques have been introduced.

Notwithstanding that the mechanisms that control creep and hot working are essentially the same, advances in creep-resistant material and in hot working processes have often proceeded independently. The title the Editors selected for the Special Issue of *Metals*—"Creep and High-Temperature Deformation of Metals and Alloys"—underline common features between them.

2. Contributions to the Special Issue

Scholars have been invited to submit research papers dealing with innovative research and literature surveys on specific aspects of creep and high-temperature deformation so that the readers could realize the common points between them. Among the submitted manuscripts, 14 papers have been published in the issue.

2.1. Creep Deformation, Damage, and Ductility

Creep deformation mechanisms are typically described in terms of secondary creep strain rate. A good description of strain rate dependences over large temperature and applied stress values, taking into account compositional or microstructural effects, corresponds to a good understanding of the phenomena taking place at a microscopical level and leading to deformation. These general features have been considered in the issue both in the works by Delandar et al. [1] and by Kassner [2]. Delandar et al. specifically refer to copper and to its deformation mechanisms when creep deformation occurs at relatively low temperatures (up to 100 °C). Its deformation behavior has been modelled and verified both by experimental data and by Dynamic Dislocation simulations. The work by Kassner [2] suggests that the five-power-law creep can be applied in the case of alloys with pure metal behavior (class M alloys) by considering a linear superposition of a dislocation hardening term and a solute strengthening term.

Creep damage phenomena of different kinds lead material to creep rupture, with more or less accumulation of plastic deformation and with different fracture modes, depending on materials and test (or service) temperature and stress conditions. Some of the published papers deal with creep damage. In many high-temperature alloys, creep damage is due to the grain boundary cavitation, and modelling of its evolution up to the final rupture can be very important. Xu et al. contributed a paper [3] in which FE simulation of grain boundary cavitation helped understanding the role played by stress redistribution, cavitation damage, and creep fracture. Creep damage evolution can also be carefully modelled by innovative combinations of experimental investigation techniques, such as small-angle neutron scattering (SANS), scanning electron microscopy, and quantitative metallography, as reported by Jazeri et al. in [4] in the case of a 304 stainless steel. The effects of prior residual stress left by welding processes on the damage at the crack tip of a 9–12% Cr steel specimen with simulated weldment was investigated experimentally by Liu et al. [5], who observed residual stress-related transition of damage forms.

The creep ductility of steels, related to the strain that can be accumulated at material rupture, is the topic of a review paper by Holdsworth [6]. A set of features involving creep ductility of high-temperature steels, including acritical analysis of creep ductility data, can be applied to predict long-term ductility exhaustion in multi-temperature and multi-cast data sets. Furthermore, the creep ductility of steels can be analyzed in cases of stress multiaxiality.

2.2. Innovative Testing Techniques of Creep Deformation

Within decades, new experimental techniques have been introduced to investigate the creep properties of materials in specific cases, for example, related to the small size of the available material from which specimens should be sampled. An example is illustrated in the work by Glee et al. [7] where, in order to avoid any influence from the substrate, miniaturized cylindrical tensile specimens of bond coatings were produced by a special grinding process, exposed to different environments and then creep tested. On the other hand, Ding et al. [8] focused on the nanoindentation technology, also suitable for small material regions characterization. The technique has been used in [8], using both a Berkovich and a spherical indenter, on a Zr-based bulk metallic glass to investigate its creep behaviour at room temperature and to evaluate at the same time the effects of testing parameters.

2.3. Creep and Hot Deformation Interacting in the Presence of Loading/Temperature Changes and Environmental Effects

Creep deformation phenomena often do not operate under the conventional constant-temperature/constant-stress force under which conventional creep characterization of materials is carried out. In fact, in industrial applications, materials can operate under different loading or temperature conditions where phenomena like creep-buckling and creep-fatigue can significantly affect the material deformation, microstructural and damage evolution, and final fracture with respect to its behaviour under conventional testing conditions. In the present issue of *Metals*, these features have been introduced and applied to representative and widely diffused steels, such as the austenitic stainless steel 304, for which buckling phenomena have been investigated by Jo et al. [9] in the specific case of tubes subjected to radial external pressure load in the temperature range of 800–1000 °C. Jürgens et al. [10] have investigated the low-cycle fatigue and relaxation phenomena for the P92 steel, a representative 9–12% Cr ferritic-martensitic steel.

2.4. Creep-Microstructure Correlations for Specific Material Classes

The strict correlations between microstructure of metals and alloys and their creep and hot deformation processes have always been of interest in the creep research field, due to practical industrial request for increasingly creep-resistant materials. This topic has also been covered in the papers included in the present Special Issue of *Metals*. One group of papers focuses on the widespread class of ferritic-martensitic steels.

The paper by Wu et al. [11] deals with the effect of heat treatment process parameters, and specifically the normalizing temperature, in grade 91, a 9–12% Cr steel, which is discussed on the basis of a careful microstructural analysis of crept samples. Other critical features for the industrial applications of these steels have been considered in the Issue, such as weld joint behaviour (Hu et al. [5] investigated the effect of residual stresses left by these processes on creep damage) or creep-fatigue and creep relaxation phenomena (Jürgens et al. [10]).

The strict correlation between microstructural features is also of utmost importance for the development and application of other high-temperature alloys. Moving to steels and Ni-based alloys characterized by an austenitic matrix, the strengthening role played by carbides, other precipitates, or dispersoids, has been experimentally investigated in the temperature range of 675–750 °C by Gobbi et al. [12] in the case of alloys VAT 32 and VAT 36. In the gamma prime-containing Ni-base superalloys, the creep resistance at high temperature is also affected by the presence of solute atoms in solid solution. In their paper, Gao et al. [13], by three-dimensional (3D) discrete dislocation dynamics simulations, proved that solute atoms such as Re and W affect dislocation glide and climb differently, and thus the back stress on dislocation motion. The different effects of these elements and their concentration as solute atoms on creep deformation resistance have also been proven.

The role played by alloy microstructure on creep deformation and creep resistance of other alloy-classes has also been experimentally investigated and modelled in scientific works included in the present Issue. Dobes et al. [14] focused on the effect of Nb to Fe-27 at. % Al alloy which, by modifying the microstructure, also acts on the creep behavior, as demonstrated experimentally in the temperature range of 650–900 °C.

3. Conclusions

The Special Issue, "Creep and High Temperature Deformation of Metals and Alloys", includes papers covering in innovative ways the relevant topics and materials in the field. The Guest Editors are aware of the quality of the contributions and of their inspiring potential for scientists and technicians who deal with materials facing creep during service. As a matter of fact, even if the specific materials, testing/modelling conditions, and microstructures have been addressed by these contributions, further innovative approaches and studies can take their cue from them.

Acknowledgments: The Guest Editors thank all who contributed effectively to the development of this Special Issue. Thanks to the authors who submitted manuscripts to share results of their research activities, and to the reviewers who agreed to read them and gave suggestions to improve their final quality. Thanks to the Editors and to Assistant Editor Kinsee Guo as well as to all the staff of the Metals Editorial Office for their management and practical support in the publication process of the Issue.

Conflicts of Interest: The authors decline conflict of interest.

References

1. Hosseinzadeh Delandar, A.; Sandström, R.; Korzhavyi, P. The Role of Glide during Creep of Copper at Low Temperatures. *Metals* **2018**, *8*, 772. [CrossRef]
2. Kassner, M.E. Application of the Taylor Equation to Five-Power-Law Creep Considering the Influence of Solutes. *Metals* **2018**, *8*, 813. [CrossRef]
3. Xu, Q.; Tu, J.; Lu, Z. Development of the FE In-House Procedure for Creep Damage Simulation at Grain Boundary Level. *Metals* **2019**, *9*, 656. [CrossRef]
4. Jazaeri, H.; Bouchard, P.J.; Hutchings, M.T.; Spindler, M.W.; Mamun, A.A.; Heenan, R.K. An Investigation into Creep Cavity Development in 316H Stainless Steel. *Metals* **2019**, *9*, 318. [CrossRef]
5. Liu, D.; Li, Y.; Xie, X.; Liang, G.; Zhao, J. Estimating the Influences of Prior Residual Stress on the Creep Rupture Mechanism for P92 Steel. *Metals* **2019**, *9*, 639. [CrossRef]

6. Holdsworth, S. Creep-Ductility of High Temperature Steels: A Review. *Metals* **2019**, *9*, 342. [CrossRef]

7. Giese, S.; Neumeier, S.; Bergholz, J.; Naumenko, D.; Quadakkers, W.J.; Vaßen, R.; Göken, M. Influence of Different Annealing Atmospheres on the Mechanical Properties of Freestanding MCrAlY Bond Coats Investigated by Micro-Tensile Creep Tests. *Metals* **2019**, *9*, 692. [CrossRef]

8. Ding, Z.Y.; Song, Y.X.; Ma, Y.; Huang, X.W.; Zhang, T.H. Nanoindentation Investigation on the Size-Dependent Creep Behavior in a Zr-Cu-Ag-Al Bulk Metallic Glass. *Metals* **2019**, *9*, 613. [CrossRef]

9. Jo, B.; Okamoto, K.; Kasahara, N. Creep Buckling of 304 Stainless-Steel Tubes Subjected to External Pressure for Nuclear Power Plant Applications. *Metals* **2019**, *9*, 536. [CrossRef]

10. Jürgens, M.; Olbricht, J.; Fedelich, B.; Skrotzki, B. Low Cycle Fatigue and Relaxation Performance of Ferritic–Martensitic Grade P92 Steel. *Metals* **2019**, *9*, 99. [CrossRef]

11. Wu, H.-W.; Wu, T.-J.; Shiue, R.-K.; Tsay, L.-W. The Effect of Normalizing Temperature on the Short-Term Creep Rupture of the Simulated HAZ in Gr.91 Steel Welds. *Metals* **2018**, *8*, 1072. [CrossRef]

12. Gobbi, V.J.; Gobbi, S.J.; Reis, D.A.P.; Ferreira, J.L.A.; Araújo, J.A.; Moreira da Silva, C.R. Creep Behaviour and Microstructural Characterization of VAT 36 and VAT 32 Superalloys. *Metals* **2018**, *8*, 877. [CrossRef]

13. Gao, S.; Yang, Z.; Grabowski, M.; Rogal, J.; Drautz, R.; Hartmaier, A. Influence of Excess Volumes Induced by Re and W on Dislocation Motion and Creep in Ni-Base Single Crystal Superalloys: A 3D Discrete Dislocation Dynamics Study. *Metals* **2019**, *9*, 637. [CrossRef]

14. Dobeš, F.; Dymáček, P.; Friák, M. The Influence of Niobium Additions on Creep Resistance of Fe-27 at. % Al Alloys. *Metals* **2019**, *9*, 739. [CrossRef]

metals

MDPI

Communication

Application of the Taylor Equation to Five-Power-Law Creep Considering the Influence of Solutes

Michael E. Kassner

Chemical Engineering and Materials Science, University of Southern California, 3650 McClintock Ave, Los Angeles, CA 90089, USA; kassner@usc.edu

Received: 18 September 2018; Accepted: 8 October 2018; Published: 11 October 2018

Abstract: This study determines the feasibility of describing the flow stress within the five-power-law creep regime, using a linear superposition of a dislocation hardening term and a significant solute strengthening term. It is assumed that the solutes are randomly distributed. It was found that by using an energy balance approach, the flow stress at high temperatures can be well-described by the classic Taylor equation with a solute strengthening term, τ_o, that is added to the $\alpha MGb\rho^{1/2}$ dislocation hardening term.

Keywords: creep; microstructural features; constitutive equations

1. Introduction

This paper addresses the theoretical validity of the application of a Taylor equation to five-power-law creep in pure alloys and class M alloys. Previous work on aluminum and stainless steel by the author [1–3] shows that the density of dislocations within the subgrain interior influences the flow stress for steady-state substructures as well as primary creep. The hardening is consistent with the Taylor relation if a linear superposition of solute hardening (τ_o, or the stress necessary to cause dislocation motion in the absence of a dislocation substructure) and dislocation hardening ($\cong \alpha MGb\rho^{1/2}$) is assumed, or

$$\tau = \tau_o + \alpha MGb\rho^{1/2} \tag{1}$$

It appears that dislocation hardening is athermal and the constant, α, is temperature independent. The value of α is consistent with the range of values observed in cases where dislocation hardening is unambiguous. M is the Taylor factor, G is the shear modulus, ρ is the Frank network dislocation density, b is the Burgers vector and τ is the applied stress. Part of the reason that the question of superposition is important is because, historically, the τ_o term is not included or is very small. The question is whether for cases where the τ_o term is large, a linear superposition in fundamentally reasonable. This endeavor complements the earlier work by the author that demonstrated, that, at least phenomenologically, the superposition (i.e., Equation (1)) is effective.

It must be mentioned that the basis for strengthening in five power-law –creep in generally attributed to the Frank dislocation network such as [4–7]. Others have held to the proposition that subgrain walls are associated with the strength of materials within the five-power-law regime [8,9]. This Communication considers the Frank network to be associated with strength and the rate-controlling process for creep.

References [1,2] show for the case of annealed (very low dislocation density) 99.999% pure aluminum, that the yield stress appears to be a significant fraction of the, eventual, steady-state flow stress. This is illustrated in Figure 1. This is also true for 304 stainless steel, a class M (pure metal behavior) alloy. It is assumed that there are no long range internal stresses as consistent with the findings of [10,11]. In both cases the yield stress of the annealed metal is roughly 0.5 at 371 °C for Al and 0.3 at 750 °C for 304 stainless steel of the steady-state flow stress. Clearly, a description of the stress at steady-state must consider both the solute and the dislocation features of the microstructure. In the author's case, the dislocation feature has been suggested to be the Frank dislocation network within the grain and subgrains of the polycrystalline aggregate. Careful work by the author and coworkers determined that the Frank network of dislocations, rather than the subgrains, is the dislocation feature associated with elevated temperature strength [1,2,6].

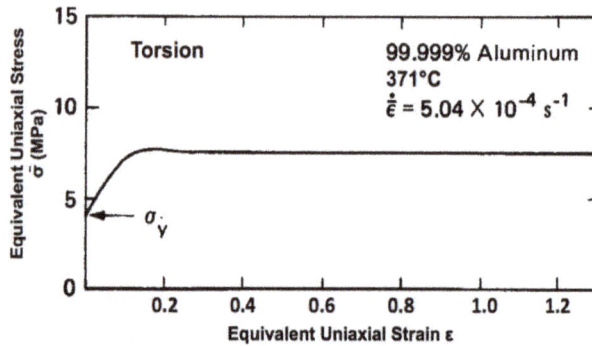

Figure 1. The stress versus strain behavior of 99.999% pure polycrystalline aluminum at 371 °C in torsion.

The yield strength in Figure 1 can only be attributed to the small amount of solute. The grain size of the aluminum in the figure is about 0.5 mm. There are a variety of possibilities to superimpose the strengthening variables (e.g., linear summation, root mean square as in [6], one hardening term exclusively controls the strength, etc.). The current work explores a simple linear superposition such as in Equation (1) that appears phenomenologically effective. The possibility that part of the annealed yield strength could be a lattice friction stress (similar to a Peierls stress) should be acknowledged.

2. Analysis and Discussion

The solutes in the present case are considered to be randomly dispersed. The Frank network coarsens with time at temperature and the link lengths increase. Eventually, some of the links are sufficient in length to operate a multiplication mechanism (e.g., Frank–Read source) and plasticity (creep) ensues. This multiplication also causes network refinement. Thus, there is both recovery (network coarsening) and hardening (network refinement) and steady-state is a balance of these two processes. The yield stress of the starting annealed material would be very low in the absence of the solute as the network strength is presumed to be very low [5]. At the yield stress, then, plastic flow is dictated by the impurities. At 99.999 (wt)% purity, roughly 10 ppm of impurities are present or roughly one in 10^5 host (solvent) atom sites are occupied by impurities. This suggests that the separation between impurities is roughly 50 atomic diameters or 13 nm. This separation is assumed constant with dislocation hardening. Hardening by solutes might occur by elastic interaction between the solute and the dislocations. From the bowing equation,

$$\tau = Gb/r \tag{2}$$

If each solute atom perfectly pins a dislocation then the yield strength is roughly $G/25$, which is much too large. The dislocation must "tear away' from the solute at a much lower stress. The solute atoms do not diffuse to "follow" the dislocation as with three-power-law creep [3]. There are a large variety of solutes that comprise the 0.001% total impurity concentration.

Once the stress has reached the yield stress, dislocation bowing can occur within the Frank network. As the dislocation bows it must (1.) perform work to tear away from the solute atoms that elastically interact with the stress fields of the dislocation and (2.) also perform work to compensate for the increase in elastic strain energy associated with increasing dislocation line length with bowing. Defining:

- l = bowed dislocation length;
- l_s = distance between solutes;
- r = radius of bowed dislocation links;
- ΔV = difference in volume between solute and solvent;
- P = hydrostatic pressure component of the dislocation stress field;
- k = constant;
- θ = radians

Then, if the loop expands by dr, by the First Law,

$$(\tau b l)dr = \theta dr\ [Gb^2/2] + dr\{d(P\Delta V)/dr\}(/l/l_s) \tag{3}$$

Again, this is just that the work done by the applied stress as the dislocation moves, is equal to the increase in elastic strain energy of dislocation line plus the work done to "tear away" the dislocation from the solutes. Equation (3) leads to,

$$\tau b l = 1/r\ [Gb^2/2] + (1/l_s)\{d(P\Delta V)/dr\} \tag{4}$$

$$\tau = [Gb/2r] + [d(P\Delta V)/dr)]/(bl_s) \tag{5}$$

for a typical r, and a simple arrangement of dislocations,

$$r = kl = k/\rho^{0.5} \tag{6}$$

$$\tau = [Gb\rho^{0.5}]/2k + [dP\Delta V)/dr]/(bl_s) \tag{7}$$

In the above, tau is the resolved shear stress on the loop. Therefore, for the applied stress or,

$$\tau_a = \tau_o + \alpha MGb\rho^{1/2} \tag{8}$$

or Equation (1), which is the classic Taylor equation, with, in this case, a τ_o that is a significant fraction of the flow stress. Thus, this article fundamentally confirms that dislocation hardening within the five-power-law creep regime, can be described by a classic Taylor equation using a linear superposition of a dislocation hardening term and a solute strengthening term. Because τ_o is a thermally activated term and $\alpha MGb\rho^{1/2}$ is the athermal term, then the constant α is expected to be of a similar value that those cases where dislocation hardening is unambiguous. In the authors earlier work on dislocation hardening in five-power-law creep [1–3], the α value is reasonable for dislocation hardening.

3. Conclusions

This study determined the theoretical feasibility of describing dislocation hardening within the five-power-law creep regime using a classic Taylor equation using a linear superposition of a dislocation hardening term and a solute strengthening term. It was assumed that the solutes are randomly distributed. This assumption and an energy balance approach demonstrated that the high temperature flow stress can be described by the classic Taylor equation with a linearly added solute strengthening term to the dislocation hardening term. The fundamental analysis complements earlier work that showed that the flow stress at steady-state can be satisfactorily described by a summation of a dislocation hardening terms consistent with the Taylor equation and a solute strengthening term.

Metals **2018**, *8*, 813

Acknowledgments: The support by the National Science Foundation under grant DMR-1401194 is greatly appreciated.

Conflicts of Interest: The authors declare no conflict of interest.

References

1. Kassner, M.E. Taylor Hardening in Five Power Law Creep of Metals and Class M Alloys. *Acta Mater.* **2004**, *52*, 1–9. [CrossRef]
2. Kassner, M.E. A Case for Taylor Hardening During Primary and Steady-State Creep in Aluminum and Type 304 Stainless Steel. *J. Mater. Sci.* **1990**, *25*, 1997–2003. [CrossRef]
3. Kassner, M.E. *Fundamentals of Creep in Metals and Alloys*, 3rd ed.; Elsevier: Amsterdam, The Netherlands, 2015; pp. 1–338.
4. Evans, H.E.; Knowles, G. A Model for Creep in Pure Metals. *Acta Metall.* **1977**, *25*, 963–975. [CrossRef]
5. Shi, L.; Northwood, D.O. Dislocation Network Models for Recovery Creep Deformation. *J. Mater. Sci.* **1993**, *28*, 5963–5974. [CrossRef]
6. Kassner, M.E.; Miller, A.K.; Sherby, O.D. The Separate Roles of Forest Dislocations and Subgrains in the Isotropic Hardening of Type 304 Stainless Steel. *Metall. Trans.* **1982**, *13A*, 1977–1986. [CrossRef]
7. Ardell, A.J.; Przystupa, M. Dislocation Link-length Statistics and Elevated Temperature Deformation of Crystal. *Mech. Mater.* **1984**, *4*, 319–332. [CrossRef]
8. Ginter, T.J.; Mohamed, F.A. The Stress Dependence of the Subgrain Size in Aluminum. *J. Mater. Sci.* **1982**, *17*, 2007–2012. [CrossRef]
9. Konig, G.; Blum, W. Comparision between the Cell Structures Produced in Aluminum by Cycling and by Monotonic Creep. *Acta Metall.* **1980**, *28*, 519–537. [CrossRef]
10. Kassner, M.E. Determination of Internal Stresses in Cyclically Deformed Cu Single Crystals Using CBED and Dislocation Dipole Separation Measurements. *Acta Mater.* **2000**, *48*, 4247–4254. [CrossRef]
11. Kassner, M.E.; Pérez-Prado, M.-T.; Long, M.; Vecchio, K.S. Dislocation Microstructures and Internal Stress Measurements by CBED on Creep Deformed Cu and Al. *Metall. Mater. Trans.* **2002**, *33A*, 311–318. [CrossRef]

metals

MDPI

Article

The Role of Glide during Creep of Copper at Low Temperatures

Arash Hosseinzadeh Delandar, Rolf Sandström * and Pavel Korzhavyi

Materials Science and Engineering, KTH Royal Institute of Technology, SE-100 44 Stockholm, Sweden; arashhd@kth.se (A.H.D.); pavelk@kth.se (P.K.)
* Correspondence: rsand@kth.se; Tel.:+46-8-7908321

Received: 13 August 2018; Accepted: 18 September 2018; Published: 27 September 2018

Abstract: Copper canister will be used in Scandinavia for final storage of spent nuclear fuel. The copper will be exposed to temperatures of up to 100 °C. The creep mechanism at near ambient temperatures has been assumed to be glide of dislocations, but this has never been verified for copper or other materials. In particular, no feasible mechanism for glide based static recovery has been proposed. To attack this classical problem, a glide mobility based on the assumption that it is controlled by the climb of the jogs on the dislocations is derived and shown that it is in agreement with observations. With dislocation dynamics (DD) simulations taking glide but not climb into account, it is demonstrated that creep based on glide alone can reach a quasi-stationary condition. This verifies that static recovery can occur just by glide. The DD simulations also show that the internal stress during creep in the loading direction is almost identical to the applied stress also directly after a load drop, which resolves further classical issues.

Keywords: creep; dislocation dynamics; glide; internal stress

1. Introduction

Copper shows creep deformation at as low a temperature as 75 °C. A number of creep strain versus time curves have been recorded at this temperature. The appearance of the creep curves is quite similar to those recorded at higher temperatures. Distinct primary, secondary, and tertiary stages are found [1,2].

Recovery creep theory is the basis of our understanding of the mechanisms during plastic deformation at elevated temperatures. A stationary condition is obtained when there is balance between work hardening and recovery. At high temperatures, recovery is based on climb of dislocations that can move in a non-conservative way. In this way, dislocations of opposite sign can attract and annihilate each other. When the climb rate is estimated with the climb mobility derived by Hirth and Lothe [3], the rate is so low at lower temperatures that it does not contribute significantly to the creep process. Any recovery must then be based on glide. It has also been assumed in general that the dislocation mobility is controlled by glide at low temperatures [1,4].

It is important to distinguish between two types of recovery: dynamic and static. The terminology for recovery varies in the literature. In this paper dynamic and static recovery are defined in the following way. Dynamic recovery occurs during deformation, where dislocations are forced together and in this way reduce the total dislocation content [5]. Static recovery is a time dependent process where dislocations of opposite sign attract each other and eventually annihilate [6]. The two types of recovery occur in parallel. In many papers, only one type of recovery is considered. Then it is usually assumed that dynamic recovery takes place during deformation at ambient temperatures and static recovery at high temperatures. However, there are important cases where both types of recovery must be taken into account. One such case is the creep of cold deformed materials [7]. For example, the extended tertiary stage that can appear in heavily cold worked material would be difficult to explain otherwise.

We can distinguish between three levels of recovery. In an ordinary tensile test at constant strain rate, the deformation stops when the load does not increase any more. In this case only dynamic recovery (strain controlled) takes place but no static recovery (time controlled). At high stresses close to the tensile strength at room temperature, logarithmic creep occurs in some alloys, for example austenitic stainless steels. Due to the logarithmic time dependence of the creep strain, the observable deformation ceases after some time. Although the mechanisms are not fully clarified [8,9], also some static recovery must be involved. However, the amount of static recovery is not sufficient to avoid a continuous increase in the dislocation density that will gradually block the creep deformation. The final level is creep of the type that occurs in copper. This process runs until rupture (i.e., continuous creep). Continuous static recovery must take place, otherwise the deformation would stop.

It has been assumed in the past that creep is controlled by glide at ambient or near ambient temperatures. As mentioned above, the simple reason is that the established expression for the climb mobility [3] has such a low value at near ambient temperatures that it gives a negligible contribution to creep and the natural alternative is glide. Although it is critical for the understanding of the creep mechanisms, it has never been verified that glide can be the controlling mechanism for creep at near ambient temperatures. There are two difficulties in verifying the role of glide. First, there has been no quantitative expression available for the glide mobility. However, such an expression will be derived in the present paper. Second, it is unclear whether static recovery can take place, which is essential for creep, just based on glide. Dislocation dynamics will be used to investigate that. Another complication is that a new expression for the climb mobility has recently been derived [10,11]. This expression takes the role of strain-induced vacancies into account. The resulting expression provides a much larger value and gives a significant contribution to the creep process. Consequently, it is of vital importance to investigate if glide on its own can give rise to continuous creep.

The present paper will also analyze another classical problem in creep. In the literature, it has often been assumed that creep deformation is controlled by an effective stress σ_{eff} acting on the dislocations

$$\sigma_{eff} = \sigma_{appl} - \sigma_i \tag{1}$$

σ_{appl} is the applied stress and σ_i is an internal stress due to the forest of dislocations or a back stress as it often is referred to as well. Attempts have been made to measure the size of σ_i in stress drop tests [12,13]. When the steady state has been reached, the applied stress is reduced. If the creep rate is zero after the stress drop, the stress change is considered to correspond to the effective stress and the internal stress can be obtained. Many stress change experiments have been made, but different laboratories tend to get conflicting results, for a survey see [13]. For example, Ahlquist and Nix [14] obtained a σ_i value that was about half of the applied stress, Cuddy [15] measured a value of 0.1–0.2 of the applied stress, whereas others have found it to be zero [16].

The purpose of the present paper is to analyze creep deformation at low temperatures in copper using dislocation dynamics. This technique has the advantage that creep deformation can be simulated by taking only glide into account. In addition, the size of the internal stress can be evaluated directly. The material that will be investigated is oxygen free copper alloyed with about 50 ppm phosphorus (Cu-OFP). The material was chosen for the reason that its fundamental dislocation mobility has been derived and experimentally verified. Cu-OFP has improved mechanical properties in comparison to oxygen free copper without phosphorus (Cu-OF). Cu-OFP has higher creep strength. Cu-OFP has also better creep ductility, in some cases dramatically better [17,18]. Cu-OFP is planned to be used in copper canisters for final disposal of spent nuclear fuel. The canisters will be placed about 500 m down in the bedrock. Copper was chosen because of its low corrosion rate in the reducing environment in the repository. Due to hydrostatic water pressure, the canisters will also be exposed to creep. The temperature in the repository will initially be close to 100 °C and will then decrease very slowly over hundreds of years.

2. The Creep Mobility

A model for the secondary creep rate of Cu-OF and Cu-OFP was formulated some time ago [1]. It has been used to describe the creep rate in an approximate way as a function of stress for temperatures between 75 and 250 °C [2,19,20]. It has also been used for evaluating the creep rupture strength [21]. For oxygen free copper without phosphorus, the model takes the form

$$\dot{\varepsilon}_{OF} = \frac{2bc_L}{m} \frac{D_{s0}b\tau_L}{k_BT} \left(\frac{\sigma}{\alpha mGb}\right)^3 e^{\frac{\sigma b^3}{k_BT}} e^{-\frac{Q}{RT}[1-(\frac{\sigma}{\sigma_{max}})^2]} = h(\sigma, T) \tag{2}$$

where σ is the applied stress and T the temperature. $D_{s0} \exp(-Q/RT)$ is the coefficient of self-diffusion, Q the activation energy for self-diffusion, b Burgers vector, G the shear modulus, m the Taylor factor, k_B Boltzmann's constant, R the gas constant, α and c_L constants, σ_{max} the maximum stress level, and τ_L the dislocation line tension. Equation (2) for the secondary creep rate is derived on the assumption of a balance between work hardening and static recovery. The influence of P is taken into account with the help of the break stress σ_{break} for the dislocations to break away from the Cottrell atmospheres of P atoms [22,23]

$$\dot{\varepsilon}_{OFP} = h(\sigma - \sigma_{break}, T)f_Q \tag{3}$$

σ_{break} is given by the following expression

$$\sigma_{break} = \frac{U_P^{max}}{b^3} \int_{y_L}^{y_R} c_P^{dyn} dy \tag{4}$$

The integration is performed around a dislocation ($\pm 20\ b$). The expression for the P content around a dislocation c_P^{dyn} can be found in [22]. The maximum interaction energy between a P solute and a dislocation is [22,24].

$$U_P^{max} = \frac{1}{\pi} \frac{(1+\nu)}{(1-\nu)} G\Omega_0 \varepsilon_P \frac{b}{r_{core}} \tag{5}$$

ν is Poisson's ratio, Ω_0 the atomic volume, ε_P the linear misfit for P atoms, and r_{core} the dislocation core radius. f_Q is a factor describing the influence of phosphorus on the activation energy. It is given by [23]

$$f_Q = e^{-U_P^{max}/RT} \tag{6}$$

The model in Equation (2) is based on a combination of the climb and glide dislocation mobilities by setting up a unified model [25]. The basic expression for the climb mobility was derived by Hirth and Lothe [3]

$$M_{climb} = \frac{D_{s0}b}{k_BT} e^{\frac{\sigma b^3}{k_BT}} e^{-\frac{Q}{RT}} \tag{7}$$

Due to the lack of basic expression for the glide mobility an empirical formula from Reference [26] was considered

$$M_{glide} = ke^{-\frac{Q_2}{RT}[1-(\sigma/\sigma_{max})^p]^q} \tag{8}$$

where k, Q_2, σ_{max}, p, and q were unknown constants. By combining Equations (7) and (8), the constants k and Q_2 could be fixed. This meant that Q_2 became the activation energy for self-diffusion. It has been verified that this assumption works well for aluminum as well, see for example Reference [12]. According to the suggestions in Reference [26], σ_{max} should be the maximum stress in the system. It has been put to the tensile strength at room temperature. $p = 2$ and $q = 1$ was chosen following Chandler's [27] work on copper. For aluminum, $p = 1$ and $q = 1$ have mainly been used, but those values cannot represent the high creep exponent at low temperatures. The resulting expression for the glide mobility is

$$M_{glide} = M_{climb}f_{enh} \tag{9}$$

where the glide enhancement factor is given by

$$f_{\text{enh}} = e^{\frac{Q}{RT}\left(\frac{\sigma}{\sigma_{\max}}\right)^2} \tag{10}$$

The formula that was established in this way, Equation (2), was postulated to represent the mobilities. The formula has now been used successfully for a number of years. It has been verified by direct comparison to creep rate and creep strength data. But it has also been used to model, for example, creep ductility [18] and creep crack growth [21].

The copper canisters for spent nuclear fuel should remain intact for thousands of years. The canisters can be exposed to creep for very long periods. Since the expression for the dislocation mobilities are used to predict the creep rate in the canisters, it is essential to have a physically-based model for the glide mobility. Such a model will now be derived.

During deformation, jogs will be formed on the dislocations. During the motion of the dislocations, the jogs will have to move by climb, which is a slow process at low temperatures. This means that the dislocations will be slowed down due to the presence of jogs. Following Hirth and Lothe [3], it will be assumed that it is the climb of the jogs that controls the speed of the gliding dislocations. This is a natural assumption since the jogs on the edge and screw dislocations that move by climb takes place at a speed that is orders of magnitude lower than if they move by glide.

When the jogs move and when dislocations intersect during plastic deformation, not only jogs are formed but also vacancies. The equilibrium excess of vacancies can be computed with the help of a model by Mecking and Estrin [28]. They estimated the number of vacancies produced mechanically in a unit volume per unit time as

$$P = 0.5\frac{\sigma\dot{\epsilon}}{Gb^3} \tag{11}$$

The quantities in this expression have been explained above. A detailed derivation of Equation (11) shows that the constant should be 0.5, not 0.1 as in Reference [28]. The corresponding annihilation rate A for the excess vacancies is given by

$$A = \frac{D_{\text{vac}}}{\lambda^2}(c - c_0) \tag{12}$$

c_0 is the equilibrium vacancy concentration and $\Delta c = c - c_0$ is the excess concentration. D_{vac} is the diffusion constant for the vacancies. λ is the spacing between vacancy sinks. We assume that this spacing corresponds to the distance between dislocations [28].

$$\lambda = 1/\sqrt{\rho} = \alpha m Gb/(\sigma - \sigma_y) \tag{13}$$

ρ is the dislocation density and σ_y the yield strength. Taylor's equation has been used in the second member.

$$\sigma = \sigma_y + \alpha m Gb\sqrt{\rho} \tag{14}$$

$\alpha \approx 0.2$ is a constant, $m = 3.06$ the Taylor factor, G the shear modulus and b Burgers' vector. Assuming that the generation and annihilation rates match, we find by combining Equations (11)–(13), the following expression for the excess vacancy concentration

$$\frac{\Delta c}{c_0} = 0.5\frac{\sqrt{2}\dot{\epsilon}(\alpha b)^2}{D_{\text{self}}}\frac{G\sigma}{(\sigma - \sigma_y)^2} \tag{15}$$

In deriving Equation (15), the following relation for the self-diffusion coefficient D_{self} has been used

$$D_{\text{self}} = c_0\Omega D_{\text{vac}} \tag{16}$$

Ω is the atomic volume. The creep rate is proportional to the self-diffusion coefficient, cf. Equation (2). In turn, the self-diffusion coefficient is proportional to the vacancy concentration.

It is now assumed that the climb rate of the jogs is proportional to the total vacancy concentration. This is the same assumption as was made in Reference [28]. The average distance between jogs is related to the dislocation density as $l_{jog} = 1/\sqrt{\rho}$. According to the Peach-Koehler formula, the force F on a dislocation is given by $F = b\,\sigma\,l$ where l is the length of the dislocation. We have assumed that the jogs control the motion of the dislocations. If then l is taken as l_{jog}, F is the force on each jog. This means that the stress on the jogs is enhanced by

$$g_\sigma = \frac{l_{jog}}{b} = \frac{1}{b\sqrt{\rho}} \tag{17}$$

The length of a jog is taken as the length of the Burgers vector. Using Taylor's equation where σ_y is the yield strength. Equation (17) can be rewritten as

$$g_\sigma = \frac{\alpha m G}{\sigma - \sigma_y} \tag{18}$$

The glide rate is obtained by multiplying the climb mobility by the excess vacancy concentration, Equation (15) and the stress enhancement factor, Equation (18). The glide enhancement factor becomes

$$g_{glide} = \frac{0.5\sqrt{2}\alpha\dot{\varepsilon}(\alpha m b)^2}{D_{self}} \frac{G^2\sigma}{(\sigma - \sigma_y)^3} \tag{19}$$

If the glide rate is controlled by the climb rate of the jogs, g_{glide} should represent the ratio between the glide mobility and the climb mobility of the dislocations. g_{glide} should then be directly comparable to f_{enh} in Equation (10). Such a comparison is shown in Figure 1.

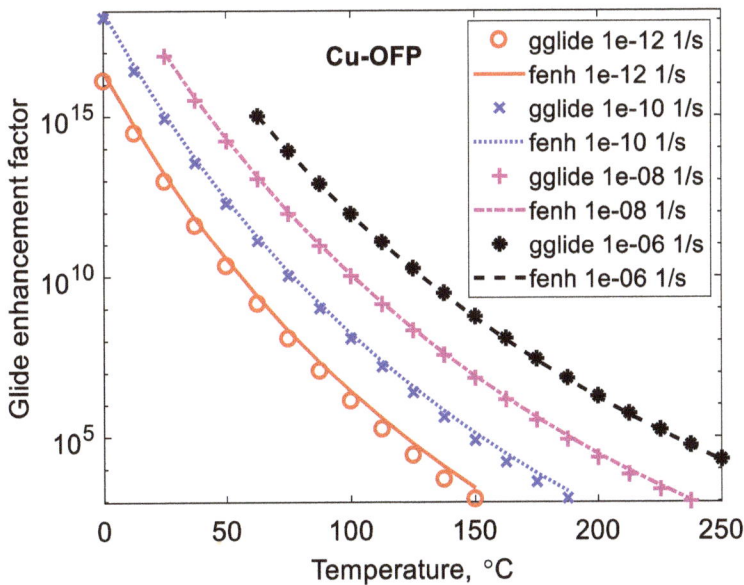

Figure 1. Comparison of the deformation induced increase in the vacancy concentration g_{glide}, Equation (19), that raises the climb rate with the expression for the climb glide enhancement factor f_{enh}, Equation (10). Results are presented for Cu-OFP at four strain rates.

The corresponding comparison at different temperatures is given in Figure 2.

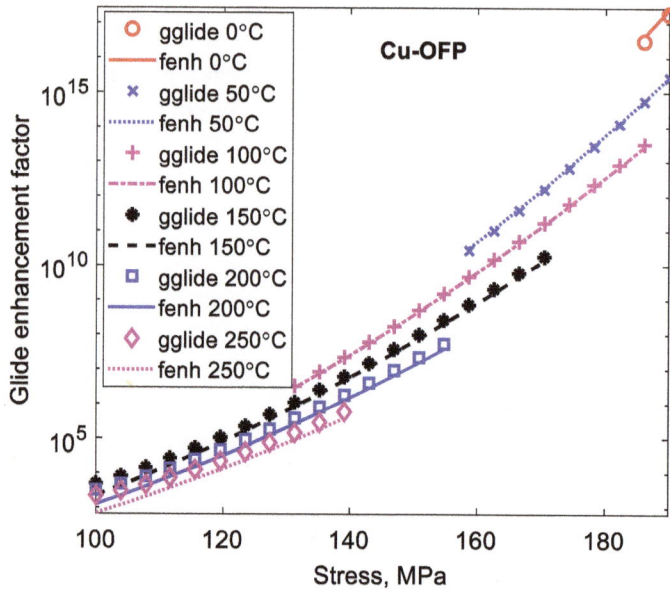

Figure 2. Comparison of the deformation induced increase in the vacancy concentration g_{climb}, Equation (19), that raises the glide rate with the expression for the climb glide enhancement factor f_{enh}, Equation (10). Results are presented for Cu-OFP at six temperatures.

It can be seen from Figures 1 and 2 that the ratio f_{enh} in Equation (10) between the glide and the climb rate is very large at temperatures below 150 °C. It increases with increasing strain rate and decreasing temperature. The stress is chosen to give the stationary creep rate marked in the figure. The predictions based on the climb rate of the jogs g_{glide} give almost identical results to those based on f_{enh} over many orders of magnitude. This demonstrates that the jog model can explain the behavior of the glide rate and confirms the validity of expression for the dislocation glide mobility, Equation (9).

3. Dislocation Dynamics Method

Creep of copper at near ambient temperatures clearly reaches a stationary stage as discussed above. The obvious way to explain this behavior is that there is balance between work hardening and recovery. The work hardening part is well understood and is fairly straightforward to model. However, practically all models for recovery are based on the interaction of just a pair of dislocations. It is far from obvious that these models can be generalized to the complex networks of dislocations that appear in alloys. To handle these complex networks, dislocation dynamics must be utilized. Two-dimensional dislocation dynamics has been used to demonstrate that a large number of gliding and climbing dislocation actually follow the recovery model that is usually attributed to Friedel [29]. However, 2D dislocation dynamics involve only parallel dislocations. For a more detailed analysis, 3D dislocation dynamics must be used.

All dislocation dynamics simulations presented here were performed using the Parallel Dislocation Simulator (ParaDis) [30] code developed at Lawrence Livermore National Laboratory. ParaDis follows a nodal-based discretization scheme to simulate collective motion of dislocations. In this approach, dislocation lines are discretized into sets of straight segments of mixed character connected with discretization nodes. Each segment is characterized by its Burgers vector, glide plane normal and line direction [31]. As dislocation lines are properly discretized, the force acting on all dislocation segments are determined. Calculation of forces is followed by evaluation of dislocation velocities according to linear force-velocity relation applied in an over-damped regime. Once all segment velocities are calculated, all dislocation segments are allowed to move based on their velocities within a certain time step; their further positions are predicted by numerical integration of the equation of motion [32]. Detailed description of the line dislocation dynamics method can be found in Reference [33].

4. Glide Controlled Creep of Copper at 75 °C

4.1. Dislocation Mobility

The glide mobility given in Equation (9) is used in the dislocation dynamics simulations. With the help of the glide mobility, the stationary creep rate can be obtained, cf. Equation (3). At an applied stress level of $\sigma_{appl} = 180$ MPa and temperature T = 75 °C, the dislocation mobility and stationary creep rate are evaluated using Equations (9) and (3) as $M(T,\sigma) = 2.89 \times 10^{-13}$ (Pa·s)$^{-1}$ and $\dot{\varepsilon} = 2.78 \times 10^{-8}$ s^{-1}, respectively. The constants used in the computations are listed in Table A1 in Appendix A. The climb mobility at the considered temperature is negligible. Thus, the predicted value represents the glide mobility. Using the mobility in Equation (9) in the equation for the creep rate (3), a direct comparison with experimental data can be made, Figure 3. Such satisfactory comparisons have been obtained under different conditions in several papers [1,2,20].

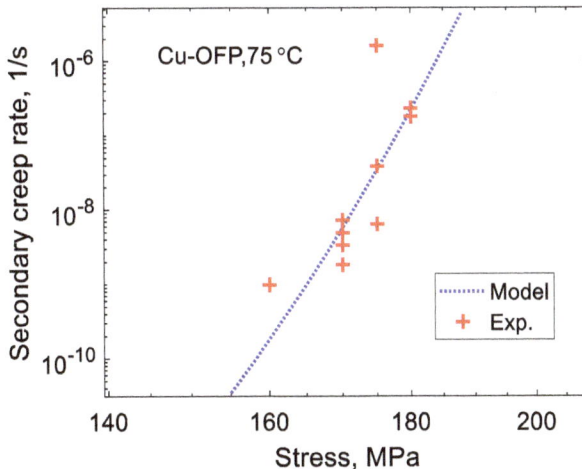

Figure 3. Predicted secondary creep rate with the dislocation mobility in Equation (9).

In derivation of the dislocation mobility in Equation (9), it is assumed that the full applied stress is acting on the dislocations. However, dislocation dynamics simulations reveal that during plastic deformation, considerable back stresses are acting on the dislocation lines. Thus, when the original value of glide mobility obtained from Equation (9) is used as input parameter, dislocation dynamics modeling yields significantly lower stationary creep rates than the value from Equations (2) and (3) due to the presence of an internal stress. To tackle this problem, the magnitude of glide mobility was increased over three orders of magnitude to balance the low effective stress and to reach the original value of stationary creep rate. Therefore, in addition to simulation of creep deformation of the single-crystal copper with the original mobility value of $M_{glide} = 2.89 \times 10^{-13}$ (Pa·s)$^{-1}$, dislocation dynamics simulations were performed with two higher glide mobilities of of $M_{glide} = 8.67 \times 10^{-10}$ (Pa·s)$^{-1}$ and $M_{glide} = 1.44 \times 10^{-9}$ (Pa·s)$^{-1}$. The increase in glide mobility corresponds to the ratio between the applied stress and the effective stress in both the analyzed cases. When the simulations match the original value of the stationary creep rate, the correct value of the effective stress and thereby also the level of the internal stress have been found.

4.2. Dislocation Dynamics Simulation

A three-dimensional dislocation dynamics method was employed to demonstrate how creep deformation of copper at 75 °C can be governed by glide motion of dislocations. To do so, plastic deformation of single-crystal copper was modelled by applying constant tensile stress of σ_{appl} = 180 MPa along the [001] direction. Initial dislocation setup consisted of 865 straight dislocations of mixed character randomly distributed inside the cubic simulation volume with an edge length of 5 μm. The simulation box was subjected to three-dimensional period boundary conditions (PBCs) to mimic the bulk material. To resemble a non-deformed crystal, initial dislocation density over the simulation volume was set to ρ_0 = 4.1 × 10^{13} m^{-2}.

The material parameters were set for copper as follows, shear modulus G = 42 GPa, Poisson's ratio ν = 0.31, and Burger vector b = 0.256 nm. Identical drag coefficient B and consequently the same glide mobility were used for both edge and screw dislocations. While screw dislocations may leave their original primary plane and cross slip to a secondary plane, climb motion of edge dislocation was hindered over the course of simulations. Therefore, edge dislocations were mostly confined to their {111} slip planes.

Simulations were performed with three different values of the glide mobility. The first set of simulations were conducted with an original mobility value of M_{glide} = 2.89 × 10^{-13} (Pa·s)$^{-1}$ and the resultant response of the single-crystal is plotted in Figure 4 as plastic strain and plastic strain rate versus simulation time. There was an almost linear increase in strain with time. Figure 4b shows that the plastic strain rate and the creep rate decreased with time and eventually leveled off at around $\dot{\varepsilon}$ = 1.0 × 10^{-11} s^{-1}. This strain rate is 3 × 10^3 times less than the value predicted by Equations (2) and (3).

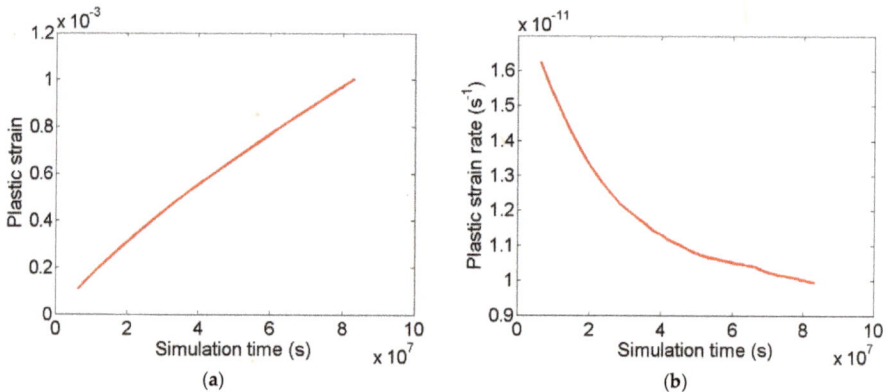

Figure 4. (**a**) Plastic strain and (**b**) plastic strain rate versus simulation time for deformation with the original mobility value M_{glide} = 2.89 × 10^{-13} (Pa·s)$^{-1}$.

The development of the dislocation density is shown in Figure 5 as a function of simulation time (a) and plastic strain (b). The dislocation density reached a fairly high value early in the simulation and then continued to grow. The general appearance of Figure 5a,b were similar, since the plastic strain approximately increased linearly with time, see Figure 4a.

To interpret these results, we considered the effective stress σ_{eff}, Equation (1). All the dislocations in the simulation generated a stress field. Each individual dislocation was exposed to a combination of these stress fields from surrounding dislocations in addition to the externally applied stress. These combined stress fields are referred to as the internal stress σ_i. The value of the internal stress can be computed from the simulation. This is analyzed in the next section. The effective stress σ_{eff} drives the motion of the dislocations. Since the computed strain rate was much less than the one evaluated from the mobility, it is natural to assume that the effective stress was much lower than the applied stress. To investigate this possibility, a new simulation was made with a mobility that was increased by a factor 3×10^3, giving a mobility of $M_{\text{glide}} = 8.67 \times 10^{-10}$ (Pa·s)$^{-1}$. The results of the simulations are shown in Figures 6 and 7.

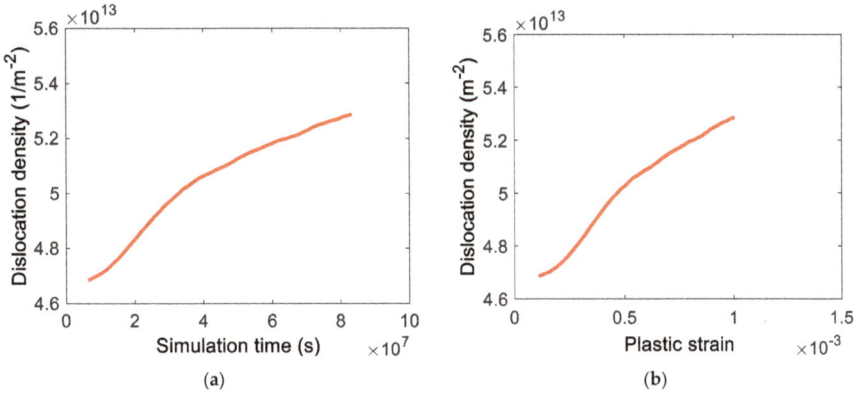

(a)

(b)

Figure 5. Dislocation density versus (**a**) simulation time and (**b**) plastic strain for deformation with original mobility value $M_{\text{glide}} = 2.89 \times 10^{-13}$ (Pa·s)$^{-1}$.

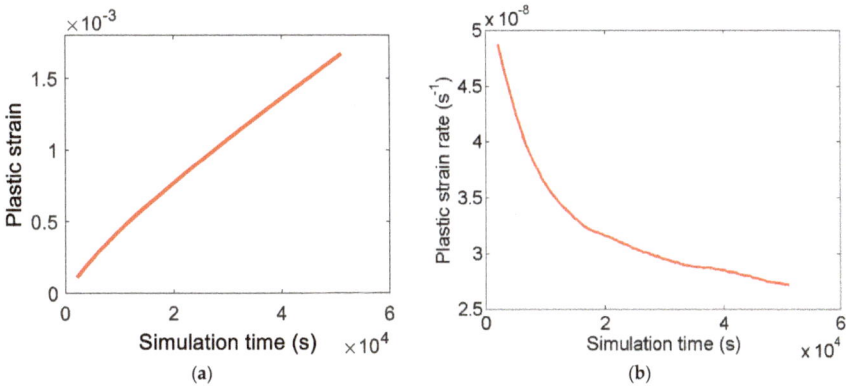

(a)

(b)

Figure 6. (**a**) Plastic strain and (**b**) plastic strain rate versus simulation time for $M_{\text{glide}} = 8.67 \times 10^{-10}$ (Pa·s)$^{-1}$.

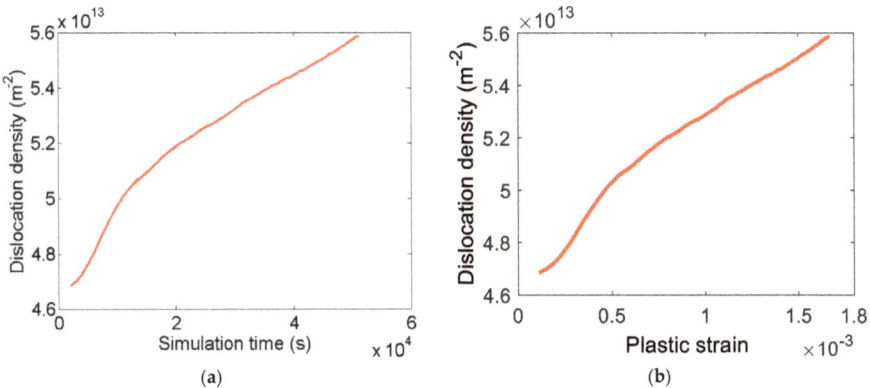

(a)

(b)

Figure 7. Dislocation density versus (**a**) simulation time and (**b**) plastic strain for deformation with the mobility value $M_{\text{glide}} = 8.67 \times 10^{-10}$ (Pa·s)$^{-1}$.

The strain increases with time in Figure 6a and the strain rate versus time curve in Figure 6b shows a decreasing creep rate with increasing time in the same way as in Figure 4a,b. The dislocation density as a function of plastic strain in Figure 6b has also a similar appearance to that in Figure 5b. The final strain rate in Figure 6b is approximately $\dot{\varepsilon} = 2.7 \times 10^{-8}$ s^{-1}. This value is almost identical to that estimated from Equations (2) and (3). Since large strains cannot be obtained in the simulation, a stationary condition was not reached. However, since the variation of the dislocation density as a function of time was modest, Figure 7a, the results can be said to represent a quasi-stationary condition. By employing a higher glide mobility in the second set of calculations, a creep rate close to the ones corresponding to observations was obtained, see Figure 3.

Figure 8a,b shows the results of the third set of simulations where deformation of the single-crystal was modeled with a glide mobility of $M_{glide} = 1.44 \times 10^{-9}$ (Pa·s)$^{-1}$, which represents a further increase of 1.67 in the mobility.

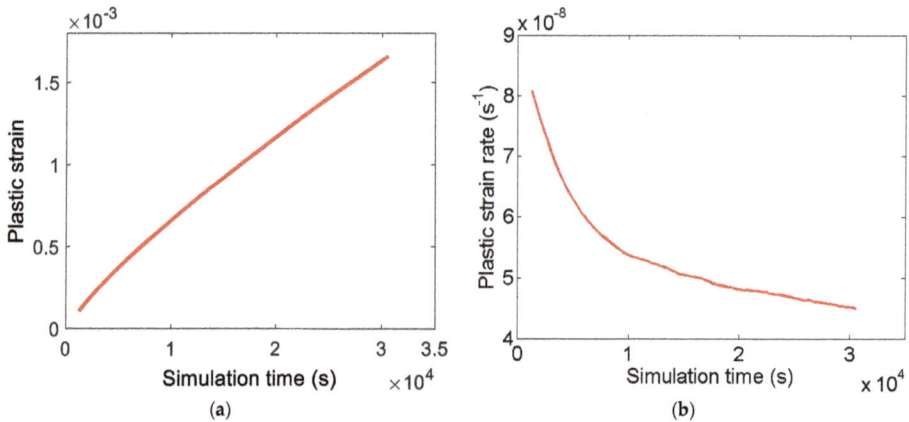

Figure 8. (a) Plastic strain and (b) plastic strain rate versus simulation time for $M_{glide} = 1.44 \times 10^{-9}$ (Pa·s)$^{-1}$.

Similar to the previous calculations with lower mobility values in Figures 4 and 6, plastic strain rate decreased during deformation of the single-crystal to finally a constant value of almost $\dot{\varepsilon} = 4.5 \times 10^{-8}$ s^{-1}. This was an increase with a factor of 1.67 in comparison with that in Figure 6 (i.e., the same as the increase in the mobility). It is obvious that the predicted creep rate was approximately linear with the mobility in the considered range. It was the purpose of the second increase to demonstrate the linearity. The dislocation density versus time and plastic strain curves are not shown, since their appearance was almost identical to the corresponding curves in Figure 7.

Dislocation dynamics simulation results shown in Figure 4 to Figure 8 demonstrate that single-crystal copper experiences creep deformation only by glide motion of dislocations under constant stress.

5. Evaluation of Effective Stresses

Effective stresses acting on the dislocation lines can be calculated directly using dislocation dynamics simulation results. The existing relation between velocity of dislocations, V, and effective stress, σ_{eff}, acting on dislocations is as follows:

$$V = M_{glide} \sigma_{eff} b \qquad (20)$$

where M_{glide} and b represent the glide mobility and the magnitude of Burgers vector, respectively. The mean value of effective stresses along each of three directions (i.e., x, y and z) is calculated using generated velocity data extracted from the dislocation dynamics results for the three sets of glide mobilities, see Table 1. The external stress is applied in the *z* direction. Effective stresses are calculated at three stages of deformation: early, intermediate, and final stages corresponding to plastic strains of 0.02%, 0.06%, and maximum computed strain. The absolute values of effective stresses are in general more than three orders of magnitude smaller than the applied stress. The amount of scatter in the results can be estimated from the average values in the *x* and *y* directions, which should be zero.

Table 1. Calculated mean value of effective stresses along *x*, *y*, and *z* directions at three stages of deformation.

M_{glide} (Pa·s)$^{-1}$	Three Stages of Deformation	$\bar{\sigma}_{effx}$(Pa)	$\bar{\sigma}_{effy}$(Pa)	$\bar{\sigma}_{effz}$(Pa)
	Early	1.22×10^5	-0.92×10^5	0.81×10^5
2.89×10^{-13}	Intermediate	0.09×10^5	-1.96×10^5	0.89×10^5
	Final	-0.61×10^5	-1.58×10^5	0.54×10^5
	Early	1.29×10^5	-1.12×10^5	1.01×10^5
8.67×10^{-10}	Intermediate	-0.86×10^5	-0.71×10^5	0.81×10^5
	Final	-0.82×10^5	-1.45×10^5	-2.39×10^5
	Early	1.32×10^5	-1.15×10^5	0.85×10^5
1.44×10^{-9}	Intermediate	0.40×10^5	-0.71×10^5	-0.41×10^5
	Final	-1.06×10^5	-0.61×10^5	-1.42×10^5

It is obvious that the scatter in the values is many times larger than the estimated effective stress in the section on glide controlled creep that was $180 \times 10^6/3000 = 0.6 \times 10^5$ Pa. From Table 1, one can conclude that the true value of the effective stress is not larger than the estimated value 0.6×10^5 Pa.

6. Effect of Change in the Applied Stress

In this section, the effect of a rapid change in the applied creep stress on the stress acting on dislocations (i.e., effective stress) is examined. To do so, the applied stress is instantly reduced from 180 MPa to 160 MPa during plastic deformation. In the first stage of deformation, the single-crystal undergoes creep deformation at constant stress level of 180 MPa and in the second stage the material deforms at a lower stress level of 160 MPa. In Figure 9, the influence of the stress change on the strain rate is illustrated. The same glide mobilities of $M_{glide} = 8.67 \times 10^{-10}$ (Pa·s)$^{-1}$ and $M_{glide} = 1.44 \times 10^{-9}$ (Pa·s)$^{-1}$ as in the section on glide controlled creep were used. The plastic strain rate exhibited an abrupt reduction as the applied stress changed from 180 to 160 MPa.

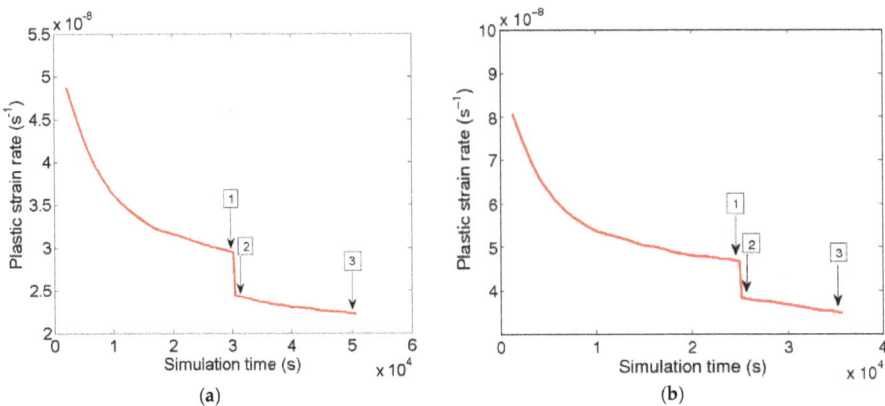

Figure 9. Abrupt change in plastic strain rate when the applied stress was reduced from 180 MPa to 160 MPa; (a) $M_{glide} = 8.67 \times 10^{-10}$ (Pa·s)$^{-1}$; (b) $M_{glide} = 1.44 \times 10^{-9}$ (Pa·s)$^{-1}$.

In Figure 10, the strain rate versus time at the lower stress 160 MPa is shown. The results in Figure 9a after the stress drop and in Figure 10 at the corresponding times were very close. Obviously, the effect of the previous deformation was weak.

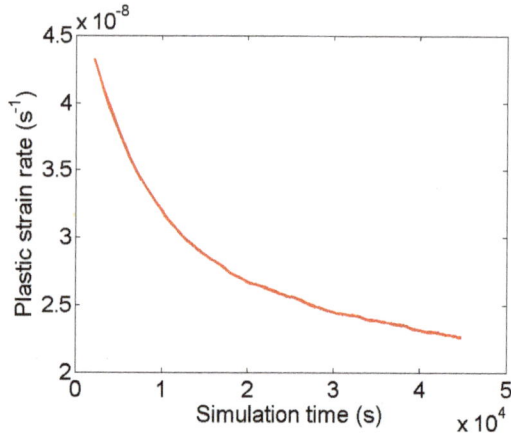

Figure 10. Plastic strain rate versus simulation time for deformation at stress level of σ_{appl} = 160 MPa and with mobility value of M_{glide} = 8.67 × 10^{-10} (Pa·s)$^{-1}$.

Evaluation of effective stresses acting on dislocations is performed at the three illustrated points in Figure 9. Point 1 corresponds to the first stage, i.e., σ_{appl} = 180 MPa, and points 2 and 3 correspond to the second stage of deformation, i.e., σ_{appl} = 160 MPa. Table 2 shows predicted mean effective stresses acting on dislocation lines along each of three x, y, and z directions at three illustrated points in Figure 9a,b.

Table 2. The calculated mean effective stresses at the three illustrated points in Figure 9.

M_{glide} (Pa·s)$^{-1}$	Illustrated Points	$\overline{\sigma}_{effx}$(Pa)	$\overline{\sigma}_{effy}$(Pa)	$\overline{\sigma}_{effz}$(Pa)
8.67 × 10^{-10}	1	−0.86 × 10^5	−0.71 × 10^5	−0.81 × 10^5
	2	−0.39 × 10^5	−0.92 × 10^5	−0.33 × 10^5
	3	−1.06 × 10^5	−0.72 × 10^5	−1.84 × 10^5
1.44 × 10^{-9}	1	−1.57 × 10^5	−1.15 × 10^5	−1.47 × 10^5
	2	−0.73 × 10^5	−0.75 × 10^5	−0.54 × 10^5
	3	−1.18 × 10^5	−0.96 × 10^5	−1.95 × 10^5

The scatter in the average effective stresses in Table 2 is of the same order as in Table 1. This means that the internal stress is at the same level as the applied stress also directly after the stress change. Thus, there is no evidence that the internal stress has any direct memory of the earlier applied stress.

7. Discussion

A fundamental expression for the climb mobility of dislocations, Equation (7), has been available for a long time [3]. A corresponding basic expression for glide has not been possible to find. Instead, a number of authors have used the empirical expression in Equation (8) or a similar one from Reference [26]. Since it involves a number of unknown constants, it cannot be used for making predictions. As described in section on the creep mobility, to solve this problem a unified climb and glide mobility was set up. This immediately fixed some of the unknown constants and the remainder of the parameters could be chosen by general arguments. The resulting semi-empirical expression has successfully been used in a number of publications for different applications.

In the present paper, the semi-empirical model was analyzed by deriving the glide rate at low temperatures. It was assumed that the glide rate is controlled by the climb rate of jogs on the dislocations. At low temperatures, the climb rate is enhanced by an excess concentration of vacancies that is generated during the creep deformation. The derived glide rate is almost identical to the results from the semi-empirical glide mobility, Figures 1 and 2, which fully verifies this expression for the glide mobility. It is possible to verify the validity of Equation (10) in an alternative way by relating it more directly to the climb rate [10,11].

Using the (verified) expression for glide mobility, creep deformation has been simulated with the help of dislocation dynamics. In these simulations, glide and cross glide have been taken into account, but not climb. Although it has not been possible to reach more than about 0.1% strain in the simulation, an apparent quasi-stationary condition with a fairly constant dislocation density was reached. The simulations show that pure glide can give rise to creep at low temperatures, which has been assumed in the literature for a number of years, see for example References [1,4].

In modelling creep deformation, a back stress or internal stress is often assumed. Many authors have considered the back stress in pure metals to be a physical quantity that can be measured with the help of stress drop tests, see for example References [14,34]. However, from the dislocation dynamics simulations one finds that the internal stress is almost identical to the applied stress. It was demonstrated that this is the case even directly after a stress drop. A direct measurement of the internal stress is then not possible. The concept of back stress can still be useful. It can be defined in many ways. But, it is not accessible to direct measurement, and many authors have realized this, see for example References [12,35]. The investigated situation does not cover the case when a stable substructure is present, which may take a longer time to relax. This is for example the case in 9% Cr-steels, where the sub-boundaries are stabilized by $M_{23}C_6$ carbides [36].

8. Conclusions

The glide rate of dislocations in copper at low temperatures is much higher than the climb rate. The glide rate has been derived based on the assumption that it is controlled by the climb of the jogs on the dislocations. The derived expression for the glide rate is in good accordance with a previously presented semi-empirical expression for the glide mobility.

This gives a fundamental understanding of the controlling creep mechanisms at near ambient temperatures, which is essential for example for the application of copper in canisters for disposal of spent nuclear waste.

Low temperature creep in copper has been simulated with dislocation dynamics (DD) taking glide but not climb into account. In the DD simulation a plastic strain of about 0.1% was reached. In spite of the low strain, the strain rate and the dislocation density reached an apparent quasi-stationary condition. It supports that creep at low temperatures can be controlled by glide alone.

Each dislocation in the DD simulations was exposed to an internal stress from surrounding dislocations. In the load direction, the internal stress was almost equal to the applied stress. In the investigated case, the difference between the applied and the internal stress, the effective stress, was only 1/3000 of the applied stress. This was taken into account in the simulation.

The effect of stress drop in the simulations was investigated. It turned out that the effective stress directly after the stress drop was as small as at constant stress. Obviously, the internal stress immediately adapts to the new stress level.

Author Contributions: A.H.D. performed the DD simulations, R.S. developed the dislocation mobility model, A.H.D., R.S. and P.K. wrote the paper.

Acknowledgments: The author would like to thank the Swedish Nuclear Fuel and Waste Management Co (SKB) for funding this work, contract 9114. Computer resources for this study at PDC Center for High Performance Computing at the KTH Royal Institute of Technology in Stockholm and National Supercomputer Center (NSC) at Linköping University were provided by the Swedish National Infrastructure for Computing (SNIC, projects 2014/11-25, 2015/16-50 and 2016/34-51).

Conflicts of Interest: The authors declare no conflict of interest.

Appendix A. Parameter Values Used in the Computations

Table A1. Values and interpretation of constants.

Parameter Description	Parameter	Value	Reference
Coefficient for self-diffusion	D_{s0}	1.31×10^{-5} m^2/s	[37]
Activation energy for self-diffusion	Q	198,000 J/mol	[37]
Burgers vector	b	2.56×10^{-10} m	
Atomic volume	Ω_0	$b^3/\sqrt{2} = 1.18 \times 10^{-29}$ m^3	
Lattice misfit for P atom	ε	0.055	[22]
Taylor factor	m	3.06	
Constant in Taylor's equation describing the influence of dislocation density on the strength	α	$(1 - \nu/2)/2\pi(1 - \nu) = 0.19$	[38,39]
Max back stress	σ_{imax}	257 MPa	[1]
Dislocation line tension	τ_L	$Gb^2/2 = 7.94 \cdot \times 10^{-16}$ MN	
Subgrain stress constant	Ksub	11	[40]
Max interaction energy between P solute and dislocation	U_P^{max}	8220 J/mol	[22]
Boltzmann's constant	k_B	1.381×10^{-23} J/grad	
Grain size	d_{grain}	100 μm	[2]
Shear modulus	G	$45,400 \times (1 - 7.1 \times 10^{-4} \times (T - 20))$, MPa, T in °C	[41]
Yield strength	σ_y	75 MPa for as hot worked in reference condition. For value at other temperatures and strain rates, see Ref.	[19]
Work hardening constant	c_L	28–31	[19]

References

1. Sandstrom, R.; Andersson, H.C.M. Creep in phosphorus alloyed copper during power-law breakdown. *J. Nucl. Mater.* **2008**, *372*, 76–88. [CrossRef]
2. Sandstrom, R. Basic model for primary and secondary creep in copper. *Acta Mater.* **2012**, *60*, 314–322. [CrossRef]
3. Hirth, J.P.; Lothe, J. *Theory of Dislocations*; Krieger: Malabar, FL, USA, 1982.
4. Blum, W.; Rosen, A.; Cegielska, A.; Martin, J.L. Two mechanisms of dislocation motion during creep. *Acta Metall.* **1989**, *37*, 2439–2453. [CrossRef]
5. Kocks, U.F. A statistical theory of flow stress and work-hardening. *Philos. Mag.* **1966**, *13*, 541–566. [CrossRef]
6. Lagneborg, R. Dislocation mechanisms in creep. *Int. Metall. Rev.* **1972**, *17*, 130–146.
7. Sandström, R. The role of cell structure during creep of cold worked copper. *Mater. Sci. Eng. A* **2016**, *674*, 318–327. [CrossRef]
8. Cottrell, A.H. Logarithmic and andrade creep. *Philos. Mag. Lett.* **1997**, *75*, 301–307. [CrossRef]
9. Nabarro, F.R.N. The time constant of logarithmic creep and relaxation. *Mater. Sci. Eng. A* **2001**, *309–310*, 227–228. [CrossRef]
10. Spigarelli, S.; Sandström, R. Basic creep modelling of aluminium. *Mater. Sci. Eng. A* **2018**, *711*, 343–349. [CrossRef]
11. Sandström, R. Fundamental Modelling of Creep Properties. In *Creep*; Tanski, T., Zieliński, A., Eds.; inTech: London, UK, 2017.

12. Biberger, M.; Gibeling, J.C. Analysis of creep transients in pure metals following stress changes. *Acta Metall. Mater.* **1995**, *43*, 3247–3260. [CrossRef]

13. Chen, B.; Flewitt, P.E.J.; Cocks, A.C.F.; Smith, D.J. A review of the changes of internal state related to high temperature creep of polycrystalline metals and alloys. *Int. Mater. Rev.* **2015**, *60*, 1–29. [CrossRef]

14. Ahlquist, C.N.; Nix, W.D. The measurement of internal stresses during creep of al and Al-Mg alloys. *Acta Metall.* **1971**, *19*, 373–385. [CrossRef]

15. Cuddy, L.J. Internal stresses and structures developed during creep. *Metall. Mater. Trans. B* **1970**, *1*, 395–401. [CrossRef]

16. Davies, P.W.; Williams, K.R. Cavity growth by grain-boundary sliding during creep of copper. *Met. Sci.* **1969**, *3*, 220–221. [CrossRef]

17. Henderson, P.J.; Sandstrom, R. Low temperature creep ductility of OFHC copper. *Mater. Sci. Eng. A* **1998**, *246*, 143–150. [CrossRef]

18. Sandström, R.; Wu, R. Influence of phosphorus on the creep ductility of copper. *J. Nucl. Mater.* **2013**, *441*, 364–371. [CrossRef]

19. Sandström, R.; Hallgren, J. The role of creep in stress strain curves for copper. *J. Nucl. Mater.* **2012**, *422*, 51–57. [CrossRef]

20. Sandström, R. Fundamental Models for Creep Properties of Steels and Copper. *Trans. Indian Inst. Met.* **2016**, *69*, 197–202. [CrossRef]

21. Wu, R.; Sandström, R.; Jin, L.Z. Creep crack growth in phosphorus alloyed oxygen free copper. *Mater. Sci. Eng. A* **2013**, *583*, 151–160. [CrossRef]

22. Sandstrom, R.; Andersson, H.C.M. The effect of phosphorus on creep in copper. *J. Nucl. Mater.* **2008**, *372*, 66–75. [CrossRef]

23. Sandström, R. Influence of phosphorus on the tensile stress strain curves in copper. *J. Nucl. Mater.* **2016**, *470*, 290–296. [CrossRef]

24. Korzhavyi, P.A.; Sandström, R. First-principles evaluation of the effect of alloying elements on the lattice parameter of a 23Cr25NiWCuCo austenitic stainless steel to model solid solution hardening contribution to the creep strength. *Mater. Sci. Eng. A* **2015**, *626*, 213–219. [CrossRef]

25. Nes, E.; Marthinsen, K. Modeling the evolution in microstructure and properties during plastic deformation of f.c.c.-metals and alloys—An approach towards a unified model. *Mater. Sci. Eng. A* **2002**, *322*, 176–193. [CrossRef]

26. Kocks, U.F.; Argon, A.S.; Ashby, M.F. Thermodynamics and kinetics of slip. *Prog. Mater. Sci.* **1975**, *19*, 291.

27. Chandler, H.D. Effect of unloading time on interrupted creep in copper. *Acta Metall. Mater.* **1994**, *42*, 2083–2087. [CrossRef]

28. Mecking, H.; Estrin, Y. The effect of vacancy generation on plastic deformation. *Scr. Metall.* **1980**, *14*, 815–819. [CrossRef]

29. Sandström, R. The role of microstructure in the prediction of creep rupture of austenitic stainless steels. In Proceedings of the ECCC Creep & Fracture Conference, Düsseldorf, Germany, 10–14 September 2017.

30. Arsenlis, A.; Cai, W.; Tang, M.; Rhee, M.; Oppelstrup, T.; Hommes, G. Enabling strain hardening simulations with dislocation dynamics. *Model. Simul. Mater. Sci. Eng.* **2007**, *15*, 553–595. [CrossRef]

31. Haghighat, S.M.H.; Eggeler, G.; Raabe, D. Effect of climb on dislocation mechanisms and creep rates in γ'-strengthened Ni base superalloy single crystals: A discrete dislocation dynamics study. *Acta Mater.* **2013**, *61*, 3709–3723. [CrossRef]

32. Delandar, A.H.; Haghighat, S.M.H.; Korzhavyi, P.; Sandström, R. Dislocation dynamics modeling of plastic deformation in single-crystal copper at high strain rates. *Int. J. Mater. Res.* **2016**, *107*, 988–995. [CrossRef]

33. Bulatov, V.V.; Cai, W. *Computer Simulations of Dislocations*; Oxford University Press: Oxford, UK, 2006.

34. Matsuo, T.; Nakajima, K.; Terada, Y.; Kikuchi, M. High temperature creep resistance of austenitic heat-resisting steels. *Mater. Sci. Eng. A* **1991**, *146*, 261–272. [CrossRef]

35. Hausselt, J.; Blum, W. Dynamic recovery during and after steady state deformation of Al-11wt%Zn. *Acta Metall.* **1976**, *24*, 1027–1039. [CrossRef]

36. Magnusson, H.; Sandstrom, R. Creep strain modeling of 9 to 12 Pct Cr steels based on microstructure evolution. *Metall. Mater. Trans. A* **2007**, *38*, 2033–2039. [CrossRef]

37. Neumann, G.; Tölle, V.; Tuijn, C. Monovacancies and divacancies in copper Reanalysis of experimental data. *Phys. B Condens. Matter* **1999**, *271*, 21–27. [CrossRef]

38. Horiuchi, R.; Otsuka, M. Mechanism of high temperature creep of aluminum-magnesium solid solution alloys. *Trans. Jpn. Inst. Met.* **1972**, *13*, 284–293. [CrossRef]
39. Orlová, A. On the relation between dislocation structure and internal stress measured in pure metals and single phase alloys in high temperature creep. *Acta Metall. Mater.* **1991**, *39*, 2805–2813. [CrossRef]
40. Wu, R.; Pettersson, N.; Martinsson, Å.; Sandström, R. Cell structure in cold worked and creep deformed phosphorus alloyed copper. *Mater. Charact.* **2014**, *90*, 21–30. [CrossRef]
41. Ledbetter, H.M.; Naimon, E.R. Elastic Properties of Metals and Alloys. II. Copper. *J. Phys. Chem. Ref. Data* **1974**, *3*, 897–935. [CrossRef]

metals

MDPI

Article

Development of the FE In-House Procedure for Creep Damage Simulation at Grain Boundary Level

Qiang Xu *, Jiada Tu and Zhongyu Lu

School of Computing and Engineering, Huddersfield University, Huddersfield HD1 3DG, UK;
u1361251@hud.ac.uk (J.T.); z.lu@hud.ac.uk (Z.L.)
* Correspondence: q.xu2@hud.ac.uk; Tel.: +44-1484-472842

Received: 20 April 2019; Accepted: 29 May 2019; Published: 5 June 2019

Abstract: A two-dimensional (2D) finite element framework for creep damage simulation at the grain boundary level was developed and reported. The rationale for the paper was that creep damage, particularly creep rupture, for most high temperature alloys is due to the cavitation at the grain boundary level, hence there is a need for depicting such phenomenon. In this specific development of the creep damage simulation framework, the material is modeled by grain and GB (grain boundary), separately, where smeared-out grain boundary element is used. The mesh for grain and grain boundary is achieved by using Neper software. This paper includes (1) the computational framework, the existing subroutines, and method applied in this procedure; (2) the numerical and programming implementation of the GB; (3) the development and validation of the creep software; and (4) the application to simulate plane stress Copper–Antimony alloy. This paper contributes to the development of finite element simulation for creep damage/rupture at a more realistic grain boundary level and contributes to a new understanding of the intrinsic relationship of stress redistribution and creep fracture.

Keywords: creep rupture; creep grain boundary; finite element method; grain boundary cavitation; creep damage; poly-crystal

---

1. Introduction

Creep damage is a serious problem limiting the lifetime of high-temperature components in many practical applications. A good understanding and accurate description of creep deformation and creep rupture is of great interest to people who research this field.

It is generally understood that under creep conditions, the higher the temperature, the quicker the deformation and the shorter the lifetime of material. On the other hand, the higher the operating temperature, the higher the thermal efficiency for power plants.

It is generally accepted that, for the majority of metals and alloys, creep rupture is due to creep cavitation at the grain boundary where cavities nucleate, grow, and coalesce [1,2].

Three different approaches are used in the modelling of creep fracture [3]: (1) at the macroscopic level, classical fracture mechanics approaches are extended to time-dependent behavior; (2) still at macroscopic level, in continuum damage models, cavitation is incorporated in an average, smeared-out manner by means of a damage parameter; and (3) the micro-mechanical models where various physical mechanisms are directly involved.

Creep continuum damage mechanics have been developed and widely used now and their applicability depends on the reliability of the development of a set of creep deformation and damage equations and the availability of a computational platform (typically finite element analysis (FEA) package). However, the most current approach in the development of creep damage constitutive equation is phenomenologically based on using macroscopic creep strain to fit models of various kinds including creep cavitation. In a comprehensive review paper, Xu et al. [4] identified that the current phenomenological approach suffers a lack of precise understanding surrounding the damage, the constitutive model thus has issues of low reliability for extrapolation beyond the stress range which has been calibrated and difficulty in generalizing a one-dimensional constitutive equation to a three-dimensional one.

Cavity nucleation [2]: cavity nucleation mechanisms are still not well understood. It has generally been observed that cavities frequently nucleate on grain boundaries, particularly on those transverse to a tensile stress; agglomeration of vacancies, dislocation pile-up, and grain boundary sliding have all been considered to promote nucleation. It has long been suggested that (transverse) grain boundaries and second phase particles are the common locations for cavities. Empirical equations of nucleation were well established for use.

Cavity growth [2]: research has suggested various cavity growth models [3] including (1) diffusion-grain boundary control; (2) diffusion-surface control, (3) grain boundary sliding, and (4) constrained diffusion cavity growth. The local (true) normal stress has been used as the driving force, hence it is more realistic; and this model has been found in good agreement with experimental observation.

Recently, synchrotron micro-tomography has been used to investigate the cavitation of high Cr steel [5–7], and continuous cavity nucleation and cavity growth models were calibrated by Xu et al. [8] and an explicit creep fracture model, based on the coalescence of grain boundary cavities was derived [8]. The applicability of Xu's creep fracture lifetime model to a stress range of 120 MPa to 180 MPa and a lifetime of 2825 to 51,406 h, has been demonstrated with 87% of accuracy [9].

The micro-mechanical model initially was focused on the single cavity to that of the failure of a polycrystalline aggregate comprising a number of grains [3], however, it is still not feasible to directly incorporate them into any engineering analysis due to the need of computational power. Hence, the need of the development of smeared-out grain boundary constitutive equations for macroscopic creep deformation and damage.

Onck and van der Giessen [3] were amongst these to propose the concept of grain and grain boundary elements, via a two-dimensional version. The material in the grain is assumed to be homogeneous and to deform by power creep law in addition to elasticity. Grain boundary processes like cavitation and sliding are accounted for by grain boundary elements that connect the grains. Results are compared with the full-field finite element analysis, the method is demonstrated to capture the essential features of creep fracture, like constrained cavitation and interlinkage of micro-cracks. It also reported that there is a gain in computational time of 600 times by using the smeared-out element.

This micro-mechanical based smeared-out grain boundary element has eventually been further developed [10,11] for the simulation of copper–antimony alloy, and the main contents are: (1) grain boundary nucleation: Dyson's empirical equation [12] has been consulted; (2) cavity annihilation: probabilistic description of crack annihilation [13] has been adopted; (3) cavity growth: constrained cavity growth model [1,14] adopted; (4) grain boundary sliding: Ashby viscosity model [15] adopted; and (5) creep fracture criterion when the cavity area fraction along grain boundary reached 0.5, experimental observation Cocks and Ashby [16]. In this model, grain boundary sliding has been considered for deformation, but not applied to the cavity nucleation. This proposed model is for 3-dimensional. A slightly simple version has been developed by [17], where the annihilation was not considered, also it is a two-dimensional version.

The grain element has been modelled by simple power law in [3] and [17], both have captured the main features of the creep damage occurring in the grain boundary; sophisticated slip-system model has also been developed and utilized in [10]. Thus, it is justified to adopt simple power law in this type research unless it has been found it is not suitable anymore.

The grain boundary element can be implemented via the cohesive zone element [10,17]; the cohesive zone element can have a small thickness or no thickness at all, such as Goodman element [18], but the mechanical properties can and have been fully represented through its formulation. The former is slightly more complicated in computation. Hence it is desirable to use Goodman element.

Both the cohesive zone element and contact element can be implemented in ABQUS or in-house software. The published work [13,17] used ABAQUS. The authors of this paper have extensive experience and already developed creep damage software for a multi-material zone version.

It is reported [17] that the mesh size of the grain boundary, assuming a perfect hexagonal shape, that eight grain boundary elements per side is required.

It is concluded that: (1) a computational platform able to model grain and grain boundary separately is of importance for research; (2) a two-dimensional version is not only the first step before the development of three-dimensional version, the two-dimensional version is still of use.

FORTRAN language on the Visual Studio 2013 platform (version 11.0.61219.00, Microsoft, Redmond, WA, USA), which is based on object-oriented programming (OOP) [19], will be used in this software development. This procedure is developed under the framework of the traditional FEM, some subroutines for the assembly and solution the stiffness matrix are from Smith et al. [20] and the non-linear displacement iteration method is from Hayhurst et al. [21]. The specific work is based on some existing subroutines and methods, combined with a series of subroutines which were developed to implement the numerical methods of GB to obtain the FE in-house procedure which can realize the 2D polycrystalline creep simulation. The main purpose of this paper is to record and present the development process, more specifically, it reports the program framework and the theoretical background. Finally, through the bi-grain case study, it benchmarked the numerical stability and accuracy of the entire system. The generation of grain boundary mesh will be achieved by the use of Neper (version 3.3.1, Romain Quey, Cornell University, Ithaca, NY, USA) [22].

In the following sections, this paper will report the specific development of creep damage simulation framework: theory, development, validation, and its application to plane stress test of copper-alloy. This paper contributes to the methodology and new insight of the intrinsic relationship of stress redistribution and failure.

2. Overview of the Procedure

The in-house procedure was developed from the FORTRAN language on the Visual Studio 2013 platform (version 11.0.61219.00, Microsoft, Redmond, WA, USA), which is based on object-oriented programming (OOP) [19] method to implement the microscopic simulation of creep. In essence, the procedure solves the creep problem, which is a kind of time-dependent and non-linear boundary problem. For simplicity, only plane strain is considered under the 2D case.

2.1. Computational Framework

The flow diagram structure of the in-house procedure is shown in Figure 1, its idea came from program P61 of Smith et al. [20] for homogeneous material and it is expanded into this non-homogeneous version, highlighted in red. These exiting techniques are adopted: the assembly and solution of the global stiffness matrix, the stiffness matrix of the plane strain element, the integration method of the constitutive equation, the nonlinear iterative method, etc. The independent development parts are marked with a red dotted line box and are summarized as follows:

1. Calculation of the GB element stiffness matrix.
2. Generation of the GB creep body loads.

In the following sections, we will introduce the existing technologies and subroutines used in this procedure, and then describe the parts of independent development, from a mathematical perspective and through program implementation.

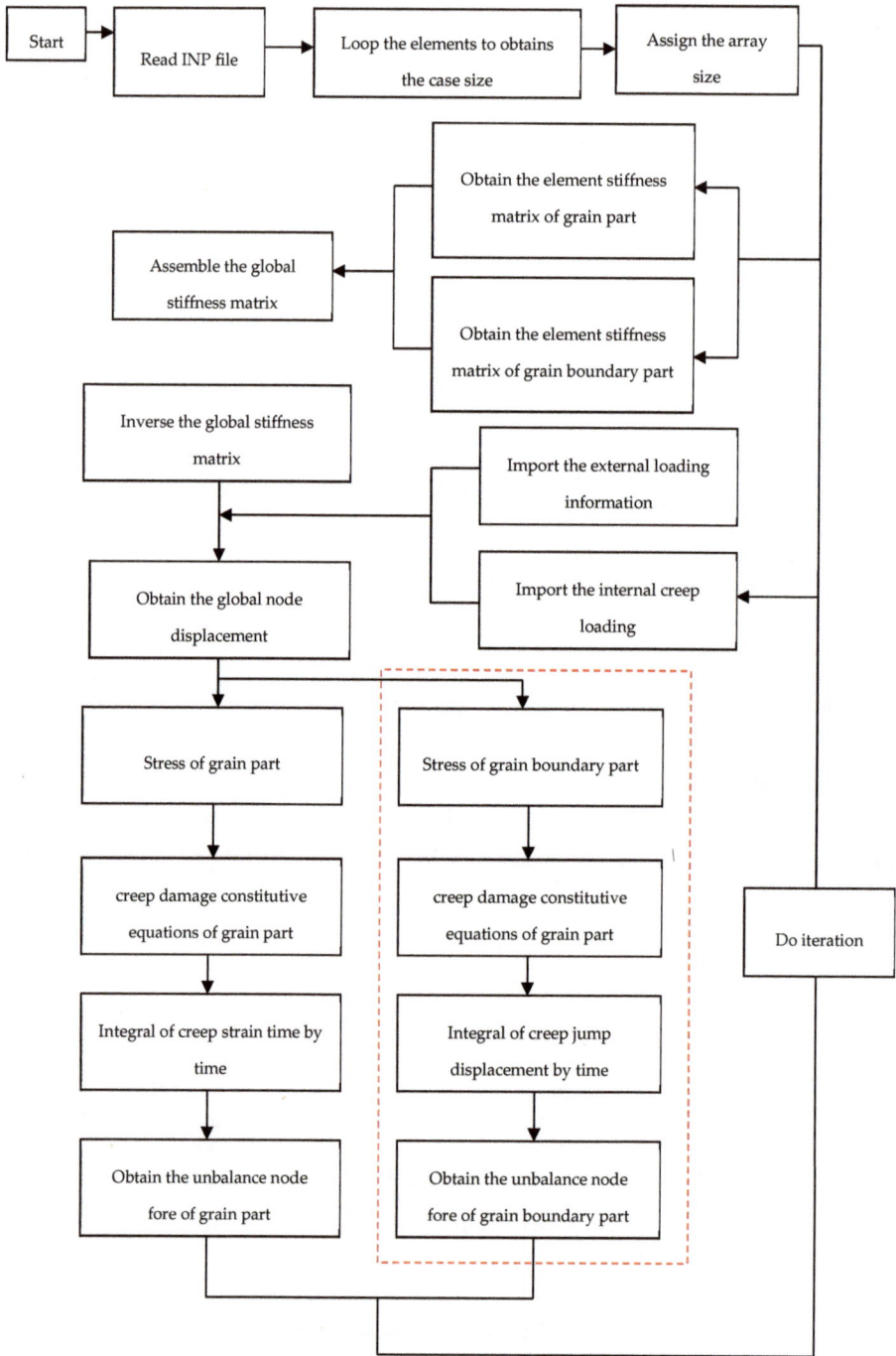

Figure 1. Flow diagram structure of the in-house procedure.

2.1.1. Packaged Block

The most relevant standard subroutines (11) were adopted from the library 'main' directly [20] and they are listed in Table 1 for completeness, where the description under the function column was simplified by current authors. These subroutines mainly implement the functions of: (1) obtaining the element stiffness matrix; (2) assembling and storing the element to the global stiffness matrix; (3) the solution of the global stiffness matrix, etc. The specific details of these subroutines are summarized in Table 1.

Table 1. List of the standard subroutines for the in-house procedure.

Name	Function
formnf	Returns the nodal freedom array.
num_to_g	This subroutine is used to obtain the element steering array from the nodal freedom array and element node number.
fkdiag	This subroutine is used to store the global stiffness matrix by the lower triangle stored method, it returns the bandwidth value.
sample	Returns the local coordinates and the weighting coefficients for the element integration by Gauss method.
deemat	Returns the stress–strain matrix.
shape_fun	Returns the shape function of each Gauss integrating points.
shape_der	Returns the shape function derivatives of each Gauss integrating points.
invert	Returns the inverse matrix onto itself.
fsparv	Return the global matrix in skyline form, which is from the lower triangular global stiffness matrix.
sparin_gauss spabac_gauss	These two subroutines are used to solve the global stiffness matrix, by forward and back-substitution of Gaussian factorized vector of pervious stiffness matrix.

2.1.2. Non-Linear Creep Iteration Method

Creep is a kind of time-related visco-plastic deformation, therefore, the solution of this problem requires non-linear iteration. Here, the displacement iteration method is chosen to solve the residual stress updating. The displacement method was proposed by Hayhurst et al. [21], which was adopted to implement the damage analysis of Weldment [23].

The iterative process for solving this creep problem is as follows. Firstly, calculate the node displacement. This involves multiplying the total load and the inverse stiffness matrix where the total load consists of the actually applied external load plus the additional load at the node due to stress redistribution. Secondly, obtain the new elastic strain for each element. This is achieved by deducting the creep strain from the total strain which, in turn, is obtained by multiplying the displacement with the [B] (Displacement-Strain) matrix. Finally, the newly updated element stress, strain, damage, etc., will be output for next iteration.

2.1.3. Generation of the GB Creep Body Loads

The conventional body force from grain element will be calculated utilizing the Gauss Legendre quadrature over the plane strain element regions [20,21], while the body force from grain boundary elements can use the analytical integration directly. They are:

For the grain part P_{CG} [20,21,23]:

$$P_{CG} = \iint [B_G]^T \cdot [D_G] \cdot \varepsilon_{CG} \, dxdy, \tag{1}$$

$$P_{CG} = \sum_{i=1}^{nip} [B_G]^T \cdot [D_G] \cdot \varepsilon_{CG} \cdot W_i, \tag{2}$$

where the $[B_G]$ is the displacement–strain matrix of grain element, $[D_G]$ is the stress–strain matrix of grain element, and ε_{CG} is the creep strain, nip is the number of the Gauss point and W_i is the weighting coefficient.

For the GB part P_{CGB}:

$$P_{CGB} = \int [[B_{GB}] \cdot [T]]^T \cdot [D_{GB}] \cdot U_{GB} \, dl, \tag{3}$$

$$P_{CGB} = [[B_{GB}] \cdot [T]]^T \cdot [D_{GB}] \cdot U_{GB} \cdot L, \tag{4}$$

where the $[B_{GB}]$ is the node-element displacement matrix of grain boundary element, $[T]$ is the local–global coordination transfer matrix of grain boundary element, $[D_{GB}]$ is the stress—relative displacement matrix of grain boundary element, U_{GB} is the creep jump displacement of the grain boundary, and L is the length of the joint element.

Therefore, the global body loads can be obtained

$$P_C = P_{CG} + P_{CGB}, \tag{5}$$

where P_C is the global body loads, P_{CG} is the global body loads of grain part, P_{CGB} is the global body loads of grain boundary part.

2.2. Numerical Implementation of GB

As mentioned before, the mesh of the GB part is implemented by the Goodman element [18]. Therefore, the core of the simulation of the GB is to obtain the element stiffness matrix of the Goodman element.

2.2.1. Goodman Element Stiffness

The Goodman element is a four-node with an eight-DOF element without thickness (as shown in Figure 2), initially, the nodal pairs (1,2) and (3,4) lie together under the unloading condition. The deformation of the element is represented by the relative displacement of the upper and lower surface.

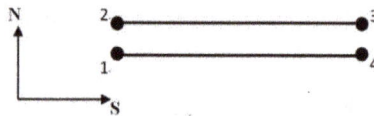

Figure 2. Goodman element schematic figure. N for normal direction. S for separate direction.

The derivation process of the element stiffness matrix is as follows [18,24]:

The displacement on separate direction for each node is v_1, v_2, v_3, v_4, while the displacement on the normal direction is u_1, u_2, u_3, u_4.

The shape functions for this four-node Goodman element with corner nodes take the form of (Equation (6))

$$N_1 = N_2 = 1 - \tfrac{2x}{L}, \\ N_3 = N_4 = 1 + \tfrac{2x}{L}, \tag{6}$$

The displacement for the upper and lower surface of the element

$$[\Phi] = \begin{bmatrix} u^u - u^l \\ v^u - v^l \end{bmatrix} = \frac{1}{2} \cdot [B] \cdot \begin{bmatrix} u_1 \\ v_1 \\ u_2 \\ v_2 \\ u_3 \\ v_3 \\ u_4 \\ v_4 \end{bmatrix}, \tag{7}$$

$$[B] = \begin{bmatrix} -N_1 & 0 & N_1 & 0 & N_3 & 0 & -N_3 & 0 \\ 0 & -N_1 & 0 & N_1 & 0 & N_3 & 0 & -N_3 \end{bmatrix}, \tag{8}$$

The vector of the forces on unit length of the Goodman element at normal and separate directions are

$$[F] = \begin{bmatrix} F_n \\ F_s \end{bmatrix}, \tag{9}$$

$$\begin{bmatrix} F_n \\ F_s \end{bmatrix} = [D] \cdot [\Phi], \tag{10}$$

Here [D] denotes the stiffness matrix and have the form $\begin{bmatrix} k_n & 0 \\ 0 & k_s \end{bmatrix}$. $[\Phi]$ denotes the element displacement (the details are shown in Equation (7)).

Based on the potential-energy theory, the element deformation energy P can be obtained as

$$P = \frac{1}{2} L \Phi^T [K] \Phi = \frac{1}{2} \int_{-\frac{L}{2}}^{\frac{L}{2}} \frac{1}{4} [\Phi]^T [B]^T [D][B][\Phi] dx, \tag{11}$$

Here, [K] is the stiffness matrix of the Goodman element per unit length.
Based on the calculation

$$[K] = \int_{-\frac{L}{2}}^{\frac{L}{2}} \frac{1}{4} [B]^T [D][B] \cdot dx, \tag{12}$$

According to the Equations (10) and (11),

$$[B]^T [D][B] = \begin{bmatrix}
N_1^2 K_S & 0 & -N_1^2 K_S & 0 & -N_1 N_3 K_S & 0 & N_1 N_3 K_S & 0 \\
0 & N_1^2 K_N & 0 & -N_1^2 K_N & 0 & -N_1 N_3 K_N & 0 & N_1 N_3 K_N \\
-N_1^2 K_S & 0 & N_1^2 K_S & 0 & N_1 N_3 K_S & 0 & -N_1 N_3 K_S & 0 \\
0 & -N_1^2 K_N & 0 & N_1^2 K_N & 0 & N_1 N_3 K_N & 0 & -N_1 N_3 K_N \\
-N_1 N_3 K_S & 0 & N_1 N_3 K_S & 0 & N_3^2 K_S & 0 & -N_3^2 K_S & 0 \\
0 & -N_1 N_3 K_N & 0 & N_1 N_3 K_N & 0 & N_3^2 K_N & 0 & -N_3^2 K_N \\
N_1 N_3 K_S & 0 & -N_1 N_3 K_S & 0 & -N_3^2 K_S & 0 & N_3^2 K_S & 0 \\
0 & N_1 N_3 K_N & 0 & -N_1 N_3 K_N & 0 & -N_3^2 K_N & 0 & N_3^2 K_S
\end{bmatrix} \tag{13}$$

The integrals of these N_1^2, N_3^2 and $N_1 N_3$ are

$$\int_{-\frac{L}{2}}^{\frac{L}{2}} N_1^2 = \tfrac{4}{3} L,$$
$$\int_{-\frac{L}{2}}^{\frac{L}{2}} N_3^2 = \tfrac{4}{3} L, \tag{14}$$
$$\int_{-\frac{L}{2}}^{\frac{L}{2}} N_1 N_3 = \tfrac{2}{3} L,$$

Thence, the [k] is

$$
K = \frac{1}{6} * \begin{bmatrix}
2K_S & 0 & -2K_S & 0 & -K_S & 0 & K_S & 0 \\
0 & 2K_N & 0 & -2K_N & 0 & -K_N & 0 & K_N \\
-2K_S & 0 & 2K_S & 0 & K_S & 0 & -K_S & 0 \\
0 & -2K_N & 0 & 2K_N & 0 & K_N & 0 & -K_N \\
-K_S & 0 & K_S & 0 & 2K_S & 0 & -2K_S & 0 \\
0 & -K_N & 0 & K_N & 0 & 2K_N & 0 & -2K_N \\
K_S & 0 & -K_S & 0 & -2K_S & 0 & 2K_S & 0 \\
0 & K_N & 0 & -K_N & 0 & -2K_N & 0 & 2K_N
\end{bmatrix}
\tag{15}
$$

2.2.2. Coordinate System Transmission

The previous consideration is a special case, which is the local coordinate that coincides with the global coordinate system. Therefore, it is necessary through the transmission matrix to relate these two coordinate systems (the geometry is shown in Figure 3).

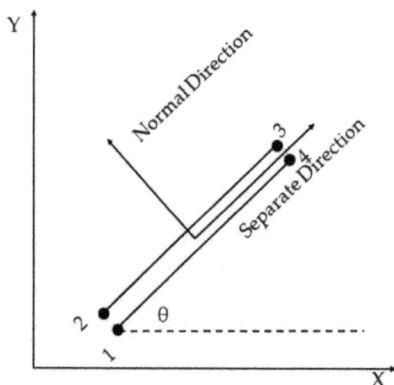

Figure 3. The schematic of the Goodman element in global coordinate system.

The rotation matrix is introduced in this procedure [24].

$$
T = \begin{bmatrix}
N & 0 & 0 & 0 \\
0 & N & 0 & 0 \\
0 & 0 & N & 0 \\
0 & 0 & 0 & N
\end{bmatrix},
\tag{16}
$$

where

$$
N = \begin{bmatrix}
\cos\theta & \sin\theta \\
-\sin\theta & \cos\theta
\end{bmatrix},
\tag{17}
$$

Based on the matrix [T], the stiffness matrix of the Goodman element can be obtained by

$$
[K]_{Glo} = [T]^T [K]_{loc} [T],
\tag{18}
$$

2.3. Programming Implementation of GB

The structured programming method is used to improve programming efficiency, and a mature overall structure and related standard subroutines were adopted, as mentioned at Section 2.2. In addition, some subroutines and structures have been developed to achieve material micro-creep simulation. These developments are based on the mathematical theory which were described in

Section 2.3, the function is to acquire the GB element stiffness matrix. The introduction and description of these subroutines are summarized in Table 2.

Table 2. Developed subroutines for GB simulation.

Name	Arguments	Description
element_inf	length, angle, coord2	Returns the element information: length ('length') and the angle θ ('angle'). The import information is the coordination of these four nodes under the global system, which store in array 'coord2'.
Loc_Gol	angle_t, angle	Returns the rotation matrix 'angle_t', for the transmission of the Goodman element form local to global coordination system. The import information is the angle θ which is generated by subroutine 'element_inf'.
new_km	km, kcoh	Returns the stiffness matrix 'km', 'km' is 8 × 8 size matrix, for Goodman element with the normal (n) and separate (s) rigidity of the Goodman element. The information is shore in 'kcoh' array. The mathematical background is Equation (22).

The specific process for obtaining the Goodman element stiffness matrix, is: firstly, initialize the variables of each Goodman element, such as the rotation angle, the length and the element stiffness matrix, then based on the mathematical background which had been mentioned in Section 2.2 to obtain the stiffness matrix of single element. Finally, the 'element-by-element' and 'lower triangular' techniques were used to assemble the global stiffness matrix and to store the global matrix as a skyline vector respectively.

2.4. Creep Constitutive Equation

2.4.1. Grain Element

Power law creep was chosen for the grain element in the current version of the software. Its uni-axial version is

$$\dot{\varepsilon} = A \cdot \exp\left(-\frac{Q}{RT}\right) \sigma^n t^m, \tag{19}$$

The multi-axial form is

$$\dot{\varepsilon} = A \cdot \exp\left(-\frac{Q}{RT}\right) \sigma^n t^m \frac{3}{2} \frac{S_{ij}}{\sigma_e}, \tag{20}$$

where n and m are the stress and time-hardening exponents respectively, the Q is the activation energy, the R is the universal gas constant and the A is the constitutive constant.

2.4.2. Grain Boundary Element

The GB displacement jump at a normal direction can be obtained by the model which is developed by Vöse et al. [25]. It takes into account nucleation, growth, coalescence, and sintering of multiple cavities, i.e.,

$$\frac{d\beta}{d\bar{t}} = \frac{3}{2}\frac{\beta}{\bar{\rho}}(\bar{\alpha}_p - \bar{\alpha}_a) + \sqrt{\bar{\rho}}\sqrt[3]{36h(\psi)\pi\beta^2}\frac{d\bar{a}}{d\bar{t}}, \tag{21}$$

$$\frac{d\bar{\rho}}{d\bar{t}} = \bar{\alpha}_p(1-f) - \bar{\alpha}_a, \tag{22}$$

$$\bar{\alpha}_a = x_3 \cdot 8\pi\bar{\rho}^2\bar{a}\frac{d\bar{a}}{d\bar{t}}, \tag{23}$$

$$\frac{d\bar{a}}{d\bar{t}} = x_1 \cdot \frac{2\bar{D}_{gb}}{h(\psi)}\frac{\left[1 - \bar{a}_{tip}(\bar{a}) \cdot (1 - x_2\omega)\right]}{\bar{a}^2 \cdot q(x_2\omega)}, \tag{24}$$

$$\omega = \sqrt[3]{\frac{9\pi\beta^2}{16h^2(\psi)}}; \bar{a} = \frac{1}{\sqrt{\bar{\rho}}}\sqrt[3]{\frac{3}{4}\frac{\beta}{h(\psi)\pi}}, \tag{25}$$

$$f = \frac{(\eta - 1)\omega}{1 - \omega}, \tag{26}$$

$$\eta = \exp\left(\left[x_4 \cdot 2\pi \overline{D}_{gb}\left(\overline{a}_{tip}(\overline{a} = 1) - \overline{a}_{tip}(\overline{a})\right)\overline{\rho}\left(\frac{d\overline{\mu}^p}{dt}\right)^{-1}\right]\right), \tag{27}$$

$$\frac{d\overline{\mu}^p}{d\overline{t}} = \frac{\beta}{\sqrt{\overline{\rho}^3}}(\overline{\alpha}_p - \overline{\alpha}_a) + \sqrt[3]{36h(\psi)\pi\beta^2}\frac{d\overline{a}}{d\overline{t}}, \tag{28}$$

where

$$q(\omega) = -2\ln\omega - (3 - \omega)(1 - \omega); \overline{a}_{tip}(\overline{a}) = 2\overline{\gamma}_s \sin\psi/\overline{a}, \tag{29}$$

In this equation, β is the damage variable, ρ is the cavity density, a is the average radius of the cavity, $\overline{\alpha}_p$ is the stress dependent nucleation rate, $\overline{\alpha}_a$ is the annihilation rate, ψ is the dihedral angle of the cavity (70°), $\overline{D}_{gb}$ is the GB diffusion coefficient, and ω is the damaged area fraction. The creep degradation of GB is calibrated by three variables: ρ, β, and a. These three parameters not only determine the failure degree of GB, but also determine the amount of the creep non-linear deformation. Therefore, ρ, β, and a are the three indicators for the benchmark.

Unlike grains, the deformation amount of GB is quantified by the absolute relative displacement of the upper and lower surface, it is assumed that this displacement is related to the stress and damage, but not rate related. The absolute relative displacement of the GB is determined by two variables, ρ and β.

$$D_C = \frac{\beta}{\sqrt{\rho}} - \frac{\beta_0}{\sqrt{\rho_0}}, \tag{30}$$

where the D_C is the relative displacement of creep, β_0 is the initial damage value (here is 10^{-4} and ρ_0 is the initial cavity density (here is 10^{-3} mm^{-2}). For the sliding part of grain boundary, the Newtonian viscous flow [15] is used

$$\frac{du_{sliding}}{dt} = \frac{\sigma_{sliding}}{\eta_{slding}}, \tag{31}$$

where $\frac{du_{sliding}}{dt}$ is the relative sliding velocity, $\sigma_{sliding}$ is the stress at the separate direction, η_{slding} is the sliding viscosity of the GB.

2.4.3. Parameters for Constitutive Equation

This bi-grain case study simulates the micro creep process of copper at 500 °C. As mentioned before, power law creep is used to describe the creep mechanism of the grain part.

The material parameters of copper power law for grain [26] are (400–700 °C)

$$A: 38.8 \text{ MPa}^{-n}S^{-m-1}, Q: 197 \text{ KJ·mol}^{-1}, n: 4.8, m: 0.$$

The three main normalized material parameters for the grain boundary creep evolution (Markus Vöse's equations) of Copper at 500 °C are listed as [25]

$$\sqrt[3]{D_{gb}}: 3.9695, \overline{\gamma}_s: 0.089, \overline{\alpha}_p: 0.24, \overline{R}: 42.$$

3. Preliminary Validation of the In-House Procedure

In order to make the logic and efficiency of the verification, a general step is employed [20,23,27]. The main idea of the step is from the uni-axial condition to the multi-axial condition, from linear elasticity to creep nonlinearity. The bi-grain case is chosen as the first step of the benchmark [17], which is to demonstrate the application of the in-house procedure under the uni-axial loading condition. In the first case, no shear sliding part occurs in the simulation due to the stress on the separate direction being zero. Based on this case, the numerical stability and accuracy of the procedure is verified to pave the way for the subsequent polycrystalline simulation. In the second case, shear sliding occurs in the simulation GB to validate the accuracy and stability of the procedure on the separate direction.

3.1. Validation of Stress Update

The bi-grain consists of two square grains and a single GB as depicted in Figure 4.

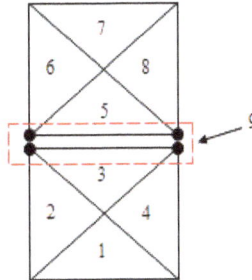

Figure 4. Schematics showing the bi-grain case study.

The model is meshed by four triangle plane strain elements for each the grain (e.g., 1–4, and 5–8, respectively), and 1 Goodman element for the GB. A uniformly distributed loading 20 MPa is applied on the top surface in the Y direction and left line fixed on the X direction and bottom line fixed on the Y direction. In this case, no shear sliding stress occurs along the GB.

3.1.1. Validation of the Elastic Stress

The theoretical stress of the grain elements at the Gauss point is 20 MPa at Y direction and the stress in the X direction and the shear stress should be zero. The theoretical stress of the GB is 20 MPa at Y direction and the stress in the X direction should be zero. By analyzing the elastic simulation results of these nine elements (eight elements for the grain part and one element for the GB part): (1) For the grain part, the simulation results is shown in good agreement with the theoretical result at Y direction. The maximum stress in X direction is 1.066×10^{-14} MPa (No. 2 grain element) and in shear direction is 4.170×10^{-15} MPa (No. 6 grain element), which is negligible as expected. (2) For the GB part, the elastic stress at normal direction is 20.000 MPa and separated direction is -2.033×10^{-15} respectively, therefore, the simulation result is expected.

3.1.2. Validation of the Non-linear Creep Iteration

The solution to the creep boundary problem is based on the non-linear iterations. Therefore, the accuracy of the elastic stress field which was generated by the iteration directly affects the integral of the creep constitutive equations. Under the uniform stress field, since the creep deformation of each element is the same, the creep body loads are equal, and there is no stress redistribution. In summary, the convergence sign of the iteration is that the elastic stress at each iteration step is to keep constant, the bi-grain case belongs to this category. The iterative process lasts for 141,564 steps, with the time step of 0.0000001 (normalized). Through the observation of the elastic stress at each iteration step, it is clear the applied stress for the non-linear iteration of each element at every step keep constant. At the end of the iteration, the cumulative maximum error value of the stress for grain part are -1.790×10^{-11} MPa (No. 7 grain element) and -1.69 MPa (No. 8 grain element) for X direction and shear direction respectively. For the GB part, the applied stress at the end of the simulation is 0 MPa at separated direction and 20.000 MPa at normal direction. Therefore, the in-house procedure has shown good convergence in the creep non-linear iteration.

3.1.3. Validation of the Integration

The accuracy verification of the non-linear iteration has been reported previously. Therefore, in this section, it reports the validation of the integration part. In the current version, the Euler forward method is adopted here for the integration of the constitutive equation (it was mentioned in Section 2.4.2).

3.1.4. The Result of the Bi-Grain Case Study

As mentioned before, there are three important indicators of the constitutive model: the cavity density ρ, the dimensionless damage variable β and the average radius cavity a. The result of these three indicators versus the normalized time is plotted in Figure 5.

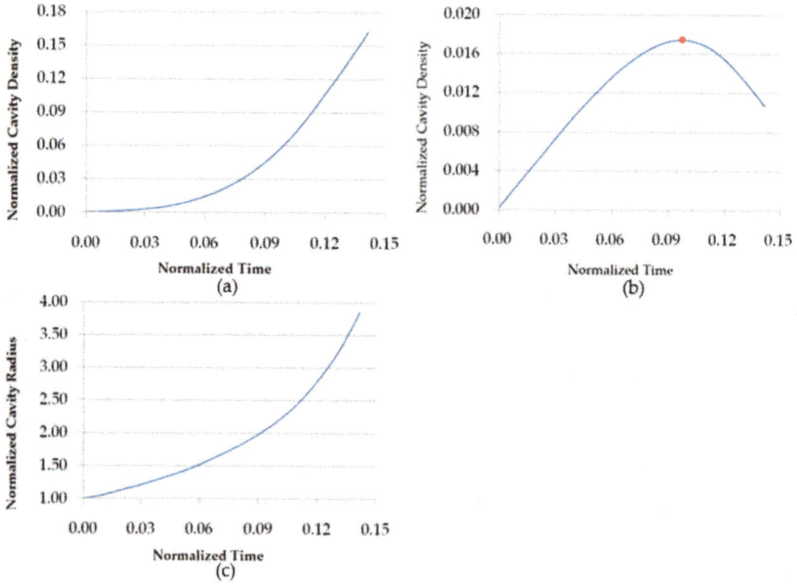

Figure 5. Cavitation evolution of these two important indicators. (**a**) The change of the damage variable β versus the normalized time. (**b**) The change of the cavity density (normalized) versus the normalized time. (**c**) The change of the average radius a (normalized) versus the normalized time.

The rupture time is 0.142 (normalized), the maximum value of the β is 0.162, the final value of ρ is 0.011. In Figure 5a, it shows the evolution of the cavity density on the GB, it also reflects the variation of the cavity on the GB. The process can be divided into two stages: from the beginning, the quantity of the cavity grows, and when it reached the upper limit 0.017 at the time point 0.097 (normalized), cavity density decrease. The main reason for the reduction of the cavity density to the end is that random nucleation occurs on the GB throughout the process. However, cavity coalescence happens when the distance between two cavities is less than the critical value (0.1 times the initial cavity radio, here is 1 μm), when the quantity of the cavities reaches a maximum, the nucleation rate is less than the coalescence rate, eventually resulting in the decrease of the cavity density. Although the cavity density is reduced, the average radius of the cavity is increased continuously (shown in Figure 5c). Finally, the total volume of the cavities is increased, and the macroscopic phenomenon is that the creep jump displacement of the GB is increased.

3.1.5. Error Analysis

The mathematical simulation result of the constitutive equations (mentioned in Section 2.4.2) have been given in the article [25], although the results obtained does not involve finite element calculations, it can still be used in the benchmark of the bi-grain case study, since it is a special uniaxial loading condition. Since the exact value of the simulation is not given, it is unavoidable that there is a corresponding reading error. The percentage error between reading values and the simulation result is shown in Table 3. Based on the bi-grain case study, the result obtained applying the in-house procedure is shown in good agreement with the reference result. Based on this result, it demonstrates the numerical stability and accuracy of the procedure at the uniform stress condition. It also proves the accuracy of the subroutines for the numerical integration of the cavitation evolution model, and the convergence of the non-linear iteration is preliminarily verified under the non-sliding condition. The next bi-grain case will be set to verify the numerical accuracy of the sliding part.

Table 3. Percentage error of these five important indicators.

Name	Reading Value [25]	Simulation Value	Percentage Error
Rupture Time (normalized)	0.15	0.142	5%
Maximum value of the β	0.165	0.162	1.53%
Final value of ρ (normalized)	0.001	0.011	5.95%
The value of the ρ at the change point (normalized)	0.017	0.017	2.11%
The time point of the change point (normalized)	0.095	0.097	1.8%

3.2. Validation of the Sliding Part

The FE model consists of two triangle grains and a single GB as depicted in Figure 6. The rotation angle of the joint element to the X axial is 135°. The model is meshed by two triangle plane strain elements (for the grain part: each grain mesh by 1triangle element) and one Goodman element (for the GB part). In order to be logical and efficient, the normal stress of the GB is set to be consistent with the previous case (20 MPa). According to the geometric relationship, a uniformly distributed loading 40 MPa is applied on the top surface in the Y direction and left line of the bottom grain element fixed on the X direction and bottom line of the grain element fixed on the Y direction. The main purpose of this case is to validate the accuracy of sliding deformation implementation and to benchmark the numerical accuracy of the Goodman element stiffness matrix in general condition (with angle). The geometric information of this FE model: the normalized length of the two short sides of the triangular grain is 1.

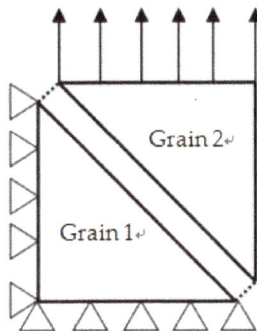

Figure 6. Schematics showing the second bi-grain case study to validate the sliding part.

3.2.1. Validation of the Elastic Stress

The theoretical stress of the grain elements at the Gauss point is 40 MPa at Y direction and the stress in the X direction and the shear stress should be zero. The theoretical stress of the GB is 20 MPa at normal direction and the stress in the separate direction should be −20 MPa. The elastic stresses for each element are shown in Table 4.

Table 4. Elastic stress for each element.

Material Zone	Element No.	Direction	Elastic Stress (Unit: MPa)	Theoretical Value (Unit: MPa)
Grain	1	X	-3.553×10^{-15}	0
		Y	40.000	40
		τ	0	0
	2	X	0	0
		Y	40.000	40
		τ	2.085×10^{-15}	0
GB	3	Separate	−20.000	−20
		Normal	20.000	20

According to the Table 4, the simulation result from the procedure has been shown in good agreement with the theoretical value.

3.2.2. Validation of the Non-Linear Creep Iteration

As mentioned in the Section 3.1.2, according to Table 5, the in-house procedure has shown good convergence in the creep non-linear solution.

Table 5. The element elastic stress field at the selected iteration step.

Iteration Step		Element No.	Direction	Elastic Stress (Unit: MPa)	Theoretical Value (Unit: MPa)
100,000	Grain 1	1	X	-9.623×10^{-9}	0
			Y	40.000	40
			τ	0	0
	Grain 2	2	X	5.359×10^{-10}	0
			Y	40.000	40
			τ	-9.812×10^{-10}	0
	GB 1	3	X	−20.000	−20
			Y	20.000	20
231,704	Grain 1	1	X	4.134×10^{-7}	0
			Y	40.000	40
			τ	0	0
	Grain 2	3	X	1.317×10^{-8}	0
			Y	40.000	40
			τ	-1.349×10^{-8}	0
	GB 1	5	X	−20.000	−20
			Y	20.000	20

3.2.3. Validation of the Sliding Part

The sliding model has been mentioned in Section 2.4.2. The magnitude $\eta_{sliding} = 3.85 \times 10^7 \text{Ns/mm}^3$ (normalized value is : 0.052) [10] of the sliding viscosity coefficient is chosen in this case. In this case, the elastic stress field keeps constant, and the sliding is a kind of linear model and the rupture time is 0.142 (normalized). Therefore, the absolute theoretical displacement of the sliding part is 2.721 (normalized). The simulation result value is 2.722. Therefore, the in-house procedure has shown good agreement with the theoretical value. Based on this result, it demonstrates the numerical stability and accuracy of the procedure at the bi-axial stress condition (normal and separate). It also proves the accuracy of the subroutines for the numerical integration of the sliding model. The convergence of the non-linear iteration with sliding deformation is preliminarily verified. The next step will be set to do a polycrystalline case study to benchmark that this procedure contains a potentially new application under the actual microstructure which considers the cavitation model and sliding model together.

4. Polycrystal Case Study

The plan strain simulation is conducted for copper–antimony alloy subjected to 10 MPa tensile stress at temperature is 823 K.

The geometry of this case is rectangle with 1 mm². The mesh for this model was generated by the Neper package [22]. Shape of the domain structure was built by the tessellation module (−T) of Neper and in a rectangular domain. The model contains 20 grains and 60 GBs. The mesh of the structure and re-mesh to generate the GB is done using the meshing module (−M) of Neper. The grains are meshed by 909 triangle plane strain elements and the GBs are meshed by 152 Goodman elements, as depicted in Figure 7 (GB is marked by the red line). The orientation of grain boundary elements is shown in Figure 8, which reveal the degree of reasonably random distribution.

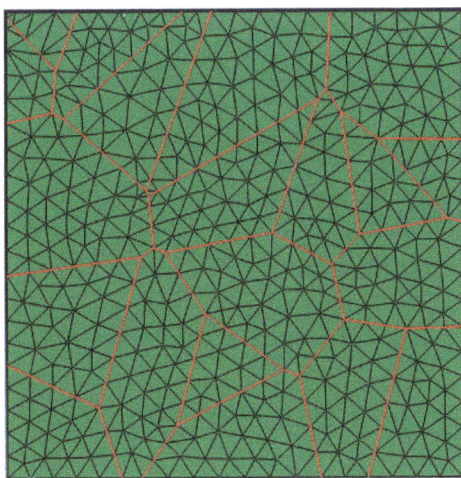

Figure 7. The mesh developed for the polycrystalline case study.

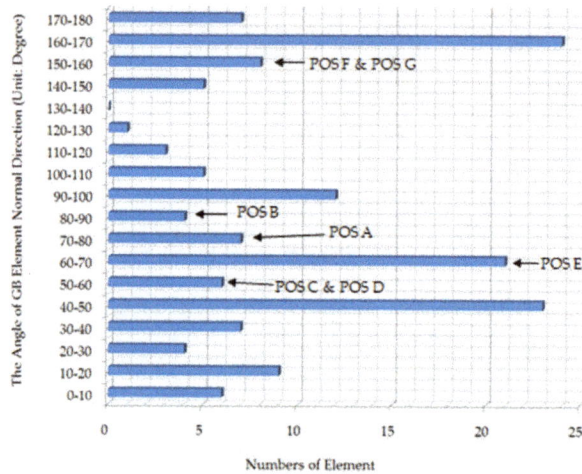

Figure 8. Orientation angle of the grain boundary element's normal direction.

In an attempt to conduct mesh convergence study, a series of much finer mesh has been produced, and it was found the computational time is too much and it is not realistic under current limitations of computational hardware to be vigorous. It is, unfortunately, to compromise to accept this size to proceed.

A uniformly distributed loading of 10 MPa was applied on the top surface in the Y direction and left line and bottom line of the domain were fixed on the X direction and the Y direction, respectively.

In this case, it simulates the creep evolution of copper–antimony alloy at 823 K [11]. The parameters for the GB cavity model is: $\bar{D}_{gb} = 10^{-14} \mathrm{mm^5 N^{-1} s^{-1}}$, $a_p = 2 \times 10^2 \mathrm{mm^{-2} s^{-1}}$ and $b_p = 1$. The parameter for the copper power law creep has been mentioned in Section 2.4.3.

4.1. Rupture Time and Creep Damage Evolution

At the time of 78.9 h, there were seven grain boundary elements that failed. If that is deemed as creep rupture time, then it agrees with the majority of all uni-axial creep tests conducted [10]. In uniaxial test, the experimental condition is: copper–antimony alloy for material, loading is 10 MPa, and temperature is 823 K; one specimen fractured at 16.6 h, one broke at 17.9 h, one broke at 58.3 h. It is worth mentioning that the simulation was conducted for plane strain case, hence it is expected a longer lifetime at the same applied stress. The sequence of fracture and its time are shown in Figure 9 and Table 6.

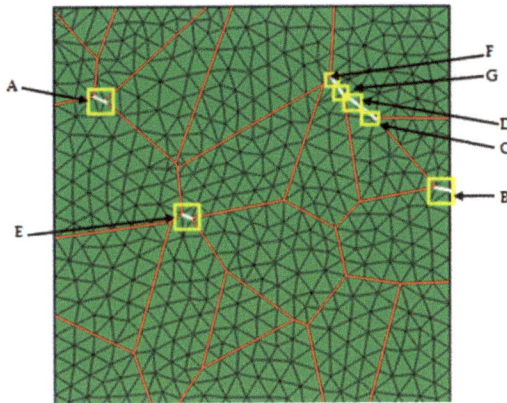

Figure 9. Sequence evolution of the failure position.

Table 6. Sequence and time of fracture of the first seven boundary elements.

Position	Orientation Angle (Normal Direction)	Element No.	Time (Unit: h)	Step
A	65.26084	48	23.55	12003387
B	76.16616	122	65.55	33246192
C	54.01357	93	68.48	34728834
D	54.01357	94	68.48	34728856
E	65.41204	111	70.69	35848560
F	146.3127	87	78.90	39987506
G	146.3128	88	78.90	39987517

The evolution of normal stress, cavitation rate, cavity density, and annihilation rate is shown in Figures 10–14, respectively.

From Figure 11, we can observe that the grain boundary A has the highest normal stress and stays relatively high until its fracture. The normal stress level in other failed elements is much lower than that is but still higher than the 10 MPa for most of the early part of the time. Hence, these elements could be subject to a higher nucleation rate and growth rate, finally resulting in failure. The initial higher normal stress clearly reveals the importance of grain boundary sliding and initial normal jumping in stress redistribution; and further creep deformation that redistributes the stress causing the decrease of high normal stress, but can still maintain itself for a longer period. Consequently, the damage developed in such a manner as confirmed by Figure 15.

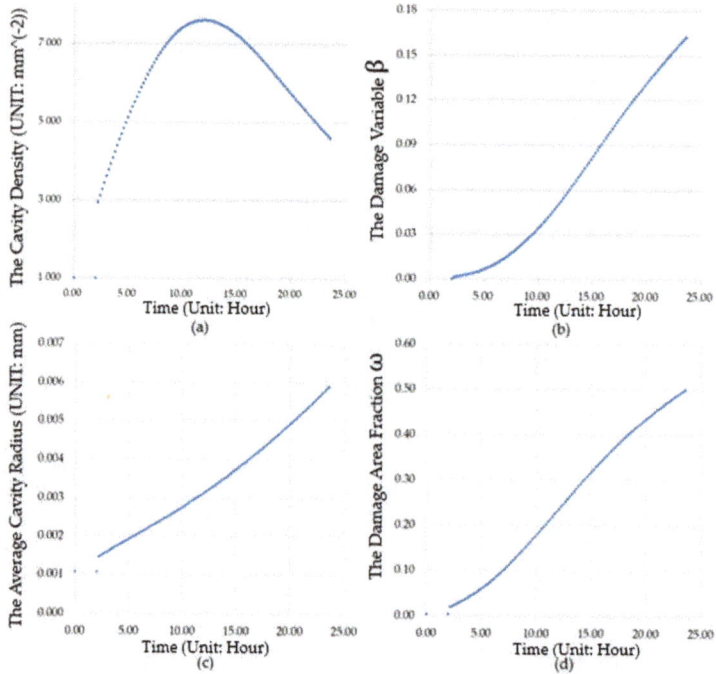

Figure 10. Evolution of the Position A. (**a**) The evolution of the cavity density versus the time. (**b**) The evolution of the average radius a versus the time. (**c**) The evolution of the average cavity radius a versus the time. (**d**) The damage area fraction versus the time.

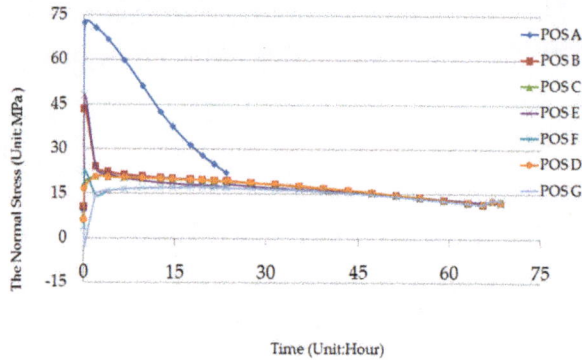

Figure 11. Evolution of the stress at normal direction.

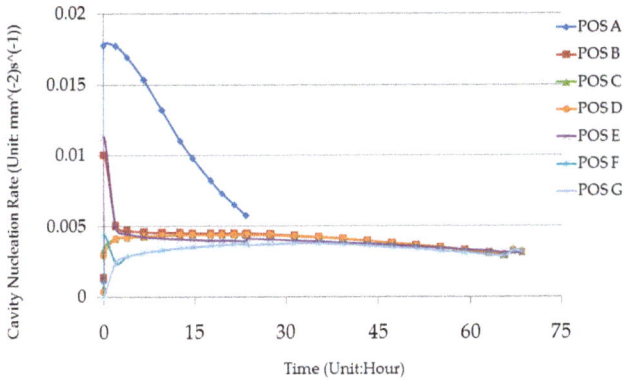

Figure 12. Evolution of the cavity nucleation ratio.

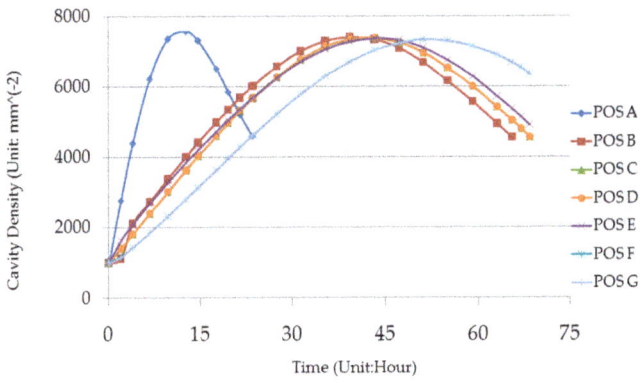

Figure 13. Evolution of the cavity density (POS B).

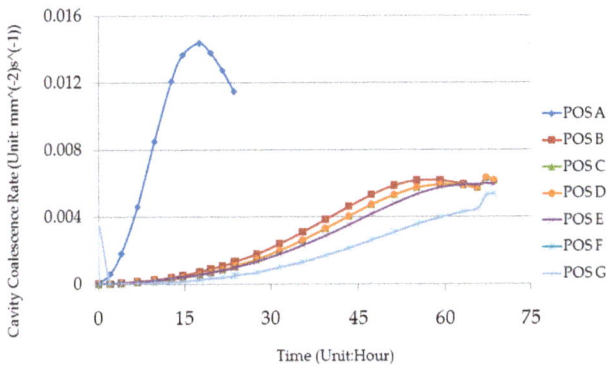

Figure 14. Evolution of the coalescence rate.

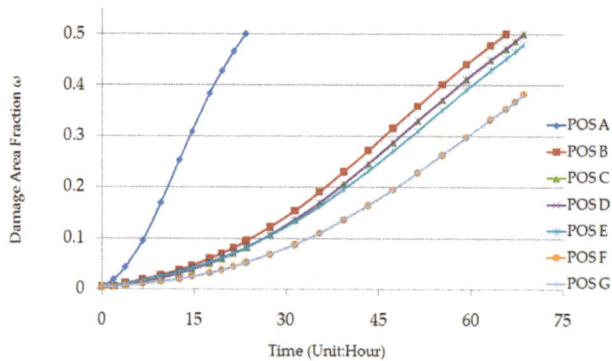

Figure 15. Damage evolution with time of the first seven failed grain boundary elements.

4.2. Stress Field Evolution

The normal stress evolution of all the grain boundary elements are shown in Figure 16a–f; and the creep damage are shown in Figure 17a–d, corresponding to the same time.

From the Figure 16c, it reveals that the damage is dominantly occurring at some slant degree to the direction of the axial stress, while the elements aligned with the direction of the stress were not damaged that much, see the range of elements in numbers 50–80. This may reveal that the normal node jump is relatively bigger than the sliding, so in this statistically under-terminated system, the normal stress has been released in those elements. This statistical trend clearly reveals the importance of grain boundary sliding and its effect on stress redistribution.

From the evolution of creep damage over time for all the elements, it also revealed another two facts: over the time, a reasonable portion of elements developed creep damage steadily and the component is about to rupture; hence it can be derived that the 152 grain boundary elements (its size and orientation distribution) did present the grain boundary fairly, though a study with finer mesh size will resolve this firmly.

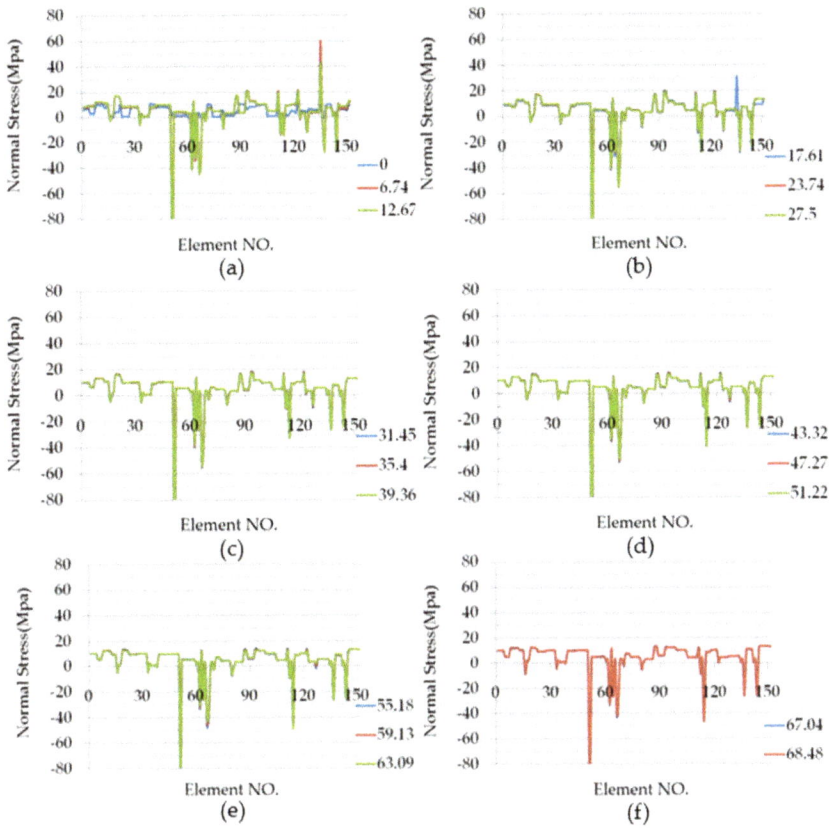

Figure 16. Evolution of normal stress of all grain boundary elements at the same time. (**a**) Time: 0 h, 6.74 h, 12.67 h. (**b**) Time: 17.61 h, 23.74 h, 27.5 h. (**c**) Time: 31.45 h, 35.4 h, 39.36 h. (**d**) Time: 43.32 h, 47.27 h, 51.22 h. (**e**) Time: 55.18 h, 59.13 h, 63.09 h. (**f**) Time: 67.04 h, 68.48 h.

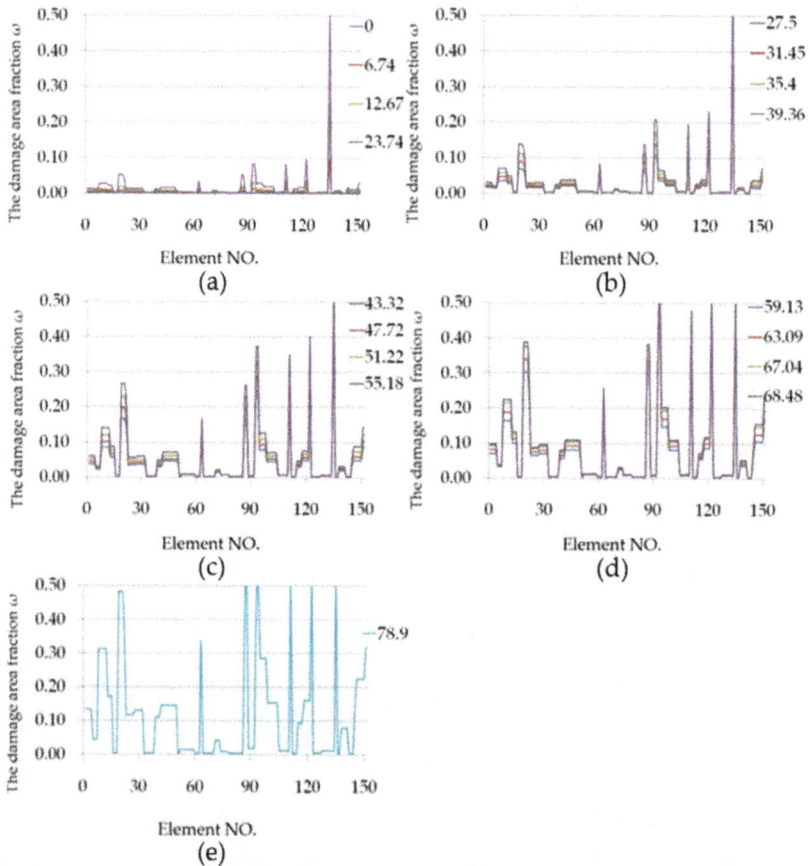

Figure 17. Damage evolution with time of the all the grain boundary elements at the same time. (a) Time: 0 h, 6.74 h, 12.67 h, 23.74 h. (b) Time: 27.5 h, 31.45 h, 35.4 h, 39.36 h. (c) Time: 43.32 h, 47.27 h, 51.22 h, 55.18 h. (d) Time: 59.13 h, 63.09 h, 67.04 h, 68.48 h. (e) Time: 78.9 h.

5. Computational Cost

The three cases above was run in PC, its CUP is: Intel (R) Core (TM) i5-3337U 1.80 GHz. The computational consumption (CPU time) of these three cases is shown in Table 7.

Table 7. CPU consumption time.

Case	Consumption CPU Time (Unit: S)
Bi-grains (without sliding)	6.36
Bi-grains (with sliding)	2.59
Polycrystal (20 Grains) (first failure)	598672 (6.9 days)
Polycrystal (20 Grains) (first seven failures)	1995000 (23.1 days)

Though the computer hardware is not high spec, it is still reasonable to claim that the computing time and then the cost is not trivial if a full three-dimensional model is running. This justifies the development and use of a two-dimensional model.

6. Conclusions and Future work

6.1. Conclusions

The development of and the preliminary application of two-dimensional finite element framework for creep damage simulation at grain boundary level was reported. It is concluded that:

1. The computational platform has been developed easily by adopting existing standard subroutines and/or algorithm. The computational time and cost are significant and higher spec computational hardware is desirable, if not necessary.
2. A simple plane strain simulating case revealed the stress concentration and its reduction.
3. The lifetime prediction still falls well within the experimental results. This is because the smeared-out grain boundary element is micro-mechanical based.
4. It also confirms that the modelling of grain is of second order of importance hence, some simplification is acceptable and justified.

6.2. Future Work

1. Conduct a parametric study to provide insight on the relative importance of various parameters throughout a lifetime;
2. Develop a three-dimensional version;
3. Develop and/or validate the micro-mechanical model using synchrotron micro-tomography cavity data.

Author Contributions: Q.X. conceived the idea to simulate the grain and grain boundary separately and preferred the in-house software development approach; Q.X. also participated in drafting, finalizing and submitting the paper; rewrote the abstract, introduction, discussion and future work for the revised version; advised what extra data/graphs (primarily Sections 4 and 5) to be produce in revised version and wrote the discussion; Q.X. finalized the paper and made submissions for all the three versions; J.T. proceeded with the theory, programming, and testing. J.T. also drafted the paper and participated in revising prior to its first submission, in addition to producing the data/diagrams, and provided some discussion regarding the results for the revised version; Z.L. participated in the initial drafting process.

Funding: This research received no external funding.

Acknowledgments: J.T. is grateful for being given this opportunity at the University of Huddersfield and for being awarded the Vice-chancellor PhD research scholarship. Q.X. is grateful for the Scholarship of Santander Student Mobility fund, The University of Huddersfield, 2017–2018.

Conflicts of Interest: The authors declare no conflict of interest.

References

1. Riedel, H. *Fracture at High Temperatures*; Springer: Berlin, Germany, 1987.
2. Kassnera, M.E.; Hayesb, T.A. Creep cavitation in metals. *Int. J. Plast.* **2003**, *19*, 1715–1748. [CrossRef]
3. Onck, P.; van der Giessen, E. Microstructurally-based modelling of intergranular creep fracture using grain elements. *Mech. Mater.* **1997**, *26*, 109–126. [CrossRef]
4. Xu, Q.; Lu, Z.; Wang, X. Damage modelling: the current state and the latest progress on the development of creep damage constitutive equations for high Cr steels. *Mater. High Temp.* **2017**, *34*, 229–237. [CrossRef]
5. Gupta, C.; Toda, H.; Schlacher, C.; Adachi, Y.; Mayr, P.; Sommitsch, C.; Uesugi, K.; Suzuki, Y.; Takeuchi, A.; Kobayashi, M. Study of creep cavitation behavior in tempered martensitic steel using synchrotron micro-tomography and serial sectioning techniques. *Mater. Sci. Eng. A* **2013**, *564*, 525–538. [CrossRef]
6. Sket, F.; Dzieciol, K.; Borbely, A.; Kaysser-Pyzalla, A.R.; Maile, K.; Scheck, R. Microtomography investigation of damage in E911 steel after long term creep. *Mater. Sci. Eng. A* **2010**, *528*, 103–111. [CrossRef]
7. Renversade, L.; Ruoff, H.; Maile, K.; Sket, F.; Borbély, A. Microtomographic assessment of damage in P91 and E911 after long-term creep. *Int. J. Mater. Res.* **2014**, *105*, 621–627. [CrossRef]

8.   Xu, Q.; Yang, X.; Lu, Z. On the development of creep damage constitutive equations: a modified hyperbolic sine law for minimum creep strain rate and stress and creep fracture criteria based on cavity area fraction along grain boundaries. *Mater. High Temp.* **2017**, *34*, 323–332. [CrossRef]
9.   Xu, Q.; Lu, Z. Modelling of Creep deformation and Creep Damage Rupture Criterion. In Proceedings of the Power Plant Operation & Flexibility, IOM, London, UK, 4–6 July 2018.
10. Vöse, M.; Otto, F.; Fedelich, B.; Eggeler, G. Micromechanical investigations and modeling of a Copper–Antimony-Alloy under creep conditions. *Mech. Mater.* **2014**, *69*, 41–62. [CrossRef]
11. Vöse, M.; Fedelich, B.; Otto, F.; Eggeler, G. Micromechanical Modeling of Creep Damage in a Copper-antimony Alloy. *Procedia Mater. Sci.* **2014**, *3*, 21–26. [CrossRef]
12. Dyson, B.F. Continuous cavity nucleation and creep fracture. *Scripta Metall.* **1983**, *17*, 31–38. [CrossRef]
13. Bower, A.F. *Applied Mechanics of Solids*; Taylor & Francis Group, LLC: Abingdon, UK, 2010.
14. Dyson, B.F. Constraints on diffusional cavity growth rates. *Met. Sci.* **1976**, *10*, 349–353. [CrossRef]
15. Ashby, M.F. Boundary defects and atomistic aspects of boundary sliding and diffusional creep. *Surf. Sci.* **1972**, *31*, 498–542. [CrossRef]
16. Cocks, A.C.F.; Ashby, M.F. On creep fracture by void growth. *Prog. Mater. Sci.* **1982**, *27*, 189–244. [CrossRef]
17. Yu, C.-H.; Huang, C.-W.; Chen, C.-S.; Gao, Y.; Hsueh, C.-H. Effects of grain boundary heterogeneities on creep fracture studied by rate-dependent cohesive model. *Eng. Fract. Mech.* **2012**, *93*, 48–64. [CrossRef]
18. Goodman, R.E.; Taylor, R.L.; Brekke, T.L. A model for the mechanics of jointed rock. *J. Soil Mech. Found. Div* **1968**, *94*, 637–659.
19. Machiels, L.; Deville, M.O. Fortran 90: An entry to object-oriented programming for the solution of partial differential equations. *ACM Trans. Math. Softw.* **1997**, *23*, 32–49. [CrossRef]
20. Smith, I.M.; Griffiths, D.V.; Margetts, L. *Programming the Finite Element Method*; John Wiley & Sons: Hoboken, NJ, USA, 2013.
21. Hayhurst, D.; Dimmer, P.; Morrison, C. Development of Continuum Damage in the Creep Rupture of Notched Bars. *Philos. Trans. R. Soc. Lond. Ser. A* **1984**, *311*, 103–129. [CrossRef]
22. Quey, R.; Dawson, P.R.; Barbe, F. Large scale 3D random polycrystals for the finite element method: Generation, meshing and remeshing. *Comp. Meth. Appl. Mech. Eng.* **2011**, *200*, 1729–1745. [CrossRef]
23. Liu, D.Z.; Xu, Q.; Lu, Z.Y.; Xu, D.L.; Xu, Q.H. The Techniques in Developing Finite Element Software for Creep Damage Analysis. *Adv. Mater. Res.* **2013**, *744*, 199–204. [CrossRef]
24. Шамровский, А.; Богданова, Е. Solution of contact problems of elasticity theory using a discrete finite-size element. *East. Eur. J. Enterp. Technol.* **2014**, *3*, 41–45.
25. Vose, M.; Fedelich, B.; Owen, J. A simplified model for creep induced grain boundary cavitation validatedby multiple cavity growth simulations. *Comput. Mater. Sci.* **2012**, *58*, 201–213. [CrossRef]
26. Li, G.; Thomas, B.G.; Stubbins, J.F. Modeling creep and fatigue of copper-alloys. *Metall. Mater. Trans. A* **2000**, *31*, 2491–2502. [CrossRef]
27. Becker, A.A.; Hyde, T.H.; Sun, W.; Andersson, P. Benchmarks for finite element analysis of creep continuum damage mechanics. *Comput. Mater. Sci.* **2002**, *25*, 34–41. [CrossRef]

Article

An Investigation into Creep Cavity Development in 316H Stainless Steel

Hedieh Jazaeri [1,*], P. John Bouchard [1], Michael T. Hutchings [1], Mike W. Spindler [2], Abdullah A. Mamun [1,3] and Richard K. Heenan [4]

[1] Department of Engineering and Innovation, The Open University, Walton Hall, Milton Keynes MK7 6AA, UK; john.bouchard@open.ac.uk (P.J.B.); michael.hutchings@open.ac.uk (M.T.H.); abdullah.mamun@bristol.ac.uk (A.A.M.)

[2] Assessment Technology Group, EDF Energy Nuclear Generation Ltd., Barnett Way, Barnwood GL4 3RS, UK; mike.spindler@edf-energy.com

[3] Department of Mechanical Engineering, University of Bristol, Bristol BS8 1TR, UK

[4] ISIS Facility, STFC Rutherford Appleton Laboratory, Didcot OX11 0QX, UK; richard.heenan@stfc.ac.uk

* Correspondence: hedieh.jazaeri@open.ac.uk; Tel.: +44-1908-653897

Received: 1 February 2019; Accepted: 6 March 2019; Published: 12 March 2019

Abstract: Creep-induced cavitation is an important failure mechanism in steel components operating at high temperature. Robust techniques are required to observe and quantify creep cavitation. In this paper, the use of two complementary analysis techniques: small-angle neutron scattering (SANS), and quantitative metallography, using scanning electron microscopy (SEM), is reported. The development of creep cavities that is accumulated under uniaxial load has been studied as a function of creep strain and life fraction, by carrying out interrupted tests on two sets of creep test specimens that are prepared from a Type-316H austenitic stainless steel reactor component. In order to examine the effects of pre-strain on creep damage formation, one set of specimens was subjected to a plastic pre-strain of 8%, and the other set had no pre-strain. Each set of specimens was subjected to different loading and temperature conditions, representative of those of current and future power plant operation. Cavities of up to 300 nm in size are quantified by using SANS, and their size distribution, as a function of determined creep strain. Cavitation increases significantly as creep strain increases throughout creep life. These results are confirmed by quantitative metallography analysis.

Keywords: creep damage; cavitation; small angle neutron scattering; scanning electron microscopy; austenitic stainless steel

---

1. Introduction

Creep cavitation [1] is an important failure mechanism in steel components operating at high temperature. The role of cavities in limiting the creep life of materials was first reported by Greenwood [2,3]. Since then much research has been conducted in this area; however, the mechanisms of cavity nucleation are still not completely understood [1], particularly in engineering alloys with complex secondary particles [4]. Although creep cavitation is often associated with the final stage of creep deformation, it actually nucleates at a relatively early stage of creep. The minimum stable nucleation size of creep cavity is not well-established, but theoretical work has indicated it to be in the range of 2–5 nm [5]. Nucleation usually occurs at the grain boundaries that are oriented normal to the principal stress direction, with the number of cavities per unit grain boundary that are approximately proportional to the creep strain [6–8]. However, cavities can also nucleate within the grains, particularly under higher applied stresses. The cavities gradually grow in size during deformation [9], as well as new cavities nucleating continuously during the entire creep life.

Classical cavitation theories associate cavity nucleation with high stress concentration at grain boundary ledges, triple junctions, and particles [10,11]. These models usually invoke the concept of a threshold stress, below which the cavity nucleation rate is negligible for practical purposes and above which the rate of nucleation is so rapid that cavities nucleate at all available sites over a short period of time. However, these models fail to capture experimental observations; such as extensive cavity nucleation during creep deformation under a much lower applied stress than the theoretical threshold stress that is predicted by these models. This discrepancy has led to the development of other cavity nucleation models based upon grain boundary sliding [12], and interactions of dislocation sub-structures with grain boundaries [13]. These models have been supported by experimental observations of creep cavitation, using various microscope-based techniques (such as Optical, Scanning Electron Microscopy (SEM), and Transmission Electron Microscopy (TEM)) [7,14–18]. Whilst these techniques can provide important information regarding the location, extent, and morphology of creep cavities, it is difficult to infer the distribution of sub-micron-sized cavities in a representative volume of the material. The latter is required to elucidate cavity nucleation and growth mechanisms in complex engineering alloys, and to understand the effects of microstructural evolution and prior deformation.

In principle, it is possible to use small-angle neutron scattering (SANS) to measure cavities in a size range of 10 to 1000 nm in a macroscopic volume (several cubic millimeters) of material, depending on the instrument used [19]. Traditional metallographic techniques are destructive, time consuming, two-dimensional, operator-dependent, and are often best suited for studying larger cavities [20]. In contrast, SANS can provide relatively quick, non-destructive volumetric measurement of cavities in cubic millimeter-sized samples of material. Together with complementary quantitative metallography, it has been previously used to measure creep-induced cavities in Type 316H stainless steel samples taken directly from power plant weldments that have exhibited cracking [17,18]. SANS can also provide quantitative volumetric measures of the size and growth evolution of creep cavities that depend upon the applied stress and temperature environment [21].

In this paper, an investigation into creep-induced cavity development in Type 316H austenitic steel under two different conditions of stress and temperature is reported. The conditions chosen are representative of those experienced in current and future power plant operations. One set of creep specimens was plastically pre-strained before creep deformation, in order to investigate plasticity-driven cavity nucleation, while the other set was deformed under creep loading conditions. The plastically pre-strained samples represent the as-manufactured condition of many components entering high-temperature power plant operation, including welded joints [22]. The magnitude and sign of plastic pre-strain can influence the minimum creep rate, strain-to-failure, and the rupture life [23]. SANS experimental work is presented by measuring creep cavitation as a function of creep strain in each of the two sets of creep specimens (where each specimen was interrupted at increasing levels of creep strain). Complementary measurements of cavitation using SEM quantitative metallography are also reported. The broader object of the study is to improve the current state of understanding of the mechanisms of cavity nucleation and growth in Type 316H stainless steel, and to help formulate new physically-based cavitation damage models.

2. The Specimens Examined

Two sets of specimens, creep-deformed under the two different conditions of pre-strain, stress, and temperature were examined in this study. The loadings of each specimen were carried out in accordance with BS EN ISO 204:2009. The specimens were extracted from a 316H stainless steel forged cylinder after long-service in a UK Advanced Gas Cooled Reactor nuclear power plant. This material has a grain size of 88 ± 9 µm, measured by the mean linear intercept method, a room temperature 0.2% proof stress of 285 MPa with the chemical composition presented in Table 1 [24]. This material had been removed from service after approximately 91,000 h exposure to temperatures of up to about 520 °C, following the discovery of reheat cracking near to cylinder to the nozzle weld. Due to the material's long exposure in service, extensive inter- and intragranular precipitation was observed from optical microscopy analysis. The observed precipitates are mainly $M_{23}C_6$ carbides [24]. The distribution of the intragranular precipitates were inhomogeneous among different grains. The observed intragranular precipitates were mostly of circular morphology, while the intergranular precipitates were relatively large, with high aspect ratios. Hong et al. [25] has reported the observation of similar morphology of precipitates in aged 316 stainless steel.

Table 1. Chemical composition (weight %) of the specimens.

C	Si	Mn	P	S	Cr	Ni	Mo	Al	Cu
0.066	0.42	1.00	0.029	0.015	17.82	11.18	2.33	0.003	0.23
Sn	V	W	Co	Pb	B	N	Nb	Ti	Fe
0.016	0.031	0.068	0.093	0.003	0.0051	0.096	0.007	0.004	Bal.

The geometry of the creep specimen used for the creep test is shown in Figure 1. In the first set of tests, a total of six specimens were used. Each specimen was subjected to an ~8% tensile plastic pre-strain at room temperature, followed by uniaxial creep deformation at 550 °C under 320 MPa start-of-test engineering stress. A temperature of 550 °C was selected to correspond with the maximum service temperature that is experienced by 316H stainless steel within UK nuclear power plants. The 0.2% yield stress of this material at 550 °C was measured to be 185 MPa. Wilshire et al. [23] reported that in the absence of recrystallization, room temperature pre-straining of 316H stainless steel modifies the creep property values, but only when the pre-strain treatment exceeds the plastic component of the initial specimen extension on loading at the creep temperature. The application of a 320 MPa stress at 550 °C was found to introduce a ~7.2% instantaneous plastic strain in the material. Therefore, a higher pre-strain of 8% was chosen, such that after pre-straining, the application of the initial stress (320 MPa) at 550 °C for the creep test would introduce a negligible inelastic instantaneous strain. This was to ensure that the creep cavity nucleation mechanism in this set of samples would be influenced by a controlled level of plastic pre-strain. The small instantaneous elastic strain upon the application of load was excluded from the total creep strain reported here.

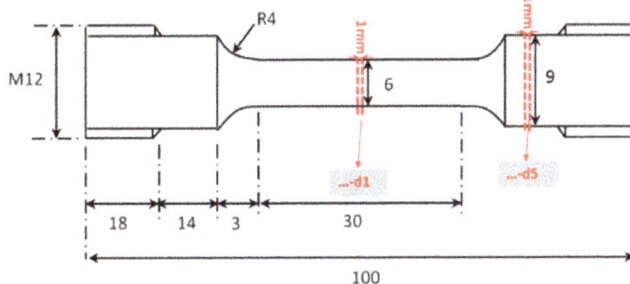

Figure 1. The creep specimen geometry in mm. The position of the disc samples, d1 and d5, removed from each interrupted creep test specimen, and used for the measurements are marked.

In the second set of tests, a total of four specimens were used. Each specimen was subjected to uniaxial creep deformation at 675 °C under 150 MPa start of test engineering stress. The aim was to trigger cavity nucleation under pure creep conditions. The 0.2% yield stress of this material at 675 °C was measured to be 165 MPa. Negligible instantaneous plastic strain was measured upon initial loading to 150 MPa for the creep tests. As for the first set, small instantaneous strains were excluded from the strain data analysis. The test temperature of 675 °C was selected, in order to accelerate the creep tests at an applied stress that is sufficiently low, to ensure that the initial plastic loading strain was negligible, whilst the material substructures and work-hardening behaviors are similar to those at 550 °C [24].

In each set of specimens the loading was interrupted at different stages of creep life, up to rupture, covering primary, secondary, and tertiary regimes [24]. The creep strain measured at each stage of the two loading conditions is shown in Figure 2, and summarized in Table 2.

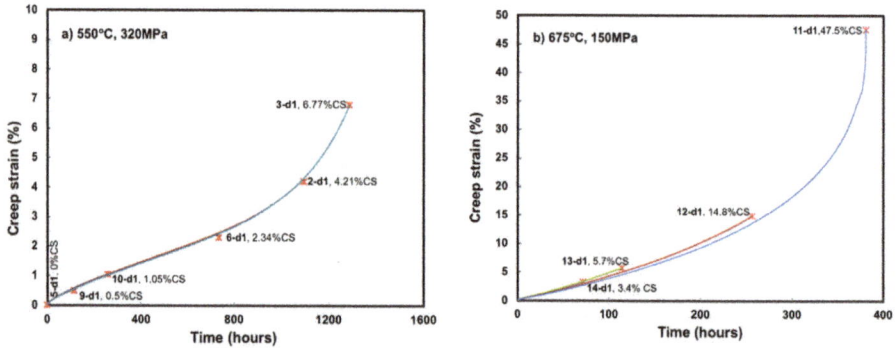

Figure 2. Creep curves for creep test conditions of: (**a**) 550 °C under 320 MPa (after 8% pre-strain); and (**b**) 675 °C under 150 MPa. The creep interruption points have been labelled with the sample ID and % of creep strain (CS) on each curve.

Table 2. The interrupted creep test specimen ID, and strain experienced by each, under the two creep test conditions. The specimen ID is that used for creep test specimens; the sample ID is that used for the discs cut from the creep test specimen.

Creep Test Condition	Specimen ID	Sample ID	Creep Strain, CS (%)
At 550 °C, 320 MPa (after 8% pre-strain)	HRA1C-5	5-d1	0
	HRA1C-9	9-d1	0.54
	HRA1C-10	10-d1	1.05
	HRA1C-4	4-d1	1.4
	HRA1C-6	6-d1	2.34
	HRA1C-2	2-d1	4.21
	HRA1C-3	3-d1	6.77 (Ruptured, t_r = 1287 h)
At 675 °C, 150 MPa	HRA1C-14	14-d1	3.3
	HRA1C-13	13-d1	5.7
	HRA1C-12	12-d1	14.8
	HRA1C-11	11-d1	47.5 (Ruptured, t_r = 381 h)

3. Analysis Techniques

Two analysis techniques were used in this study, SANS and quantitative metallography (QM) using a Zeiss Supra 55VP Field Emission Gun (FEG) scanning electron microscope (ZEISS, Germany), the details of which are described below. Two 1 mm-thick disc samples were extracted from each interrupted test specimen at the positions shown in Figure 1. Samples of 6 mm diameter, marked as -d1, were extracted from the mid-length positions of each creep specimen, and of 9 mm diameter, marked as -d5, extracted from near the end of each creep specimen, outside but near to the grip. Separate samples were prepared for each analysis technique. The samples from the two sets of creep specimens are summarized in Table 2. These have pre-numbers; for example, 5-d1, associated with creep specimens, which were originally designated by HRA1C, followed by the number. So, for example the samples, 5-d1 and 5-d5 were taken from specimen HRA1C-5.

3.1. Small-Angle Neutron Scattering

The SANS measurements were carried out on the SANS2D small-angle scattering instrument at the ISIS Pulsed Neutron Source at the Rutherford Appleton Laboratory, Didcot, UK [26]. A scattering vector, Q, ranging from 0.0015 to 0.11 Å^{-1} (0.015 to 1.1 nm^{-1}) was achieved, utilizing an incident wavelength range of 4.2 to 12.5 Å (0.42 to 1.25 nm), and an instrument set up of incident and scattered flight paths of 12 m. Two position sensitive detectors were used; the furthest, at 12 m from the sample, was of 1 m^2 area, and its center was offset vertically by 150 mm and horizontally by 200 mm. With this set up, scattering from inhomogeneities in a matrix, such as carbides and cavities, of size up to about 3000 Å (300 nm) can be measured. The upper limit on size is governed by the minimum value of the scattering vector, Q, available on the instrument, which gives a cut-off on the size distribution measured. Each disc sample examined was carefully positioned at the center of the incident beam, and the scattering measured from a circular gauge area of 4 mm diameter, corresponding to a volume of ~12.5 mm^3, at its center. The acquisition time for measurement on each disc was about 60 min. The neutron transmission of each disc was also measured in order to enable absorption corrections to be made. Each raw scattering data set was corrected for the detector efficiencies, sample transmission, and empty beam scattering to give the cross-section using the instrument-specific software [27]. The macroscopic cross section, in cm^{-1}, from each disc as a function of scattering vector was placed on an absolute scale using the scattering from a standard sample, a solid blend of hydrogenous and perdeuterated polystyrene, in accordance with established procedures [28].

The general analysis of SANS data is discussed by Hutchings and Windsor [29]. SANS measures an inverse transform of the defect size distribution in real space. Ideally, one might carry out this inverse transformation to obtain a size distribution, but this was simply not possible without ideal counting statistics and a complete range of all of the scattering vectors Q. In this study, in order to interpret the measured cross-section over the fully measured range of Q for each disc, the computer routine MAXE was used, which uses the maximum entropy algorithm. The advantage of this approach is that it obviates the need to assume any prior model of the size distribution of inhomogeneous scatterers, and it covers the full range of measured wave vector, in contrast to the Guinier or Porod approaches, which cover a limited range of Q. The technique is a means of carrying out the inverse transform of the SANS intensity with good statistical credibility. The MAXE routine was originally developed at Southampton University and Harwell [30], and written for a mainframe computer in FORTRAN code. It has been modified to work on a PC, and recently reprogrammed in C^{++} at the Open University with more convenient inputs and outputs. The version used, MAXE V4 (2015), was written by Mike J.H. Fox for the Mathematics, Computing and Technology Department at the Open University. A detailed description of the use of the program, which is unchanged in principle from the original, has been reported elsewhere [31]. In the analysis, a simplified model of a size distribution of spherical defects of diameter D is assumed. This is the best model for giving an approximation to a sample of a steel, with a complex distribution of precipitates and cavities of different shapes and orientations. It should be noted that the matrix and defects have no magnetic scattering.

The most probable fractional volume distribution $C(D)$ for defects, where $C(D)\delta D$ gives the volume fraction of defects with diameters in the range D to $D + \delta D$, is determined using the maximum entropy algorithm. This algorithm is known to give the best result for determining $C(D)$, making no prior assumptions regarding its form. In order to obtain an absolute distribution, $C(D)$, the scattering contrast factor for each defect: carbide; cavity etc., must be determined. This requires the nature of each defect to be identified which often entails the use of complementary techniques. In the present case, electron microscopy was essential to determine their nature.

The absolute macroscopic scattering cross-section, and the resulting size distribution of defects such as carbides, in the material taken from the same reactor component from which the creep test specimens reported here were fabricated, has been previously measured. It was measured on a section of material from positions where no cavitation was expected. The cross-section and size distribution from such a 'far-field' region, has been reported by Jazaeri et al. [32]. It was used as the cavity-free reference for the present analysis. Ideally, the scattering cross sections for such a 'far-field' reference position should be subtracted from that for each measured disc, in order to determine the scattering from creep-induced cavities. However, it was found that the resulting statistical uncertainty from the counting times used gave relatively large uncertainties in the resulting size distributions from the MAXE analysis method. The MAXE analysis method was therefore used to give a defect distribution, which we term the 'relative' size distribution, $V(D)$, from each set of scattering data separately, using a contrast factor of unity. A typical fit to the macroscopic cross section over the full Q range measured from sample 3-d1 is shown in Figure 3. The fit is good over most of the cross-sectional range of four orders of magnitude, but it falls just within the larger uncertainties at the highest Q. The resulting 'relative' defect distribution from the 'far-field' reference position was then subtracted from that for each disc, to isolate the size distribution from any cavities which might have developed. The absolute defect size distribution $C(D)$ is then determined by dividing by the contrast factor for the assumed nature of the defects in the present case for the cavities. The contrast factor involves the difference in scattering length density between the matrix and the inhomogeneity. For cavities, it is particularly strong as the open cavities have zero scattering length density. This approach was used to determine the size distribution of the cavities in each disc sample, arising from the stage of creep deformation listed in Table 2. It should be noted that it assumes, reasonably, that the 'far field' reference scattering is appropriate to all of the disc samples.

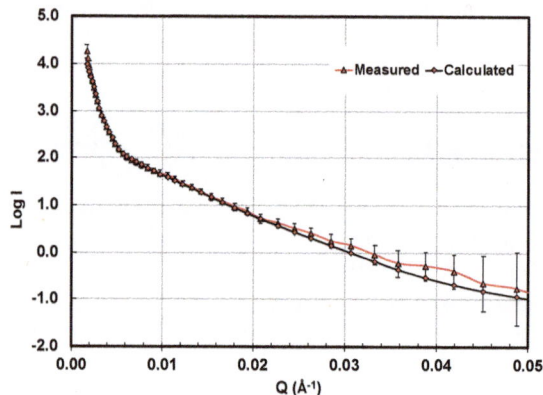

Figure 3. Variation of $\log_{10}$ of the measured absolute macroscopic cross-section, I, in cm^{-1}, with wave vector Q in inverse angstroms (triangles) for sample 3-d1. The calculated curve (diamonds) is the result of the MAXE fit to these data. $1 \text{ Å}^{-1} = 10 \text{ nm}^{-1}$.

3.2. Quantitative Metallography

Discs taken from each creep test sample were mounted in conductive Bakelite, in order to prepare them for examination by quantitative metallography. The preparation procedure included grinding with SiC papers (down to 4000 grit size) and polishing them down to 0.25 μm level with diamond suspension. The final preparatory stage involved an etching procedure, where samples were immersed in Murakami's reagent (10g K_3Fe $(CN)_6$, 10g KOH, 100 mL water) for 60 s. Murakami's reagent was found to be the optimum solution for the sample preparation of ex-service 316H austenitic stainless steel material, as it highlights the grain boundary carbides without having a significant impact on the grain boundaries themselves [17]. A Zeiss Supra 55VP FEGSEM instrument was used to examine the samples in both backscattered (BS) and secondary (SE) imaging mode, using an accelerating voltage of 5–10 kV, and an aperture size of 30 μm. Further details of the measurement technique are given in Section 4.2 below.

4. Results

4.1. Small-Angle Neutron Scattering

The best estimate, given by the MAXE analysis, of the fractional size distribution of cavities, $C(D)$, measured in the 6 mm diameter discs taken from the mid-length of the creep test specimens, under both sets of creep test conditions of stress and temperature, are presented in Figure 4. For both cases, it was seen that there are essentially two cavity size distributions. The first, with cavities of up to about 100 nm in diameter is more sharply peaked, and the other broader distribution spans the size range 100–300 nm. It should be noted that the minimum scattering vector available on SANS2D cuts off the distribution above 300 nm, so that there is no information on cavities that are larger than 300 nm. There is a clear systematic increase in the fractional size distribution of cavities, with an increase in creep strain for the population of smaller cavities of diameters of less than 100 nm, whereas the variation in the size distribution for the larger cavities is less pronounced, especially for the samples that are subjected to a temperature of 550 °C and a stress of 320 MPa.

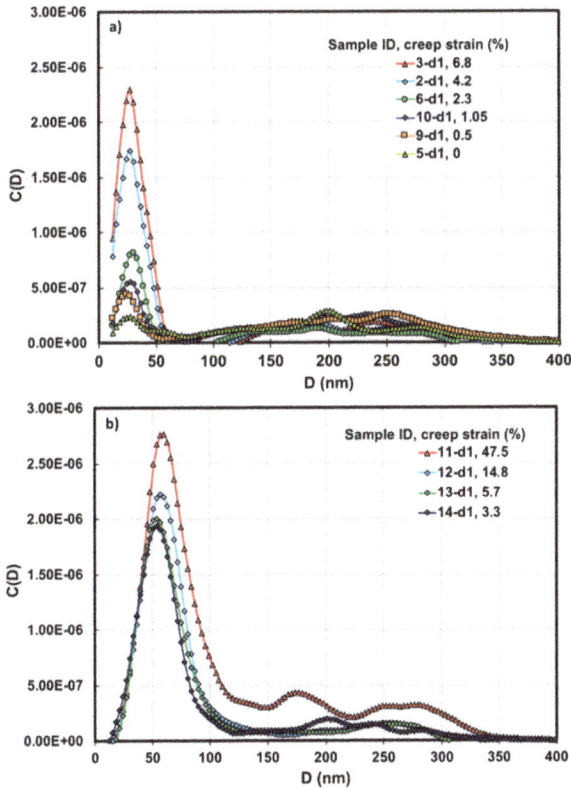

Figure 4. Small-angle neutron scattering (SANS) results, showing the increase in fractional size distribution, $C(D)$, of cavities through creep life at (**a**) 550 °C under 320 MPa (after 8% pre-strain); and at (**b**) 675 °C under 150 MPa.

In order to best represent the size distribution $C(D)$ for the smaller cavities, this has been fitted by a Gaussian function, as shown in Figure 5. The fitted function gives, for each strain level, a mean diameter of the smaller cavities, the peak value of $C(D)$, and the value of $C(D)$ integrated over the peak. An estimate of the corresponding values for the broader distribution between diameters of ~100 to

300 nm was calculated by a method of moments and direct summation. In fitting to the lower peak for the distributions from all samples, the fitted size range was taken up to 75 nm. For the broader peaks calculations, the ranges for the calculations for the samples tested at 550 °C and 320 MPa were taken to range from 75 to 300 nm, and from 110 to 300 nm for the samples tested at 675 °C and 150 MPa.

The mean diameter of these two peaks in the distribution of cavities, as a function of creep strain, is shown in Figure 6. Here the abscissa is the creep strain at each stage, expressed as a fraction of that measured closest to failure ($\varepsilon/\varepsilon_f$).

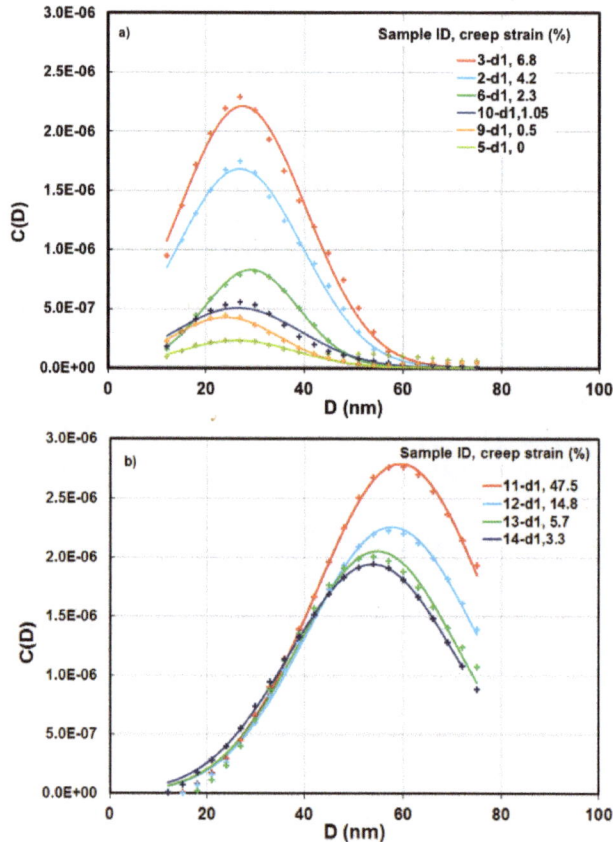

Figure 5. Gaussian functions (lines) that give the best fits to the lower peak in the distribution $C(D)$ determined from MAXE (points). Distribution from samples tested at (**a**) 550 °C with an applied stress of 320 MPa (after 8% pre-strain); and (**b**) 675 °C with an applied stress of 150 MPa.

The number density $N_d(D)$ of defects with diameters in the range D to D + δD, can be calculated from the relation:

$$N_d(D) = C(D)/V_{sph}(D),$$

where $V_{sph}(D)$ (= $\pi D^3/6$) is the volume, and $N_d(D)$ is the number density of a spherical defect of diameter D. In order to estimate the number density of cavities in the diameter range of each peak in the size distribution, the integral over $C(D)$ in each peak has been determined, and the mean diameter of the corresponding peak, shown in Figure 6, used in the above equation. This is clearly an approximation, but the broader peak in the distribution gives a comparable result to a summation over the individual $N_d(D)$ calculated at each diameter. The results are presented as a function of $\varepsilon/\varepsilon_f$

in Figure 7. It is seen that, generally, the population of smaller cavities has a higher number density compared with the population of larger cavities. Also, it is seen that the number density of the smaller cavities, less than 75 nm in size, increases significantly with life fraction, for the samples which have undergone creep at 550 °C and 320 MPa. In contrast, the number density of cavities of less than 110 nm diameter remains relatively constant over the life-to-rupture, for the samples tested at 675 °C and 150 MPa, whereas the larger-sized cavity population (110 to 300 nm) increases slightly in size with creep.

It is interesting to compare the data from the measurements on the discs of different diameters, d1 and d5, extracted from different parts of each creep test specimen. These specimens were subjected to different stress levels that were inversely proportional to their area at constant temperature. That is: 320 MPa compared with 142 MPa at 550 °C, and 150 MPa compared with 67 MPa at 675 °C. Figure 8 shows the 'relative' size distribution, $V(D)$, for the two samples from HRA1C-3, that is 3-d1 and 3-d5, and that for the far-field position. It is clear that there are far fewer defects of size less than 50 nm in size in disc 3-d5, compared with disc 3-d1, and as expected there are far fewer defects in the far field region across most of the size range measured. The difference in distribution of the smaller defects between samples d1 and d5 can be attributed unambiguously to cavity formation arising from the difference in applied stress during creep.

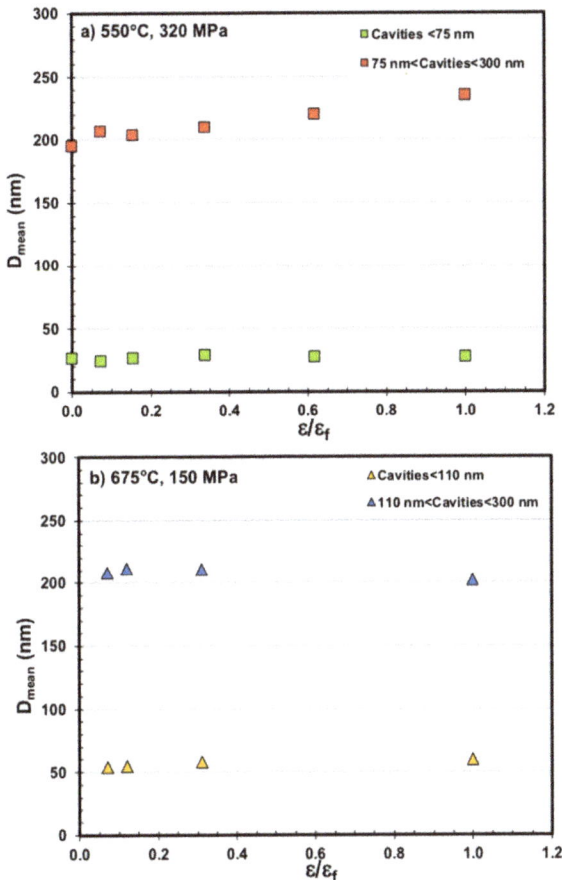

Figure 6. The mean cavity diameter, D_{mean}, of two populations of cavities as a function of $\varepsilon/\varepsilon_f$, measured by SANS: (**a**) 550 °C under 320 MPa (after 8% pre-strain); and (**b**) 675 °C under 150 MPa.

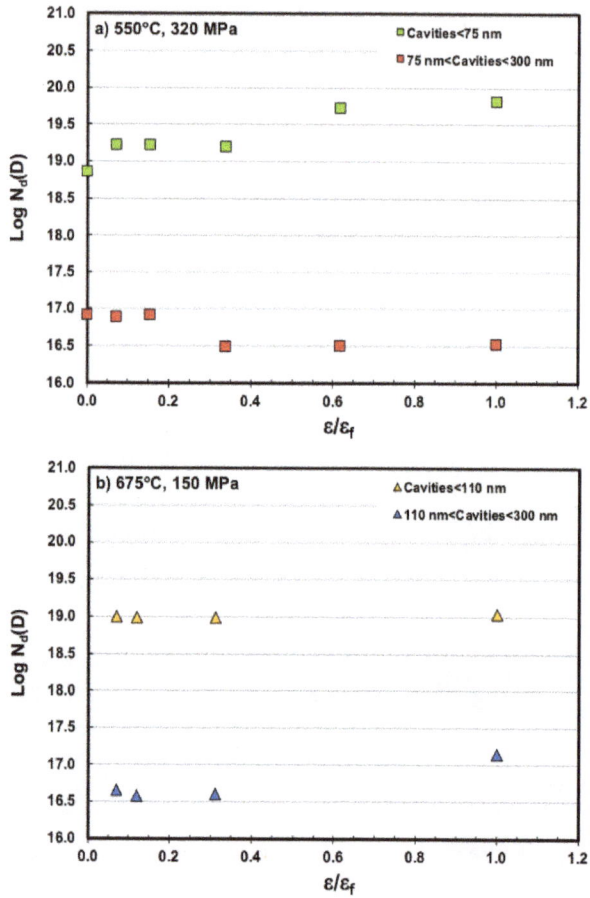

Figure 7. The number density of two populations of cavities as a function of $\varepsilon/\varepsilon_f$, measured by SANS. (a) 550 °C with an applied stress of 320 MPa (after 8% pre-strain); (b) 675 °C with an applied stress of 150 MPa.

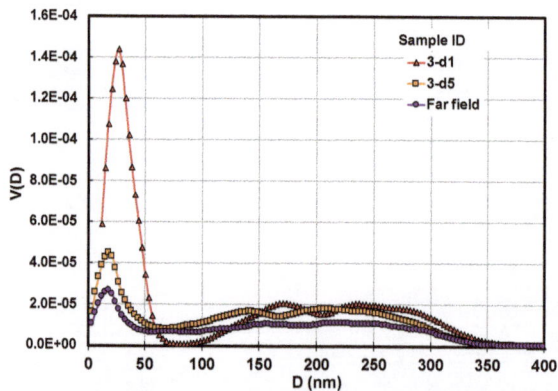

Figure 8. The "relative" size distribution, $V(D)$, of defects in sample 3-d1, from the mid position, 3-d5, from the end position of the specimen and the 'far field' reference is also shown.

By the subtraction of $V(D)$ for the d5 samples from that of the d1 samples, and using the contrast factor for cavities, the change in size distribution of cavities, $C(D)$, as a result of the difference in applied stress alone, can be determined. These are shown for the two sets of creep conditions in Figure 9.

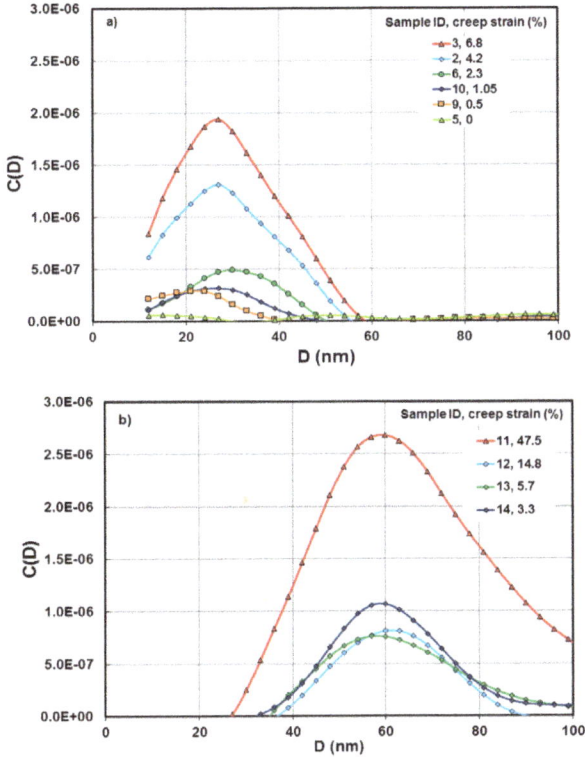

Figure 9. The increase in the fractional size distribution, $C(D)$, of cavities through creep life, determined from SANS, from samples d1 and d5. (**a**) Due to an applied stress difference of 178 MPa at 550 °C; and (**b**) an applied stress difference of 83 MPa at 675 °C.

4.2. Quantitative Metallography

A typical cavitated microstructure, for specimen HRA1C under the creep condition of 550 °C and an applied stress of 320 MPa, and at creep strain of 6.77%, is shown in Figure 10. Cavities (A) are mainly surrounding intergranular precipitates (B), and intragranular precipitates (C) are seen as dark spots. A recent study on 316H austenitic stainless steel has shown that both the intergranular and intragranular precipitates are mainly $M_{23}C_6$ carbides; however, the intragranular precipitates are associated with a long service history, and they form at later stages of creep [15]. Intergranular precipitates do not explicitly lead to cavitation at the grain boundaries [33]. It has been shown that the cavities nucleate at intergranular $M_{23}C_6$ carbides in high residual stress regions [34].

Image analysis was carried, out in order to quantify the variation in the size and area fraction of the creep cavities. For this purpose, sequential backscattered electron images (BSE), acquired by SEM at an accelerating voltage of 10 kV and magnification of 20 k, were examined, covering a total area of over 4700 mm^2. The BSE images were analyzed using ImageJ software (ImageJ1.x, National Institute of Mental Health, Bethesda, Rockville, MD, USA) [35]. As previously reported [17], by defining an appropriate contrast threshold, it is possible to separate the creep cavities from the surrounding carbides shown in Figure 10. Cavities can be separated from fine intragranular carbides according to their shape differences, using a circularity factor [4π(area)/(perimeter)^2]. This is equal to 1 for perfectly circular precipitates. Cavities were separated from intragranular carbides using a circularity factor of 0 to 0.7. To make a direct comparison between the mean sizes of the cavities measured by quantitative metallurgy and SANS, cavities up to a 300 nm diameter of cross-sectional area of 0.071 μm^2, were quantified by QM. The data were separated for the two populations of cavities identified by SANS. Cavities with a cross-sectional area of less than 0.0078 μm^2, corresponding to a diameter of 100 nm, were quantified separately from those having cross-sectional areas between 0.0078 and 0.071 μm^2, corresponding to diameters of between 100 and 300 nm. The percentage area fraction of the cavity population less than 100 nm in size, measured by QM, is shown in Figure 11. It is seen that the area fraction increases through creep life for both of the test conditions, but the area fraction is higher for 675 °C and 150 MPa than that which measured for 550 °C and 320 MPa. The same trends are seen in the volumetric SANS data for cavities of less than 100 nm in size; see Figure 4a,b.

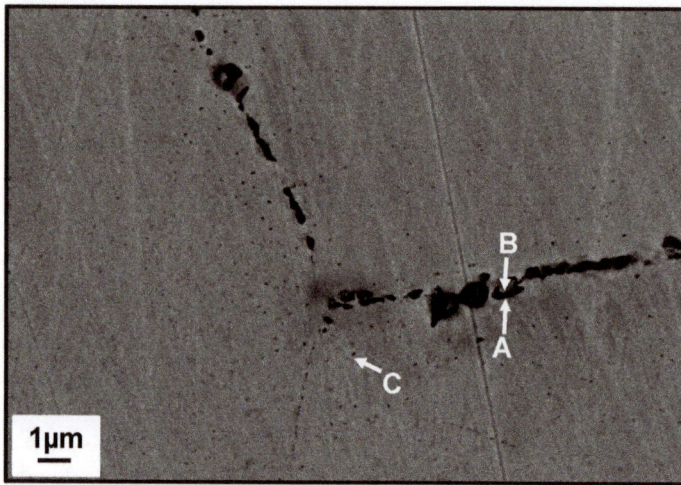

Figure 10. Backscatter image showing the microstructure of a gauge area of the sample 3-d1 under conditions of 550 °C, with an applied stress of 320 MPa (after 8% pre-strain). Cavities (A) are mainly surrounding intergranular carbides (B). A population of fine intragranular carbides (C) can also be seen, appearing as dark spots adjacent to the grain boundaries.

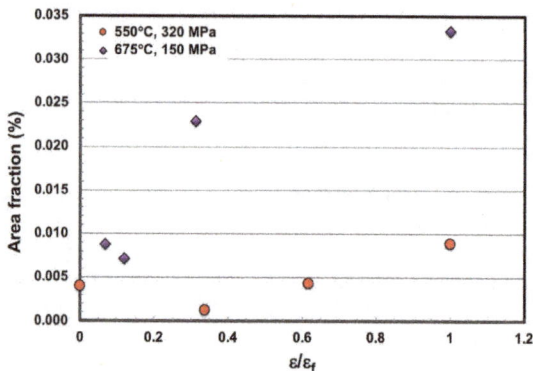

Figure 11. The variation in the area fraction (%) of cavities less than 100 nm diameter in size, through creep life, measured by quantitative metallography.

The mean cavity size associated with the two cavity populations, measured under both creep conditions by SANS and QM techniques is presented in Figure 12. It should be noted that the QM result is based on measuring cavities in a relatively small area (4700 μm^2) compared to that in SANS, which is covering a gauge volume of 9 mm^3 (9 $\times$ 10^9 μm^3). More statistically, valid QM data can only be acquired by automating the SEM imaging in order to cover a larger area. Also, in this study, the smallest cavity quantified by QM had an area of 0.0002 μm^2, or a diameter of 16 nm. Therefore, cavities smaller than 16 nm in size were not measured by QM. This might have skewed the data and resulted some discrepancy in comparing the measurements, corresponding to the largest D. Nevertheless, a good agreement between cavity sizes measured by the two techniques is evident.

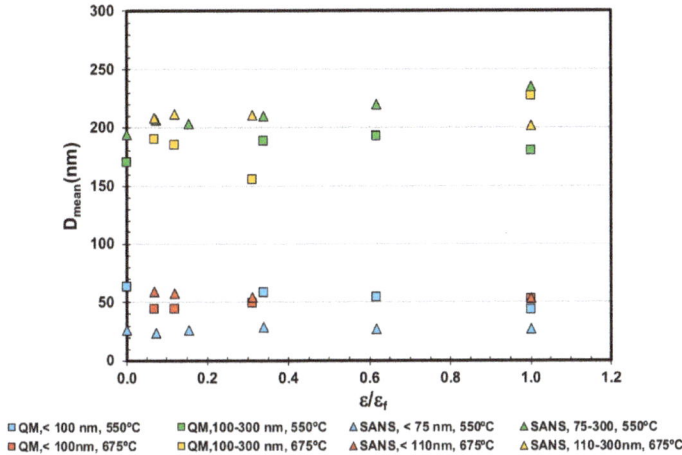

Figure 12. The mean sizes of two populations of cavities, measured by QM and SANS, for creep conditions of 550 °C, 320 MPa (after 8% pre-strain); and 675 °C, 150 MPa.

5. Discussion

The mean diameter of the population of smaller cavities observed from SANS for the tests at 550 °C under 320 MPa was found to be 39 nm, with a standard deviation of 16.4, and that for the tests at 675 °C under 150 MPa was found to be 63 nm, with a standard deviation of 30.2. Theoretically, the minimum stable cavity radius in steel is given by $2\gamma/\sigma$, where γ is the surface tension and σ is the applied direct stress. For an applied stress of 320 MPa, and a surface tension of 2.41 Nm^{-1} for stainless steel [36], a minimum cavity diameter of 30 nm is expected. The decrease in surface tension as a result of increasing the temperature from 550 to 675 °C is negligible [37] and the effect of applied stress on the minimum stable cavity radius is more significant. Therefore, at the lower applied load of 150 MPa, larger cavities, about 64 nm in diameter, are expected to be stable. These values are in reasonable agreement with the mean sizes of the small cavities, as measured by SANS.

The large difference in the mean diameters of the creep cavities observed between the two sets of specimens could possibly be attributed to the difference in cavity nucleation mechanisms under the different loading conditions. In the specimens tested at 550 °C and 320 MPa, cavity nucleation might be promoted from the initial plastic pre-straining from the coalescing of piled-up dislocations along a slip plane or against a hard particle. The plastic pre-straining may therefore result in grain boundary void development at the start of the creep test. In Figure 4a, a small but noticeable volume fraction of cavities can be observed for sample 5-d1 with 0% creep strain. Early void nucleation driven by plastic pre-strain may therefore control cavity development mechanisms in this set of samples. Whereas, in the specimens tested at 675 °C and 150 MPa, the nucleation could be promoted by vacancy accumulation.

A significant increase in the number density of smaller cavities during creep life was observed in the samples tested at 550 °C and 320 MPa, whereas that for the samples tested at 675 °C and 150 MPa showed little change (see Figure 7). This suggests that the growth mechanisms of the creep cavities during creep life are also different under these two test conditions. At 550 °C and 320 MPa cavity nucleation continues to occur during creep life, whereas at 675 °C and 150 MPa, cavity initiation and growth both play a key role. The growth of smaller cavities is seen as a shift to a larger size in the peak position of the distribution shown in Figure 4b, and a small increase in the mean size shown in Figure 6b.

Dyson [38] showed that tensile creep specimens that were pre-strained at room temperature appeared to have a predisposition for creep cavitation. He also experimentally showed a direct correlation between increasing the pre-strain and creep cavity density. In Figure 7, it can be observed that although the number density of the smaller cavities in the pre-strained specimen increased significantly over the creep life, which for the larger cavities did not. This suggests that although pre-strain promotes cavity nucleation, it does not necessarily affect the cavity growth rate, as previously reported [39]. However, the limitation of the SANS technique in measuring cavities of only up to 300 nm size must be mentioned here, as the measurement of much larger size cavities may prove otherwise. Finally, the increase in cavity density due to plastic pre-compression reduces the uniaxial creep ductility of the samples. A reciprocal relation of the creep ductility with the rate of cavity production with strain in type 347 stainless steel was reported previously [38]. It can be observed in Figure 7 that the rate of increase in the cavity number density, in particular for smaller sized cavities, with strain in the pre-strained specimens is much higher than those deformed under pure creep at higher temperature. This may explain the much lower creep ductility for the former.

In this work, a systematic increase in the fractional size distribution of cavities formed under increasing creep strain has been found in disc samples cut from creep test specimens having undergone two sets of creep conditions. A previous study using SANS, of a section of a cracked weldment taken from the same component as the specimens examined in this work, has shown a similar increase in the fractional size distribution of cavities at positions approaching the crack along lines normal to the crack line, and along lines parallel to the crack line approaching the crack mouth [32].

By measurements of disc samples from different locations of the test specimen having different diameters, the increase in creep-induced cavity size distribution with applied stress for the two sets of specimens has been determined unambiguously, without the need for a reference sample. This is the case for the pre-strained specimen tested at 550 °C with a stress difference of 178 MPa, and that tested at 675 °C with a stress difference of 83 MPa, except for the highest creep strain of the latter.

The present work shows how SANS can be employed to characterize the volumetric size distributions of creep-induced cavities in uniaxial test samples, and how measurements on sets of interrupted tests can be used to track the evolution of cavities during creep life. In particular, the observations suggest that plastic pre-strain promotes early cavity nucleation. This is significant, because most laboratory creep deformation tests on Type 316H stainless steel are undertaken by using applied stresses that are greater than the material yield stress, in order to complete the tests within a practical timescale. Thus, creep deformation and damage models based on conventional uniaxial tests innately conflate plastic strain and creep mechanisms. Future work is required to quantify the role of plastic pre-strain on physics and mechanics of cavity nucleation and growth, in order to improve published models, and to develop more realistic lifetime assessment methods for high-temperature power plant applications.

6. Conclusions

Small-angle neutron scattering and quantitative metallography are complementary techniques that have been effectively applied to measure creep cavitation up to 300 nm in size, in two sets of 316H austenitic stainless steel specimens tested up to rupture at 550 °C and 320 MPa, and at 675 °C and 150 MPa. The former set specimens had been subjected to a plastic pre-strain of 8%. Two populations of cavities were observed to develop during creep life: a population of smaller cavities under ~100 nm in size, and a population of larger cavities ~100–300 nm in size. For both test conditions, the volume fraction distribution of smaller cavities exhibited a pronounced increase with increasing creep strain; therefore, a continuous nucleation of creep cavitation has been observed throughout those creep lives, for both the pre-strained and purely creep deformed samples. A significant increase in the number density of smaller cavities, less than 100 nm in size, was observed to develop during creep life in samples that were tested at 550 °C and 320 MPa. However, samples tested at 675 °C and 150 MPa showed little change in the number density of smaller cavities during creep life, which suggests that cavity growth is the dominant mechanism of creep cavity development under these conditions. In the plastic pre-strained set of specimens, the number density of only the smaller sized cavities was observed to increase with creep life fraction, while that of the larger sized cavities were almost unchanged. This suggests that plastic pre-straining acts as a predisposition for creep cavity nucleation in this material. Future work is required to quantify the role of plastic pre-strain on physics and mechanics of cavity nucleation and growth, in order to improve published models, and to develop more realistic lifetime assessment methods for high-temperature power plant applications.

Author Contributions: Data curation: H.J.; M.T.H.; A.A.M. and R.K.H.; Formal analysis: H.J.; M.T.H.; and A.A.M.; Funding acquisition: P.J.B. and H.J.; Investigation: H.J.; M.T.H. and A.A.M.; Methodology: H.J. and M.T.H.; Supervision: P.J.B.; Validation: H.J.; M.T.H. and M.W.S. Writing—Original Draft, H.J. and Writing—Review & Editing, H.J.; P.J.B.; M.T.H.; A.A.M. and M.W.S.

Acknowledgments: This research was funded by EDF Energy, grant number 4840543478. The APC was funded by the Science, Technology, Engineering, and Mathematics (STEM) faculty at The Open University. The help of Sarah Rogers is acknowledged in undertaking the SANS experiments.

Conflicts of Interest: The authors declare no conflict of interest.

References

1. Kassner, M.E.; Hayes, T.A. Creep cavitation in metals. *Int. J. Plast.* **2003**, *19*, 1715–1748. [CrossRef]
2. Greenwood, J.N. Intercrystalline cracking of metals. *J. Iron Steel Inst.* **1954**, *176*, 268–269.
3. Greenwood, J.N.; Miller, D.R.; Suiter, J.W. Intergranular cavitation in stressed metals. *Acta Metall.* **1954**, *2*, 250–258. [CrossRef]
4. Chen, B.; Flewitt, P.E.J.; Smith, D.J.; Jones, C.P. An improved method to identify grain boundary creep cavitation in 316H austenitic stainless steel. *Ultramicroscopy* **2011**, *111*, 309–313. [CrossRef] [PubMed]
5. Kassner, M.E. *Fundamentals of Creep in Metals and Alloys*; Elsevier Science: Amsterdam, The Netherlands, 2008; pp. 225–226. [CrossRef]
6. Chen, R.T.; Weertman, J.R. Grain boundary cavitation in internally oxidized copper. *Mater. Sci. Eng.* **1984**, *64*, 15–25. [CrossRef]
7. Chen, I.W.; Argon, A.S. Creep cavitation in 304 stainless steel. *Acta Metall.* **1981**, *29*, 1321–1333. [CrossRef]
8. Riedel, H. Cavity nucleation at particles on sliding grain boundaries. A shear crack model for grain boundary sliding in creeping polycrystals. *Acta Metall.* **1984**, *32*, 313–321. [CrossRef]
9. Boettner, R.C.; Robertson, W.D. A study of the growth of voids in copper during the creep process by measurement of the accompanying change in density. *Trans. Met. Soc.* **1961**, *221*, 613–622.
10. Evans, H.E. *Mechanisms of Creep Fracture*; Elsevier Applied Science Publishers Ltd.: London, UK, 1984.
11. Nix, W.D. Introduction to the viewpoint set on creep cavitation. *Scr. Metall.* **1983**, *17*, 1–4. [CrossRef]
12. Sandström, R.; He, J. Grain boundary sliding. In *Survey of Creep Cavitation in fcc Metals*; IntechOpen: London, UK, 2017. [CrossRef]
13. He, J.; Sandström, R. Formation of creep cavities in austenitic stainless steels. *J. Mater. Sci.* **2016**, *51*, 6674–6685. [CrossRef]
14. Bouchard, P.J.; Withers, P.J.; McDonald, S.A.; Heenan, R.K. Quantification of creep cavitation damage around a crack in a stainless steel pressure vessel. *Acta Mater.* **2004**, *52*, 23–34. [CrossRef]

15. Burnett, T.L.; Geurts, R.; Jazaeri, H.; Northover, S.M.; McDonald, S.A.; Haigh, S.J.; Bouchard, P.J.; Withers, P.J. Multiscale 3D analysis of creep cavities in AISI type 316 stainless steel. *Mater. Sci. Technol.* **2015**, *31*, 522–534. [CrossRef]

16. Dyson, B.F.; Loveday, M.S.; Rodgers, M.J. Grain boundary cavitation under various states of applied stress. *Proc. R. Soc. Lond. A* **1976**, *349*, 245–259. [CrossRef]

17. Jazaeri, H.; Bouchard, P.J.; Hutchings, M.T.; Lindner, P. Study of creep cavitation in stainless steel weldment. *Mater. Sci. Technol.* **2014**, *30*, 38–42. [CrossRef]

18. Jazaeri, H.; Bouchard, P.J.; Hutchings, M.T.; Lindner, P. Study of creep cavitation in a stainless steel weldment using small angle neutron scattering and scanning electron microscopy. In Proceedings of the ASME 2014 Pressure Vessels & Piping Conference, Materials and Fabrication, PVP2014-28641, Anaheim, CA, USA, 20–24 July 2014. [CrossRef]

19. Rustichelli, F. Applications of small neutron scattering in material science and rechnology. *Metall. Sci. Technol.* **1993**, *11*, 118–141.

20. Bouchard, P.J.; Fiori, F.; Treimer, W. Characterisation of creep cavitation damage in a stainless steel pressure vessel using small angle neutron scattering. *Appl. Phys. A Mater. Sci. Process.* **2002**, *74*, S1689–S1691. [CrossRef]

21. Hales, R. The role of cavity growth mechanisms in determining creep-rupture under multiaxial stresses. *Fatigue Fract. Eng. Mater. Struct.* **1994**, *17*, 579–591. [CrossRef]

22. Acar, M.; Gungor, S.; Bouchard, P.J.; Fitzpatrick, M.E. Effect of prior cold work on the mechanical properties of weldments. In Proceedings of the 2010 SEM Annual Conference and Exposition on Experimental and Applied Mechanics, Indianapolis, IN, USA, 7–10 June 2010; Volume 6, pp. 817–826.

23. Wilshire, B.; Willis, M. Mechanisms of strain accumulation and damage development during creep of prestrained 316 stainless steels. *Metall. Mater. Trans. A* **2004**, *35*, 563–571. [CrossRef]

24. Githinji, D.N. *Characterisation of Plastic and Creep Strains from Lattice Orientation Measurements*; The Open University: Milton Keynes, UK, 2013.

25. Hong, H.U.; Nam, S.W. The occurrence of grain boundary serration and its effect on the M23C6 carbide characteristics in an AISI 316 stainless steel. *Mater. Sci. Eng. A* **2002**, *332*, 255–261. [CrossRef]

26. Heenan, R.K.; Rogers, S.E.; Turner, D.; Terry, A.E.; Treadgold, J.; King, S.M. Small angle neutron scattering using Sans2d. *Neutron News* **2011**, *22*, 19–21. [CrossRef]

27. Akeroyd, F.; Ansell, S.; Antony, S.; Arnold, O.; Bekasovs, A.; Bilheux, J.; Borreguero, J.; Brown, K.; Buts, A.; Campbell, S.; et al. Mantid: Manipulation and Analysis Toolkit for Instrument Data. Available online: http://dx.doi.org/10.5286/SOFTWARE/MANTID (accessed on 31 December 2013).

28. Heenan, R.K.; Penfold, J.; King, S.M. SANS at pulsed neutron sources: present and future prospects. *J. Appl. Cryst.* **1997**, *30*, 1140–1147. [CrossRef]

29. Hutchings, M.T.; Windsor, C.G. 25. Industrial Applications. In *Methods in Experimental Physics*; Sköld, K., Price, D.L., Eds.; Academic Press: Cambridge, MA, USA, 1987; Volume 23, pp. 405–482. [CrossRef]

30. Potton, J.A.; Daniell, G.J.; Rainford, B.D. A new method for the determination of particle size distribution from Small-Angle Neutron Scattering Measurements. *J. Appl. Cryst.* **1988**, *21*, 891–897. [CrossRef]

31. Hutchings, M.T. *The Use of Small Angle Neutron Scattering for Mapping Creep Cavitation Damage in an Ex-Service Steam Header*; The Open University: Milton Keynes, UK, 2012.

32. Jazaeri, H.; Bouchard, P.J.; Hutchings, M.T.; Mamun, A.A.; Heenan, R.K. Application of small angle neutron scattering to study creep cavitation in stainless steel weldments. *Mater. Sci. Technol.* **2015**, *31*, 535–539. [CrossRef]

33. Slater, T.J.A.; Bradley, R.S.; Bertali, G.; Geurts, R.; Northover, S.M.; Burke, M.G.; Haigh, S.J.; Burnett, T.L.; Withers, P.J. Multiscale correlative tomography: an investigation of creep cavitation in 316 stainless steel. *Sci. Rep.* **2017**, *7*, 7332. [CrossRef] [PubMed]

34. Pommier, H.; Busso, E.P.; Morgeneyer, T.F.; Pineau, A. Intergranular damage during stress relaxation in AISI 316L-type austenitic stainless steels: Effect of carbon, nitrogen and phosphorus contents. *Acta Mater.* **2016**, *103*, 893–908. [CrossRef]

35. Rasband, W.S. *ImageJ1.x*; U.S. National Institutes of Health: Bethesda, MD, USA. Available online: https://imagej.nih.gov/ij/ (accessed on 31 December 2013).
36. Yu, J.; Lin, X.; Wang, J.; Chen, J.; Huang, W. First-principles study of the relaxation and energy of bcc-Fe, fcc-Fe and AISI-304 stainless steel surfaces. *Appl. Surf. Sci.* **2009**, *255*, 9032–9039. [CrossRef]
37. Kristyan, S.; Giber, J. Temperature dependence of the surface free energies of solid chemical elements. *Surf. Sci.* **1988**, *201*, L532–L538. [CrossRef]
38. Dyson, B.F. Continuous cavity nucleation and creep fracture. *Scr. Metall.* **1983**, *17*, 31–37. [CrossRef]
39. Dyson, B.F.; Rodgers, M.J. Prestrain, cavitation, and creep ductility. *Met. Sci.* **1974**, *8*, 261–266. [CrossRef]

metals

MDPI

Article

Estimating the Influences of Prior Residual Stress on the Creep Rupture Mechanism for P92 Steel

Dezheng Liu [1,*], Yan Li [1], Xiangdong Xie [2,*], Guijie Liang [1] and Jing Zhao [1]

[1]   Department of Mechanical Engineering, Hubei University of Arts and Science, Xiangyang 441053, China; wustliyan@163.com (Y.L.); guijie-liang@hotmail.com (G.L.); zjjysu@163.com (J.Z.)
[2]   School of Urban Construction, Yangtze University, Jingzhou 201800, China
*   Correspondence: liudezheng126@126.com (D.L.); xdxie@yangtzeu.edu.cn (X.X.);
     Tel.: +86-1869-620-3128 (D.L); +86-1560-861-9775 (X.X.)

Received: 13 May 2019; Accepted: 31 May 2019; Published: 2 June 2019

Abstract: Creep damage is one of the main failure mechanisms of high Cr heat-resistant steel in power plants. Due to the complex changes of stress, strain, and damage at the tip of a creep crack with time, it is difficult to accurately evaluate the effects of residual stress on the creep rupture mechanism. In this study, two levels of residual stress were introduced in P92 high Cr alloy specimens using the local out-of-plane compression approach. The specimens were then subjected to thermal exposure at the temperature of 650 °C for accelerated creep tests. The chemical composition of P92 specimens was obtained using an FLS980-stm Edinburgh fluorescence spectrometer. Then, the constitutive coupling relation between the temperature and material intrinsic flow stress was established based on the Gibbs free energy principle. The effects of prior residual stress on the creep rupture mechanism were investigated by the finite element method (FEM) and experimental method. A comparison of the experimental and simulated results demonstrates that the effect of prior residual stress on the propagation of micro-cracks and the creep rupture time is significant. In sum, the transgranular fracture and the intergranular fracture can be observed in micrographs when the value of prior residual stress exceeds and is less than the material intrinsic flow stress, respectively.

Keywords: residual stress; creep rupture mechanism; P92 steel; FEM; Gibbs free energy principle

1. Introduction

Creep deformation and failure in high-temperature structures are serious problems in industry and are becoming even more serious under the current increasing pressures of energy, economics, and sustainability [1–3]. P92 high Cr alloy steels are widely used as high-temperature construction materials in power plants due to their high creep strength, good molding property to be processed, and superior heat properties [4,5]. However, such components employed in power plants are continually exposed to high temperatures and high steam pressures, and creep crack growth could occur within these high-temperature regimes, causing the failure of these components [6]. In practical engineering, it is difficult to accurately evaluate the creep rupture mechanism of P92 steels under multi-axial stress states due to the intricate variations of the stress, strain, and damage at the tip of a creep crack with time [7]. For example, residual stresses can be invariably introduced into structural components during the fabrication processes and thermal operations and non-uniform plastic deformation [8,9], and the residual stresses can be superimposed by any applied loading and produce a complex stress state acting on in-service components, causing creep crack growth and rupture. Therefore, it is essential to investigate the influences of residual stress on the creep rupture mechanism for high Cr alloy steel.

In recent years, the creep rupture behavior of high Cr alloy steel has been studied extensively. For example, recent studies [10–15] have reported the prediction of the creep damage and lifetime for P92 steels based on the modification of existing creep damage models. However, there are few detailed studies of the effect of prior residual stress on the creep rupture mechanism of P92 steels. The difficulty in describing the creep rupture behavior of P92 steel is due to the lack of an accurate rupture mechanism for depicting the microscopic crack characteristics [16]. Creep fracture is usually caused by the growth of nucleation and mutual connection of micro-cavities and micro-cracks, and creep cracks grow from the cusp and ultimately weaken the cross section to the point where failure occurs [13]. Due to the fact that residual stresses can be generated during the fabrication processes for many mechanical or thermal operations, creep crack initiation and growth in components can be driven by these residual stresses [8]. Especially under multi-axial stress states, the growth of the crack-like defects, such as micro-cracks and creep voids, can be accelerated by the residual stresses combined with external applied loading. Furthermore, the residual stress can contribute to the amplitudes of crack tip stress fields and the cracks will be nucleated when the accumulated creep strains under the action of the stress field around the crack tip are sufficient to exhaust the creep ductility of the material [9]. Therefore, the residual stress may significantly affect the load carrying capacity and resistance to creep and rupture of the structural components made of P92 steel.

To better understand the creep rupture mechanism of residual stress and initial crack positions and to accurately analyze creep failure and life assessments, it is essential to quantitatively investigate the effect of prior residual stress levels on creep crack growth and rupture. There are two main ways to introduce residual stresses into measured specimens. One way is to extract a specimen from the structure containing residual stresses and then introduce a flaw in the region of the residual stress, and the other one is to directly introduce residual stresses into specimens through the application of either mechanical or thermal techniques [17]. The local out-of-plane compression approach [17], which is the use of mechanical techniques and the most well-known residual stress generation method, is adopted in this study. Furthermore, the numerical methods based on the damage model combined with experimental tests can provide an efficient way to analyse the effect of stress levels on creep crack growth and rupture. It should be noted that some researchers [18,19] have attempted to describe the process of high-temperature deformation of metallic materials through the use of material intrinsic flow stress. In reference [18], the thermal deformation behavior and the hyperbolic constitutive equation of T122 steel were investigated by the determination of deformed activation energy and material intrinsic flow stress. In reference [19], the relationship between residual stress and material intrinsic flow stress has been used to determine the formation of weld solidification cracks. On the basis of references [18,19], the method for exploring the effects of prior residual stress on the creep rupture mechanism through the comparison of residual stress and material intrinsic flow stress was conducted in this study.

This paper is organized as follows. Reviews of previous literature on the investigation of the relationship between the residual stress and creep rupture behavior for P92 steel were presented previously. In Section 2, the methodologies of creep crack growth test are conducted, involving methods of introducing different residual stresses for creep tests and establishing the constitutive coupling relation between the temperature and material intrinsic flow stress, followed by the construction of the corresponding finite element (FE) model. The results and discussions of the influences of prior residual stress on the creep rupture mechanism for P92 steel are presented in Sections 3 and 4, respectively. Conclusions are drawn in Section 5.

2. Materials and Methods

2.1. Specimen Preparation and Creep Test

The specimens of P92 steel were provided by Angang Steel Company (Anshan, China) Limited. In order to introduce the multi-axial stress states into the specimens for the creep test, the annular breach in the middle of each specimen was machined by the lathe. The geometry of the notched specimen is shown in Figure 1. The width at both ends of the specimen is 30 mm, and the width at the middle of the specimen is 20 mm. The annular breach radius is set to 0.7 mm and the thickness of the specimen is 2 mm. The notch sharpness is set to 28.6, which is the ratio of the width of the middle segment to the radius of the notch. The chemical composition of the P92 specimen was measured using an FLS980-stm Edinburgh fluorescence spectrometer (Edinburgh Instruments Ltd, Livingston, UK) and the chemical composition is presented in Table 1. To analyze the evolution of the creep fracture surface, there are five samples under different residual stress levels, respectively.

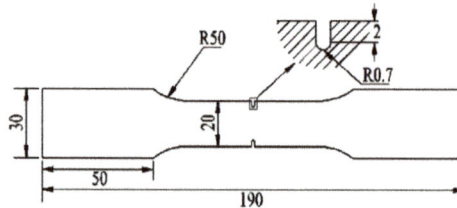

Figure 1. Schematic diagram of the specimen (mm).

Table 1. Chemical compositions of P92 steel (wt.%).

Material	C	Mn	Si	Cr	Mo	S	P	Nb	V	Al	Ni	W	N	B	Bal.
P92 steel	0.1	0.45	0.35	8.95	0.96	0.01	0.018	0.08	0.215	0.04	0.12	1.47	0.043	0.001	Fe

The Gibbs free energy interface thermodynamic method [20] was adopted to establish the constitutive coupling relation between the temperature and material intrinsic flow stress. The basic equation for the Gibbs energy of a multi-component solution phase can be expressed as

$$G_m = \sum_i x_i G_i^0 + RT \sum_i x_i ln x_i + \sum_i \sum_{j>i} x_i x_j \sum_V \Omega_V \left(x_i - x_j \right)^V \tag{1}$$

where $\sum_i x_i G_i^0$ is the Gibbs energy of the pure components, $RT \sum_i x_i ln x_i$ is the ideal entropy, and $\sum_i \sum_{j>i} x_i x_j \sum_V \Omega_V \left(x_i - x_j \right)^V$ accounts for pairwise interactions of the species. According to the Gibbs free energy principle [20], the equilibrium pressure of the system is achieved when the Gibbs's free energy attains a minimum. The solute mass fraction in the solid [21] can be expressed as

$$\tau(X,T) = \frac{1}{\beta 2^{\frac{N}{8}} D \Delta T^q} \int_0^x \frac{dx}{x^{\frac{2(1-x)}{3}} \cdot (1-x)^{\frac{2x}{3}}} \tag{2}$$

where τ is the solute mass fraction in the solid for different amounts of undercooling, β is an empirical coefficient for describing the kinetics of isothermal austenite to pearlite reaction, N is the size of a grain, D is an effective diffusion coefficient, ΔT is the undercooling, q is an exponent that depends on the effective diffusion mechanism, and x is the fraction transformed.

In a high-temperature environment, the various microstructural constituents can significantly affect the mechanical properties and the distribution of residual stresses in components. Therefore, it is important to be able to predict the final microstructure distribution for a given thermal history. Equations (1) and (2) can be used to describe the characterization of the kinetics of austenite decomposition to equilibrium or non-equilibrium phases in alloy steel. The kinetics of isothermal austenite to ferrite, pearlite, and bainite reactions can also be described through the use of the Gibbs free energy interface thermodynamic method. However, the various phase transformations that generate the required microstructure link the process parameters to the final properties. For example, the empirical coefficient in Equation (2) is only to be used in the kinetics of isothermal austenite to pearlite reaction [21].

According to reference [20], the relationship between the stress and strain can be determined by the yield strength, hardening exponent, and reference strain, as follows:

$$\sigma = \frac{\sigma_{0.2}}{\varepsilon_0^n} \cdot \varepsilon^n \tag{3}$$

where $\sigma_{0.2}$ is the yield strength, n is the hardening exponent, and ε_0 is a reference strain. The determination of the value of the parameters in Equations (1)–(3) has been reported in references [21,22]. In this study, the value of the parameters used to derive the relationship between the temperature and the material intrinsic flow stress is shown in Table 2.

Table 2. The value of the parameters used to derive the material intrinsic flow stress.

Parameter	β	N (um)	D (m^2/s)	ΔT (°C)	q	$\sigma_{0.2}$ (MPa)	n	ε_0
Value	0.3054	20	2.09×10^{-12}	123.1	3	440	0.129	1

Using Equations (1)–(3) and Tables 1 and 2, a constitutive relation between the temperature and material intrinsic flow stress can be obtained using JmatPro software (JmatPro 7.0, Sente Software Ltd, Surrey, UK).

Figure 2 shows the material intrinsic flow stress of P92 steel in the temperatures range from 0 °C to 1000 °C. According to Figure 2, the material intrinsic flow stress is about 145 MPa when the temperature is 650 °C. The residual stresses introduced into the specimens were devised as two levels; here, the value of material intrinsic flow stress is taken as the reference standard. One level of residual stress is above the material intrinsic flow stress, while the other one is below the material intrinsic flow stress.

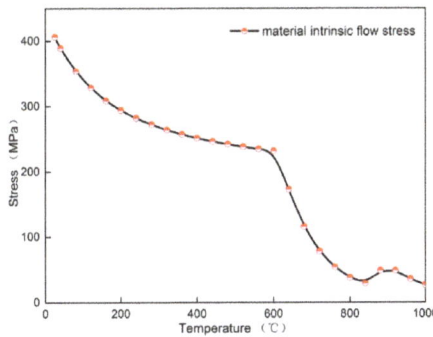

Figure 2. Material intrinsic flow stress vs. temperature.

The local out-of-plane compression approach [17] was used to introduce residual stress states in specimens. According to reference [17], the tensile residual stresses ahead of the annular breach can be generated by loading in compression beyond the yield and then unloading. Furthermore, the tensile residual stress can be changed through altering loading point displacement. The specimen surface is compressed until a specified displacement to produce some plastic deformation around the annular breach. In this study, the specimens with dimensions of 190 mm × 30 mm × 2 mm were used to introduce residual stress states through the use of the SUNS WAW-2000 universal material hydraulic experiment machine (Shenzhen Suns Technology Stock Co., Ltd, Shenzhen, China). The notched specimen was loaded in compression to introduce a tensile residual stress field over a significant distance ahead of the annular breach. Pre-strained specimens were produced by uniaxially straining large tensile specimens to −3.82, +2.15, −1.78, and +1.03 pct when the different plastic deformation levels ahead of the annular breach were produced. The specimen was then unloaded by moving the tools back to their original position, and a residual stress field was generated due to the strain incompatibility between the elastic and plastic regions ahead of the annular breach.. In conjunction with the FERMI customized X-ray diffractometer thermal stage, the value of residual stress can be measured at the temperature of 650 °C through the use of a BRUKER D8 advance X ray diffractometer (Bruker Corporation, Ettlingen, Germany), as shown in Figure 3. Two levels of residual stress were introduced into the notched P92 specimens through the different loading point displacements.

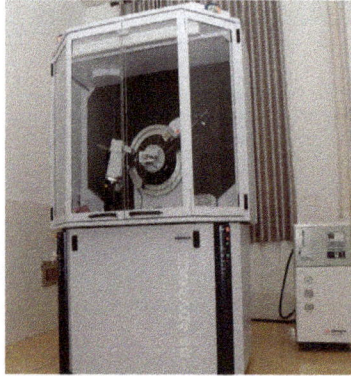

Figure 3. BRUKER D8 advance X ray diffractometer.

The parameters of the pre-compression to introduce the residual stress field into the P92 specimens are shown in Table 3. After the compression experiment of the specimens, creep tests for two groups of specimens with different residual stress levels were conducted at the temperature of 650 °C. In order to obtain creep crack initiation and propagation properties of specimens by an accelerated creep test, an external pulling stress of 110 MPa was applied to the specimens. The RDL50 electronic high-temperature creep and lasting intension testing machine (JiLin Guanteng Automation Technology Co., Ltd, Changchun, China) was adopted in the accelerated creep test, as shown in Figure 4.

Table 3. Parameters of the pre-compression to introduce the residual stress field.

Residual Stress Level	Pre-Compression Load (KN)	Loading Speed (mm/min)	Compressing Magnitude (%)	Stretching Magnitude (%)	Maximum Residual Stress Value (MPa)	Material Intrinsic Flow Stress Value (MPa)
High level	42.8	0.16	3.82	2.15	182	145
Low level	20.6	0.12	1.78	1.03	97	145

Figure 4. Accelerated creep tests for P92 specimens.

In accelerated creep tests, the temperatures were kept within ±1 °C during the test through the use of a thermocouple to measure the center of the specimen and the position of the upper and lower pull rods. Meanwhile, the displacement of loading line and the length of the crack were measured and recorded. The displacement of the loading line was measured by an extensometer with a measurement accuracy of 1 micron and the creep crack length was measured by the direct current potential drop method [17]. After the creep test, the specimen was dissected into two parts along with its axial direction of thickness by the wire-electrode cutting method. One part was used to observe the morphology of the crack growth surface, and the other one was used to observe the change of the microstructures.

2.2. Creep Damage and Crack Growth Model

The continuum damage mechanics (CDM) based creep damage model has been widely used to describe all three stages of creep because it can be easily implemented with the finite element program in the analysis of creep damage behavior, which can significantly improve the efficiency in terms of both time and economy. In applications, a damage parameter is defined through the use of the CDM-based approach that ranges from zero (no damage) to a critical damage value (full damage), and is then measured throughout the creep processes. Creep failure time is defined as the time taken for the continuum damage level to move from no damage to full damage [23]. Physically-based CDM for creep provides a suitable framework for quantifying the shapes of the creep curve caused by several microstructural damage mechanisms.

The computational capability for creep damage analysis relies on the availability of a set of creep damage constitutive equations. In this study, Hayhurst creep damage constitutive equations [24], which were developed based on the CDM approach, were used to investigate the influences of residual stress on the creep rupture mechanism for P92 steel. Hayhurst creep damage constitutive equations are shown in the following:

$$\dot{\varepsilon}_{ij} = \frac{3s_{ij}}{2\sigma_e} Asinh\left[\frac{B\sigma_e(1-H)}{(1-\varnothing)(1-\omega)}\right] \tag{4}$$

$$\dot{H} = \left(\frac{h\dot{\varepsilon}_e}{\sigma_e}\right)\left(1 - \left(\frac{H}{H^*}\right)\right) \tag{5}$$

$$\dot{\varnothing} = \left(\frac{K_c}{3}\right)(1-\varnothing)^4 \tag{6}$$

$$\dot{\omega} = CN\dot{\varepsilon}_e(\sigma_1/\sigma_e)^v \tag{7}$$

where $\dot{\varepsilon}_{ij}$ is the creep strain rate under the multi-axial stress state, ε_{ij} is the multi-axial creep rate, σ_1 is the maximum principal stress, ω is the creep damage, $\dot{\omega}$ is the creep damage rate, H is the strain hardening, $\dot{H}$ is the strain hardening rate, and n represents the stress exponents. $N = 1$ when $\sigma_1 > 0$ and $N = 0$ when $\sigma_1 < 0$. The multi-axial parameters A, B, C, h, H^* and k_c are constants to be determined from the uniaxial creep behaviour [25]. Variable $\varnothing$ increases from its initial value of zero towards a theoretical upper limit of unity. The parameter v is the multi-axial stress sensitivity index, where $\sigma_e = \left(3S_{ij}S_{ij}/2\right)^{1/2}$ is the effective stress, $S_{ij} = \sigma_{ij} - \delta_{ij}\sigma_{kk}/3$ is the stress deviator, and $\varepsilon_e = \left(2\varepsilon_{ij}\varepsilon_{ij}/3\right)^{1/2}$ is the effective creep strain. The creep damage parameter ω is defined ranging from 0% (no damage) to 100% (full damage) and the parameter is then monitored throughout the creep time. The creep rupture time is defined as the time taken for the continuum damage level to reach 99.9–100% in most elements in a given zone [24].

The behavior of creep crack growth can be evaluated in terms of the range of stress intensity factor, and the formula for describing the behavior of creep crack growth [26] can be written as

$$\Delta K = \frac{\Delta P}{B^{1/2}W^{1/2}}F(a/W) \tag{8}$$

$$\Delta P = (1-R)P_{max} \tag{9}$$

$$F(a/W) = \left[\frac{2+a/W}{(1-a/W)^{3/2}} \right] (0.886 + 4.64(a/W) - 13.32(a/W)^2 + 14.72(a/W)^3 - 5.6(a/W)^4) \quad (10)$$

where ΔK is the stress intensity factor; ΔP is the range of applied loads; $F(a/W)$ represents the behavior of creep crack growth; R is the ratio of maximum stress and normal stress; a is the length of the crack; and W and B are the width and thickness of the specimen, respectively.

The fracture parameter of creep [26] can be written as

$$(C_t)_{avg} = \frac{\Delta P \Delta V_c}{B^{1/2} W t_h} \frac{F\prime}{F} \quad (11)$$

$$\frac{F\prime}{F} = \left[\left(\frac{1}{2+a/W} \right) + \left(\frac{3}{2(1-a/W)} \right) \right] + \left[\frac{4.64 - 26.64(a/W) + 44.16(a/W)^2 - 22.4(a/W)^3}{0.886 + 4.64(a/W) - 13.32(a/W)^2 + 14.72(a/W)^3 - 5.6(a/W)^4} \right] \quad (12)$$

where $(C_t)_{avg}$ is the fracture parameter, ΔV_c is the load line displacement caused by creep deformation, and t_h is the loading time. In the process of creep crack growth, the total load line displacement is mainly composed of the displacement caused by creep deformation and the displacement caused by elastic deformation. Therefore, the displacement caused by creep deformation can be written as

$$\Delta V_c = \Delta V - \Delta V_e \quad (13)$$

$$\Delta V_e = \frac{t_h (da/dt)_{avg}}{P} B \frac{2(\Delta K)^2}{E\prime} \quad (14)$$

$$(da/dt)_{avg} = (1/t_h) \cdot (da/dN) \quad (15)$$

where ΔV is the total load line displacement, ΔVe is the load line displacement caused by elastic deformation, $(da/dt)_{avg}$ is the average crack growth rate, and $E\prime$ is the effective modulus of elasticity. $E\prime = E/(1-v^2)$ under the plane strain state and $E\prime = E$ under the plane stress state. According to reference [27], the fracture behavior can be defined as brittle fracture when $\Delta V_e > \Delta V_c$, and the fracture behavior can be defined as ductile fracture when $\Delta V_e < \Delta V_c$.

The above mentioned equations have been used to simulate the creep damage behaviors of P92 steel under a multi-axial stress state using ABAQUS finite element software (ABAQUS 6.10, Dassault Systemes Simulia Corporation, Johnston, RI, USA). A user programmable function UMAT was written by FORTRAN to calculate the values of stress, strain, and damage in notched specimens. According to reference [9], the elasticity modulus and Poisson's ratio of P92 steel at 650 °C are 125 GPa and 0.3, respectively. According to references [25,28], the creep parameters of P92 steel at the temperature of 650 °C are shown in Table 4.

Table 4. Creep parameters of P92 steel at the temperature of 650 °C.

Material Parameter	A (MPa/h)	B (MPa⁻¹)	C	H (MPa)	H*	K_c (MPa⁻³/h)	E (GPa)	v
Value	2.21618×10^{-9}	3.473×10^{-3}	9.85×10^{-2}	2.43×10^6	0.5929	9.227×10^{-4}	125	0.3

2.3. FE Analysis Model

The FE analysis model was built using the HyperWorks software (HyperWorks 11.0, Altair Corporation, Troy, MI, USA) for simulating the creep rupture process for the P92 notched specimen. The geometry and dimensions of the notched tension specimen were taken to be the same as those of the experiments, as shown in Figure 1. The simulated temperature was set to 650 °C, and the elasticity modulus and Poisson's ratio of the FE model were set to 125 GPa and 0.3, respectively. The FE mesh is shown in Figure 5, and an external tensile stress of 110 MPa was applied to the specimen.

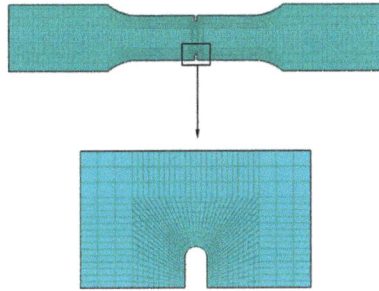

Figure 5. FE model of the notched P92 specimen.

The brick mesh (type C3D8R) and prismatic mesh (type C3D6) were used to model the evolution of creep rupture. The global size of the mesh was 1 mm, and the FE mesh was refined in the vicinity of the notch tip to obtain accurate results and to eliminate mesh dependency effects in the analyses. According to the average grain size of P92 steel [29], the smallest mesh size was set to 20 um. In this FE model, the number of elements and nodes were 18,677 and 24,136, respectively. Subsequently, the FEA model was imported into the ABAQUS software. The model was loaded by a uniform tensile stress at one end of the specimen and was fixed by an encastre constraint at the other end of the specimen, and the accumulated creep damage variables such as stress, strain, and damage could be calculated in ABAQUS using the corresponding user defined subroutine Creep UMAT. A flowchart used to link the different software between Hyperworks and ABAQUS is shown in Figure 6.

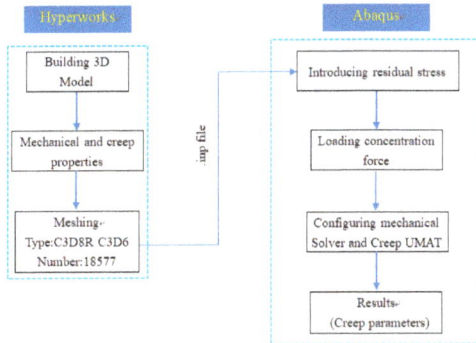

Figure 6. The flowchart used to link Hyperworks and ABAQUS.

For creep damage problems in FEM, the resulting equations are highly non-linear and stiff in nature. The nature of creep damage analysis is time dependant and the field variables such as stress, strain, and creep damage need to be updated where an integration scheme needs to be implemented. The stability and accuracy of the FE solution critically depends on the selection of the time step size associated with an appropriate integration method [30]. In this study, the well-known Runge-Kutta-Merson method [30] was adopted to control the time step because it can minimize the requirement of extra storage and reduce the amount of round-off error in creep damage analysis.

The FE analysis procedure was carried out in four stages: (1) the pre-compression was applied on the notched P92 specimens so as to produce the initial residual stress field; (2) a creep deformation modeling step at 650 °C was applied directly in order to predict relaxation, creep damage, and crack initiation under the different residual stress fields; (3) a primary load was applied in the FE models to investigate the combined effects of the residual stress with a primary load on the creep damage and crack initiation; (4) a creep loading step was applied and the integration time step size was controlled by the Runge-Kutta-Merson method.

3. Results

The creep strain and displacement of the whole rupture evolution of the P92 specimen from a constant pressure (110 MPa) test at a constant temperature of 650 °C were simulated through the implantation of Equations (4)–(15) in ABAQUS software by using the user subroutine UMAT. Figures 7 and 8 show the evolution of creep strain with the high and low residual stress level, respectively. When a specimen is stressed, its lattice spacing is altered. In the process of creep crack growth, the total load line displacement is mainly composed of the displacement caused by creep deformation and elastic deformation. The elastic strain in the material is determined by the change in crystal lattice spacing using Bragg's Law [31] and the load line displacement caused by creep deformation can be calculated by ABAQUS software.

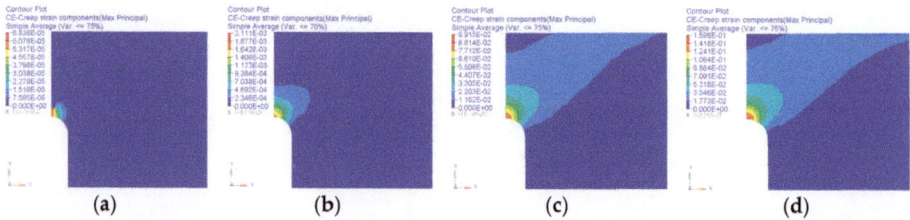

Figure 7. The evolution of creep strain at different lifetime fractions of the specimen with the high residual stress level. (a) 0%, (b) 50%, (c) 80%, and (d) 100%.

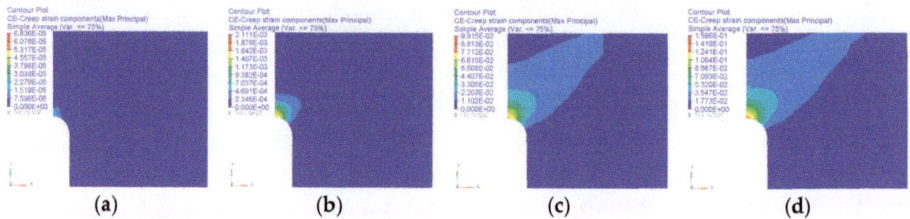

Figure 8. The evolution of creep strain at different lifetime fractions of the specimen with the low residual stress level. (a) 0%, (b) 50%, (c) 80%, and (d) 100%.

The effect of residual stress on creep deformation can be examined by comparing the predicted creep strain of the specimen with the introduced prior residual stress level higher than the material intrinsic flow stress level to that of the specimen containing the prior residual stress level less than the material intrinsic flow stress level. Figures 7 and 8 show the predicted evolution in equivalent creep strain with creep time from a constant pressure (110 MPa) test at a constant temperature of 650 °C under two different pre-residual stress levels. According to Figures 7 and 8, the strain ahead of the notch is increasing with the increase of creep time, and the maximum strain always appears near the notch tip. By extracting the creep displacement from the FE simulated results, the elastic displacement can be computed by subtracting the creep displacement from the total load line displacement measured. The evolution of load line displacement under different residual stress levels can be drawn in Figure 9.

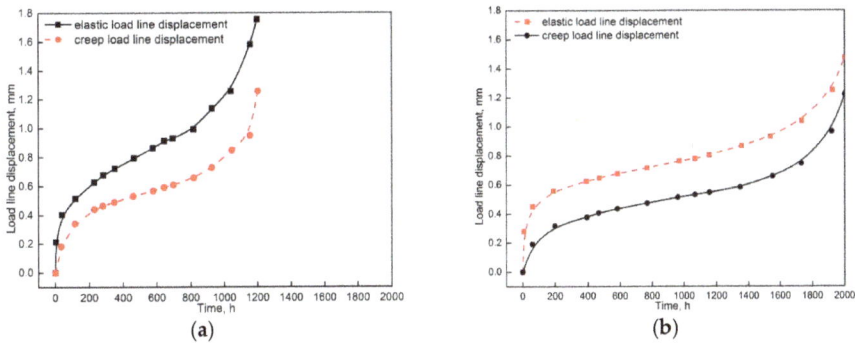

Figure 9. The evolution of the load line displacement caused by elastic deformation and creep deformation under different residual stress levels. (**a**) High residual stress level and (**b**) low residual stress level.

Figure 9 shows the evolution of the simulated load line displacement caused by elastic deformation and creep deformation under the different residual stress levels. It can be seen from Figure 9a that the simulated load line displacement caused by creep deformation is obviously higher than that of elastic deformation under the introduced residual stress level higher than the material intrinsic flow stress level. According to Figure 9b, the simulated load line displacement caused by creep deformation is obviously smaller than that of elastic deformation under the introduced residual stress value level less than the material intrinsic flow stress level.

To investigate the effects of the residual stress on creep rupture behavior, the evolution of the creep fracture surface under the different residual stress levels was studied. Figure 10 shows the microscopic morphology of creep crack distribution at the lifetime fraction of 90% under two different prior residual stress levels. There is an obvious difference among the overall morphologies of crack distribution at different prior residual stress levels. The rough and zig-zag growth path of the crack propagation can be observed in Figure 10a. By contrast, the relative smooth and straight growth path of the crack propagation can be observed in Figure 10b.

Figure 10. Creep crack distribution at the lifetime fraction of 90%. (**a**) The introduced prior residual stress level higher than the material intrinsic flow stress level and (**b**) the introduced residual stress value level less than the material intrinsic flow stress level.

Figure 11 shows the SEM images of the creep fracture surface near the annular breach of the samples along with the axial direction of thickness with different residual stress levels. There is an obvious difference among the overall morphologies at different lifetime fractions of the specimen. As shown in Figure 11a, the specimen subjected to pure local out-of-plane compression loading shows a smooth and cleavage fracture surface. With an increase of time, the microstructure of the fracture surface has obviously changed when the specimen is subjected to a constant tensile stress of 110 MPa at the temperature of 650 °C. It can be seen from Figure 11b that the fracture surface tends to be rough and some small voids were formed on the fracture surface. As shown in Figure 11c,d, dimples are visible on the fracture surface. Figure 11c shows that a high tongue-shaped cleavage fracture occurred on the fracture surface, a ligament with a width between 5 um and 10 um appeared around the tip of the micro-crack, and a quasi-cleavage surface with obvious plastic deformation was observed. According to Figure 11d, the width of the ligament was increased up to about 20 um and some obvious cleavage steps occurred on the cleavage fracture sector. Moreover, the wedge shape cavities can be clearly observed, cavities nucleate at inclusions, and the combination of time and plasticity promotes their growth.

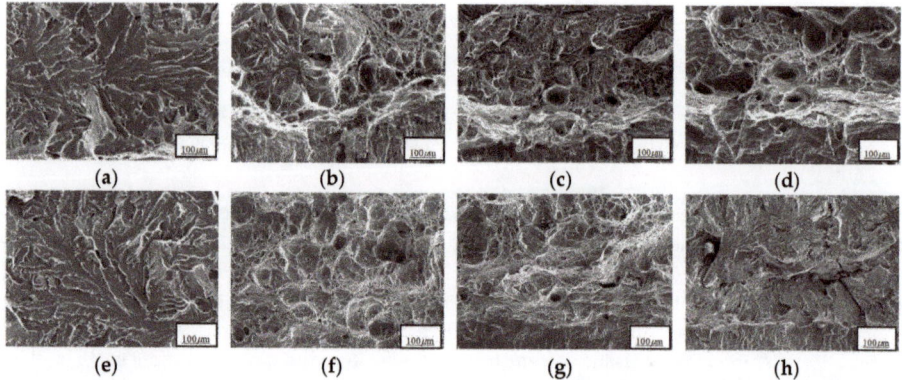

Figure 11. Creep fracture surface at different lifetime fractions of the specimen with different residual stress levels: high residual stress level (**a–d**): (**a**) 0%, (**b**) 50%, (**c**) 80%, and (**d**) 100%; and low residual stress level (**e–h**): (**e**) 0%, (**f**) 50%, (**g**) 80%, and (**h**) 100%.

As shown in Figure 11e–h, an obvious difference among the overall morphologies at different lifetime fractions of the specimen with the introduced prior residual stress value level less than the material intrinsic flow stress level was also observed. It can be seen from Figure 11e that dimples occurred on the quasi-cleavage surface. The evolution of lath boundaries from a low-angle to high-angle is apparent when the specimen is subjected to the creep loading, as shown in Figure 11f. At the lifetime fraction of 50%, voids can be obviously observed and the size of these voids is apparently smaller than that of the value of the prior residual stress that exceeds the material intrinsic flow stress through a comparison of Figure 11c,g. According to Figure 11f–h, the spheroidal-shape creep voids can be observed, the growth of the void by creep becomes slow, and many small voids are nucleated as a result of the fracture. Furthermore, the coalescence of discrete voids on prior grain boundaries were oriented approximately to the tensile stress axis. The width of the ligament was about 30 um and non-plastic lacerations were observed from Figure 11h.

4. Discussion

Using the FE analysis model described in Section 2.3, the creep strain and displacement of the whole rupture history of the P92 specimen under the two different residual stress levels were calculated and the predicted creep strain evolutions (Figures 7 and 8) indicate that larger creep strain accumulated in the vicinity of the notch tip during the stress relaxation process. However, the difference in the maximum creep strain value between the two different residual stress levels is obvious. With redistribution of the stress and strain during crack growth, the higher stress triaxiality could accelerate strain accumulation because the creep deformation was constrained by the hard crack growth. This means that residual stress could significantly affect the creep properties. According to the formula for describing the behavior of creep crack growth [27,28] in Section 2.2, the simulated load line displacement caused by elastic deformation and creep deformation under the different residual stress levels (Figure 9) indicated that the ductile fracture is predominant when the value of the prior residual stress exceeds the material intrinsic flow stress, while the brittle fracture is predominant when the value of the prior residual stress is less than the material intrinsic flow stress.

The cavitation process, which includes creep voids' nucleation, growth, and linkage, is the most important factor that weakens the material and results in rupture. Using the experimental method described in Section 2.1, the evolution of the creep fracture surface under the different residual stress levels was investigated. SEM images of the creep fracture surface in Figure 11 indicate that the creep rupture process under the two different residual stress levels was obviously different. Figure 11a–d shows the process of void growth associated with a wedge-type cavity at the triple point of the grain boundary, which indicates the degradation of creep ductility [32]. According to references [32,33], evolution of the creep fracture surface under the high residual stress level in Figure 11a–d proves that transgranular cracking has resulted in ductile failure and the plasticity-controlled cavity growth is the predominant mechanism. Figure 11c,f–h shows that spheroidal-shape creep voids can be observed and the growth of the void by creep occurs slowly. Many small voids are nucleated as a result of fracture, and the small spheroidal-shape creep voids can be effectively explained by the brittle mechanism [34]. According to references [34,35], evolution of the creep fracture surface under the low residual stress level in Figure 11e–h proves that intergranular dimple mixed cracking has resulted in brittle failure and nucleation-controlled constrained cavity growth is the predominant mechanism.

Through the FE simulation of the evolution of creep deformation variables, the load line displacement caused by creep deformation elastic deformation can be obtained. According to the formula employed for describing the behavior of creep crack growth [27,28], the fracture behavior can be defined as brittle and ductile fracture when the elastic load line displacement is larger and less than the creep load line displacement, respectively. Thus, the failure type of the P92 specimen under different prior residual stress levels can be determined by comparing the simulated load line displacement caused by elastic deformation and the simulated load line displacement caused creep deformation. Furthermore, the SEM images of the creep fracture surface show that failure types were observed as ductile failure when the introduced residual stress level was higher than the material intrinsic flow stress level and brittle failure when the introduced residual stress value level was less than the material intrinsic flow stress level. It can be observed that the FE simulated results agree well with experimental results. Based on the Gibbs free energy principle [20], the constitutive coupling relation between the temperature and material intrinsic flow stress can be established. Unlike the modification of current creep damage models based on creep test results to predict the life of high-temperature components in previous studies [10–15], the main innovative method in this study is the estimation of the influences of residual stress on the creep rupture mechanism through a comparison of stress levels between the prior residual stress and the material intrinsic flow stress. FE simulations and experimental observations have proven that the new method for assessing the creep rupture mechanism under different prior residual stress levels presented in this study is reliable.

5. Conclusions

In summary, this study has investigated the effect of prior residual stress on the creep rupture mechanism for P92 steel. The different prior residual stress levels were introduced into the P92 specimens, were are then subjected to thermal exposure at the temperature of 650 °C for accelerated creep tests. At the same time, the constitutive coupling relation between the temperature and material intrinsic flow stress was established based on the Gibbs free energy principle through the extraction of the chemical composition of the P92 specimen. The evolution of the creep rupture process under the different prior residual stress levels was studied by FEM and experimental observations. FE investigations for determining the creep failure type in the P92 specimen under different prior residual stress levels were conducted through the use of the existing formula for describing the behavior of creep crack growth, and a comparison of the creep failure mechanism between the FE simulations and experimental results was then conducted. The FE simulated results agree well with experimental results. It was found that when the introduced prior residual stress level is higher than the material intrinsic flow stress level, the transgranular cracking results in ductile failure and the plasticity-controlled cavity growth is the predominant mechanism. Additionally, the intergranular dimple mixed cracking results in brittle failure and nucleation-controlled constrained cavity growth is the predominant mechanism when the introduced prior residual stress level is less than the material intrinsic flow stress level. Moreover, it also demonstrates that material intrinsic flow stress can be used to assess the creep rupture mechanism under the different prior residual stress levels.

Author Contributions: Conceptualization, D.L.; methodology, D.L. and J.Z.; software, D.L. and Y.L.; validation, Y.L. and G.L.; formal analysis, X.X. and Y.L.; writing, D.L.; supervision, X.X.

Funding: This research was funded by the Hubei Superior and Distinctive Discipline Group of "Mechatronics and Automobiles" (No. XKQ2019009) and Hubei Key Laboratory of Power System Design and Test for Electrical Vehicle.

Conflicts of Interest: The authors declare no conflict of interest.

References

1. Skelton, R.P. Deformation, diffusion and ductility during creep—continuous void nucleation and creep-fatigue damage. *Mater. High Temp.* **2017**, *34*, 121–133. [CrossRef]
2. Auerkari, P.; Salonen, J.; Holmstrom, S.; Laukkanen, A.; Rantala, J.; Nikkarila, R. Creep damage and long term life modelling of an X20 steam line component. *Eng. Fail. Anal.* **2013**, *35*, 508–515. [CrossRef]
3. Khodamorad, S.H.; Fatmehsari, D.H.; Rezaie, H.R.; Sadeghipour, A. Analysis of ethylene cracking furnace tubes. *Eng. Fail. Anal.*. **2012**, *21*, 1–8. [CrossRef]
4. Lee, K.H.; Suh, J.Y.; Hong, S.M.; Huh, J.Y.; Jung, W.S. Microstructural evolution and creep-rupture life estimation of high-Cr martensitic heat-resistant steels. *Mater. Charact.* **2015**, *106*, 266–272. [CrossRef]
5. Wang, S.; Peng, D.; Chang, L.; Hui, X. Enhanced mechanical properties induced by refined heat treatment for 9Cr–0.5Mo–1.8W martensitic heat resistant steel. *Mater. Des.* **2013**, *50*, 174–180. [CrossRef]
6. Falat, L.; Výrostková, A.; Homolova, V.; Svoboda, M. Creep deformation and failure of E911/E911 and P92/P92 similar weld-joints. *Eng. Fail. Anal.* **2009**, *16*, 2114–2120. [CrossRef]
7. Watanabe, T.; Tabuchi, M.; Yamazaki, M.; Hongo, H.; Tanabe, T. Creep damage evaluation of 9Cr-1Mo-V-Nb steel welded joints showing type IV fracture. *Int. J. Press. Vessels Pip.* **2006**, *83*, 63–71. [CrossRef]
8. Chen, L.Y.; Wang, G.Z.; Tan, J.P.; Xuan, F.Z.; Tu, S.T. Effects of residual stress on creep damage and crack initiation in notched CT specimens of a Cr–Mo–V steel. *Eng. Fract. Mech.* **2013**, *97*, 80–91. [CrossRef]
9. Zhao, L.; Jing, H.; Xu, L.; Han, Y.; Xiu, J. Effect of residual stress on creep crack growth behavior in ASME P92 steel. *Eng. Fract. Mech.* **2013**, *110*, 233–248. [CrossRef]
10. Tezuka, H.; Sakurai, T. A trigger of Type IV damage and a new heat treatment procedure to suppress it. Microstructural investigations of long-term ex-service Cr-Mo steel pipe elbows. *Int. J. Press. Vessels Pip.* **2005**, *82*, 165–174. [CrossRef]
11. Kimura, K.; Sawada, K.; Kushima, H.; Kubo, K. Effect of stress on the creep deformation of ASME Grade P92/T92 steels. *Int. J. Mater. Res.* **2008**, *99*, 395–401. [CrossRef]
12. Dean, J.; Bradbury, A.; Aldrich-Smith, G.; Clyne, T.W. A procedure for extracting primary and secondary creep parameters from nanoindentation data. *Mech. Mater.* **2013**, *65*, 124–134. [CrossRef]
13. Sugiura, R.; Yokobori, A.T.; Sato, K.; Tabuchi, M.; Kobayashi, K.; Yatomi, M. Creep crack initiation and growth behavior in weldments of high Cr steels. *Strength Fract. Complexity* **2014**, *8*, 125–133.
14. Hyde, T.H.; Saber, M.; Sun, W. Testing and modelling of creep crack growth in compact tension specimens from a P91 weld at 650 °C. *Eng. Fract. Mech.* **2010**, *77*, 2946–2957. [CrossRef]
15. Basirat, M.; Shrestha, T.; Potirniche, G.P.; Charit, I.; Rink, K. A study of the creep behavior of modified 9Cr–1Mo steel using continuum-damage modeling. *Int. J. Plast.* **2012**, *37*, 95–107. [CrossRef]
16. Rouse, J.P.; Cortellino, F.; Sun, W.; Hyde, T.H.; Shingledecker, J. Small punch creep testing: review on modelling and data interpretation. *Mater. Sci. Technol.* **2013**, *29*, 1328–1345. [CrossRef]
17. Mahmoudi, A.H.; Truman, C.E.; Smith, D.J. Using local out-of-plane compression (LOPC) to study the effects of residual stress on apparent fracture toughness. *Eng. Fract. Mech.* **2008**, *75*, 1516–1534. [CrossRef]

18. Cao, J.R.; Liu, Z.D.; Cheng, S.C.; Yang, G.; Xie, J.X. Influences of strain rate and deformation temperature on flow stress and critical dynamic recrystallization of heat resistant steel T122. *Acta Metall. Sinica* **2007**, *43*, 35–40.

19. Liu, D.Z.; Li, Y.; Liu, H.S.; Wang, Z.R.; Wang, Y. Numerical Investigations on Residual Stress in Laser Penetration Welding Process of Ultrafine-Grained Steel. *Adv. Mater. Sci. Eng.* **2018**, *2018*, 8609325. [CrossRef]

20. Torres, J.D.; Faria, E.A.; Prado, A.G. Thermodynamic studies of the interaction at the solid/liquid interface between metal ions and cellulose modified with ethylenediamine. *J. Hazard. Mater.* **2006**, *129*, 239–243. [CrossRef]

21. Kirkaldy, J.S.; Vanugopalan, D. Prediction of microstructure and hardenability in low alloy steels. In Proceedings of the International Conference on Phase Transformation in Ferrous Alloys; Philadelphia, PA, USA, 4–6 October 1983, Marder, A.R., Goldstein, J.I., Eds.; TMS-AIME: Warrendale, PA, USA, 1984; pp. 125–148.

22. Ye, Q.B.; Xie, Q.; Liu, Z.Y.; Wang, G.D. Effect of ultrafast cooling on preventing abnormal microstructural banding at centerline of steel plate. *J. Northeastern Univ.* **2017**, *38*, 1696–1702.

23. Rice, J.; Tracey, D. Computational fracture mechanics. In *Numerical and Computer Methods in Structural Mechanics*; Fenves, S.J., Ed.; Academic Press: New York, NY, USA, 1973; pp. 585–623.

24. Hayhurst, D.R.; Hayhurst, R.J.; Vakili-Tahami, F. Continuum damage mechanics predictions of creep damage initiation and growth in ferritic steel weldments in a medium bore branched pipe under constant pressure at 590 °C using a five-material weld model. *Proc. Math. Phys. Eng. Sci.* **2005**, *461*, 2303–2326. [CrossRef]

25. Zhao, L.; Jing, H.Y.; Xu, L.Y.; An, J.C.; Xiao, G.C. Numerical investigation of factors affecting creep damage accumulation in ASME P92 steel welded joint. *Mater. Des.* **2012**, *34*, 566–575. [CrossRef]

26. Rui, B.; Hao, L.; Lu, S.; Yue, C.; Fei, B. Experimental investigation of creep crack growth behavior in nickel base superally by constant displacement loading method at elevated temperature. *Mater. Sci. Eng. A* **2016**, *665*, 161–170.

27. Zhou, H.; Mehmanparast, A.; Davies, C.M.; Nikbin, K.M. Evaluation of fracture mechanics parameters for bimaterial compact tension specimens. *Mater. Res. Innovations* **2013**, *17*, 318–322. [CrossRef]

28. Zhao, L. Research on life assessment method considering constraint effect for P92 pipe with defects at elevated temperature. Ph.D. Thesis, Tianjin University, Tianjin, China, 2012.

29. Yatomi, M.; O'Dowd, N.P.; Nikbin, K.M.; Webster, G.A. Theoretical and numerical modelling of creep crack growth in a carbon–manganese steel. *Eng. Fract. Mech.* **2006**, *73*, 1158–1175. [CrossRef]

30. Ling, X.; Tu, S.T.; Gong, J.M. Application of Runge–Kutta–Merson algorithm for creep damage analysis. *Int. J. Press. Vessels Pip.* **2000**, *77*, 243–248. [CrossRef]

31. Humphreys, C.J. The significance of bragg's law in electron diffraction and microscopy, and bragg's second law. *Acta Crystallogr.* **2013**, *69*, 45–50. [CrossRef]

32. Nix, W.D.; Matlock, D.K.; Dimelfi, R.J. A model for creep fracture based on the plastic growth of cavities at the tips of grain boundary wedge cracks. *Acta Metall.* **1977**, *25*, 495–503. [CrossRef]

33. Hayhurst, D.R.; Lin, J.; Hayhurst, R.J. Failure in notched tension bars due to high-temperature creep: Interaction between nucleation controlled cavity growth and continuum cavity growth. *Int. J. Solids Struct.* **2008**, *45*, 2233–2250. [CrossRef]

34. Crichton, G.C.; Karlsson, P.W.; Pedersen, A. Partial discharges in ellipsoidal and spheroidal voids. *IEEE Trans. Electr. Insul.* **1989**, *24*, 335–342. [CrossRef]

35. Anderson, P.M.; Rice, J.R. Constrained creep cavitation of grain boundary facets. *Acta Metall.* **1985**, *33*, 409–422. [CrossRef]

metals

MDPI

Review
Creep-Ductility of High Temperature Steels: A Review

Stuart Holdsworth [1,2]

[1] Inspire Centre for Mechanical Integrity, 8600 Dübendorf, Switzerland; stuart.holdsworth@empa.ch;
    Tel.: +41-58-765-4732
[2] Empa: Swiss Federal Laboratories for Materials Science & Technology, 8600 Dübendorf, Switzerland

Received: 27 February 2019; Accepted: 14 March 2019; Published: 18 March 2019

Abstract: A number of measures of the creep-ductility of high temperature steels are reviewed with an ultimate focus on intrinsic creep-ductility. It is assumed that there will be a future requirement for the determination of long duration creep ductility values for design and product standards in the same way as there is currently for creep strength values. The determination of such information will require specialist modelling techniques to be applied to the complex nature of multi-temperature, multi-heat (multi-cast) data collations, and possible solutions are considered. In service, the exhaustion of creep-ductility is most likely to occur at stress concentrations, and for this, a knowledge of the multiaxial creep-ductility is required, and its relationship to uniaxial creep-ductility. Some practical applications requiring a knowledge of creep-ductility are reviewed.

Keywords: creep ductility; intrinsic ductility; modelling; multiaxiality

1. Introduction

The following review concerns the creep ductility of high temperature steels. At elevated temperatures, metallic structures deform with time under load. Ultimately, the accumulation of such plastic deformation leads to fracture by a creep-rupture mechanism (Figure 1). A list of the symbols and terminology adopted in this paper is given in the Nomenclature section.

Creep strain is typically accumulated in three stages. During the primary creep stage, creep rates decrease with time due to strain hardening until a minimum ($\dot{\varepsilon}_{c,min}$, or steady secondary) creep rate is achieved [1,2]. During the secondary creep stage, there is a balance between strain hardening and thermal softening (dynamic recovery) [3,4], and the creep rate remains approximately constant ($\dot{\varepsilon}_{c,min}$). During the tertiary stage, creep rates increase due to thermal softening (microstructural change), damage accumulation, and/or associated net section reduction, until rupture occurs [5]. Creep curve profiles during the three stages are determined by the respective magnitudes of temperature and stress, and can be strongly influenced by relatively small differences in chemical composition, e.g., Reference [6].

Above T_c, dislocations can more readily cross-slip and climb, the latter being a diffusion controlled mechanism. With increasing temperature, more slip systems become operative, resulting in easier cross-slip and climb responsible for dislocation rearrangement into sub-boundaries by polygonisation. Damage accumulation during tertiary creep may be a consequence of plastic hole growth [7], diffusion controlled cavity growth [8,9], or constrained cavity growth [10], as the magnitude of stress decreases. In general, creep ductility is determined by the superposition of strains accumulated due to void nucleation and growth [11].

Traditionally, creep ductility is defined in terms of rupture elongation, A_u (Equation (1a)) or reduction of area at rupture, Z_u (Equation (1b)) [12].

$$A_u = \frac{L_{r,u} - L_{r,o}}{L_{r,o}} \tag{1a}$$

$$Z_u = \frac{S_o - S_u}{S_o} \tag{1b}$$

In practice, rupture elongation and reduction of area at rupture are the most widely reported creep ductility parameters, although their magnitudes are known to be dependent on specimen geometry. In such circumstances, like-with-like comparisons should only strictly be made when A_u and Z_u are determined using proportional specimens with specific dimensional configurations, i.e., in terms of $L_{r,o} = k\sqrt{S_o}$, where k is usually ≥ 5 (often 5.65) for stress-rupture specimens and ideally ~11 for high precision creep strain specimens [12,13].

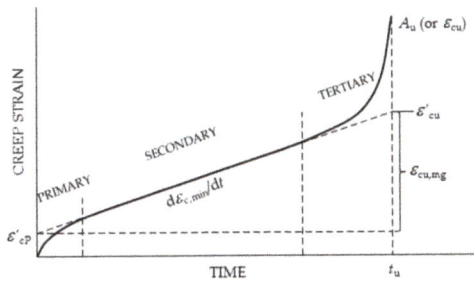

Figure 1. Variation of creep strain with time to rupture in constant load tests (creep rupture strain and elongation at rupture are only the same when there is insignificant instantaneous plastic strain on loading).

It is important to recognize that constant load tests of the type depicted in Figure 1 may actually be conducted with or without strain measurement (i.e., respectively in creep-rupture or stress-rupture tests). As a generality (and in particular for lower cost medium-to-long duration power generation industry material characterization in the mid-to-late 1900s), stress-rupture tests were mainly conducted without strain measurement to only give t_u (σ_o), A_u and Z_u. There was therefore no intermediate strain (or strain rate) data for the majority of creep tests performed for industry in this era (e.g., Reference [14]). For low alloy ferritic and higher alloy martensitic steels, for which the instantaneous plastic strain on loading (ε_o) is negligible in creep tests typically conducted for engineering assessments, A_u and creep strain at rupture (ε_{cu}) are identical. This is not necessarily the case for austenitic steels in higher load tests at lower temperatures for which the accumulated instantaneous plastic strain on loading can be significant, and A_u comprises both ε_o and ε_{cu}.

Of the two most common direct measures of creep ductility, Z_u is often the preferred quantity because its magnitude is considered to be less dependent on specimen geometry (k), although this may be more related to the method of determination being more repeatable and reproducible. Z_u may be alternatively expressed as a local strain to rupture in the region of necking, i.e.,

$$A_{u,loc} = \ln\left(\frac{1}{1 - Z_u}\right) = \varepsilon_{cu,loc} \tag{2}$$

With all their limitations A_u and Z_u are the most widely reported (and sometimes the only available) measures of creep ductility in large international data collations (e.g., Reference [14]), and therefore cannot be ignored.

Increasingly, there is a tendency to determine the uniform elongation at rupture, $\varepsilon_{cu,u}$, as a means of eliminating the geometry dependent component of creep ductility associated with necking (e.g., Reference [15]), Equation (3), with the meaning of the symbols shown in Figure 2a.

$$A_{u,u} = \left(\frac{d_o}{(d_{u1} - d_{u2})/2} \right)^2 - 1 \tag{3}$$

The evidence for a number of engineering steels indicates that the magnitude of $A_{u,u}$ is comparable with A_u for $A_u \leq {\sim}5\%$, and approximately constant at $A_{u,u} \geq {\sim}10\%$ for $A_u \geq {\sim}20\%$ [15], Figure 2b.

Other creep ductility concepts are introduced later in the paper, most notably intrinsic creep ductility. The intrinsic ductility, ε'_{cu}, in Figure 1, is regarded by many (e.g., References [16,17]) as being the most practical indicator of creep ductility.

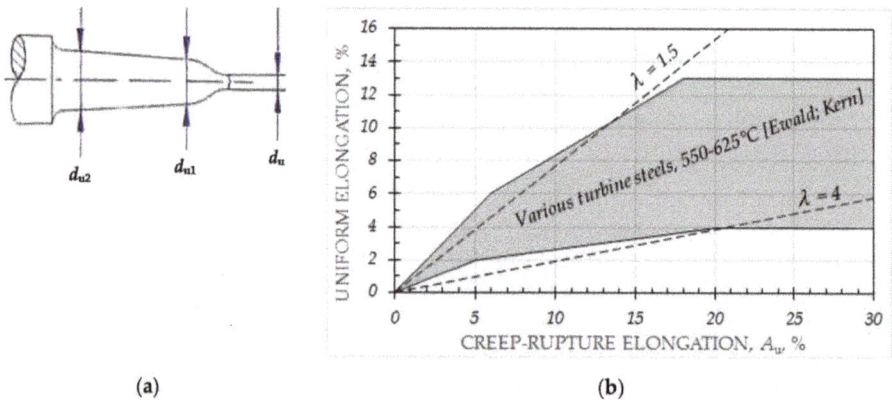

(a)   (b)

Figure 2. (**a**) Quantities required for the determination of uniform elongation at rupture, (**b**) Schematic of relationship between uniform elongation ($A_{u,u}$) and rupture elongation (A_u, ε_{cu}) for various turbine steels at temperatures in the range of 550–625 °C [15] (the significance of the λ lines is discussed later).

In the following text, the uniform elongation at rupture ($A_{u,u}$) is assumed to provide a reasonable approximation of the intrinsic creep ductility (ε'_{cu}, assuming insignificant instantaneous plastic strain on loading). This is the basis for the λ line constructions in Figure 2b (i.e., $\lambda \sim A_u/A_{u,u} \sim \varepsilon_{cu}/\varepsilon_{cu,u}$).

Values of creep ductility are increasingly being used to form the basis of high temperature component long time creep strain limits, in particular those with stress concentrating features, the interest being in service duty involving both steady and cyclic loading. As will be seen, there is a requirement for multi-temperature creep ductility models for a given alloy based on observations for a number of heats (or casts), and typically from a number of sources) [18]. While there are models becoming available for this purpose, their implementation is complicated by the inherent variability in the source experimental observations and heat (or cast) sensitive variations in underlying mechanism changes after long times.

2. Creep Ductility

For many high temperature steels, the variation of creep ductility with time and temperature can be complex, due in part to associated mechanism changes (e.g., Figure 3a). For example, for ferritic steels at a constant temperature above T_c, four creep regimes can be exhibited. In Regime-I (at high stresses), ductility is high, with ductile-transgranular rupture resulting from the formation of voids due to particle/matrix decohesion (and plastic hole growth) [19]. Regime-II is a transition region in which the ductility drops due to the increasing incidence of grain boundary cavitation, but still accompanied by relatively high levels of matrix deformation (and diffusion controlled cavity growth). In Regime-III, rupture is by the nucleation and subsequent diffusive growth of grain boundary cavities (constrained cavity growth). Ultimately, in Regime-IV, over-ageing of the microstructure reduces the rate of cavity nucleation and/or growth leading to a progressive increase in ductility. In practice, times to the start and end of each mechanism regime typically vary with temperature (e.g., Figure 3b).

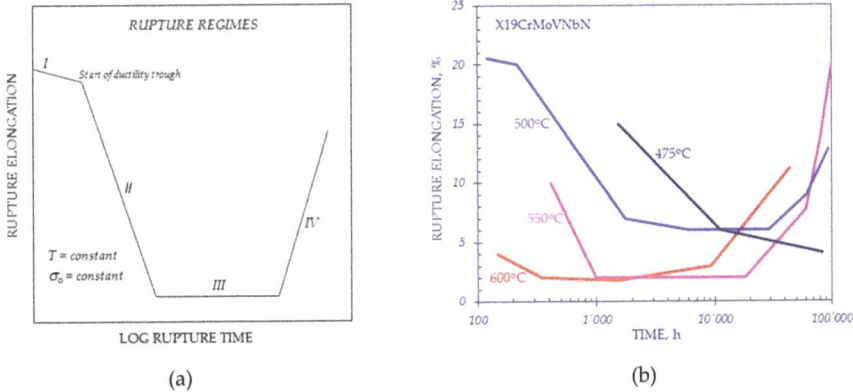

(a)
(b)

Figure 3. (a) Schematic representation of sequence of creep-rupture regimes for ferritic steels, and (b) experimental evidence of the influence of temperature on kinetics of creep-rupture regime development (data mainly available for regimes II, III and IV).

The multi-temperature creep ductility characteristics of alloys exhibiting a number of mechanism regimes is clear, although the evidence in Figure 3b does not reflect the within-heat (within-cast) test-to-test variability typically experienced with creep ductility observations. Consequently, the level of complexity increases yet again for multi-heat, multi-temperature creep ductility data collations, further complicating the process of model fitting.

A number of models have already been evaluated for predicting the behaviour depicted in Figure 3b (e.g., Reference [20]). These include adaptations of the algebraic models used to represent creep-rupture strength (e.g., Reference [21]):

$$\varepsilon_{cu} = \min\left(\varepsilon_{cu,U}, \max\left[\varepsilon_{cu,L}, a_1 \cdot T_K{}^{b_1} \cdot \sigma^{c_1} \cdot \exp(d_1/T_K + e_1\sigma/T_K)\right]\right) \tag{4}$$

and the stress modified ductility exhaustion model [22], i.e.,

$$\varepsilon_{cu} = \min\left(\varepsilon_{cu,U}, \max\left[\varepsilon_{cu,L}, C_1 \cdot e^{Q*/T_K} \cdot (\dot{\varepsilon}_{c,min})^{n_1} \cdot \sigma^{m_1}\right]\right) \tag{5}$$

Regimes I, II and III of the ductility (time) diagram in Figure 3a may be alternatively expressed in terms of creep ductility as a function of creep strain rate (Figure 4). This provides a graphical representation of Equation (5) (for a given stress), and the plastic hole growth, diffusion controlled cavity growth and constrained cavity growth damage accumulation regimes referred to earlier.

Figure 4 indicates creep ductility responses associated with two types of cavity growth mechanism. In regime II where cavity growth is diffusion controlled [8], creep ductility reduces with reducing strain rate, whereas, in regime III where cavity growth is constrained [10], creep ductility is independent of creep strain rate.

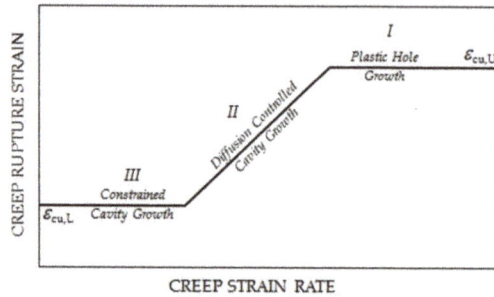

Figure 4. Variation of creep rupture strain with creep strain rate and the influence of damage mechanism on creep ductility (the indicated regime numbers coincide with those in Figure 3a).

While Equation (5) can effectively model the multi-temperature creep ductility behavior of a single heat (cast) of material, its effectiveness is more limited when the multi-temperature data collation to be assessed is comprised of observations from a number of heats. In order to account for heat-to-heat variability, Binda modified Equation (5) to include a material pedigree parameter, i.e., $f(MP)$ [23]. For 1CrMoV rotor steel, Binda first derived a number of independent and simple-interaction $f(mp)$ functions to represent the influences of individual chemical elements and heat treatment procedures on creep ductility. These were then combined into the single $f(MP)$ parameter contained in Equation (6), i.e.,

$$\varepsilon_{cu} = \min\left(\varepsilon_{cu,U}, \max\left[\varepsilon_{cu,L}, C_1 \cdot f(MP) \cdot e^{Q*/T_K} \cdot \left(\dot{\varepsilon}_{c,min} \right)^{n_1} \cdot \sigma^{m_1} \right] \right) \tag{6}$$

The ability of Equation (6) to predict creep rupture strains for 1CrMoV steel heats with different material pedigree characteristics is demonstrated by the black lines for the stresses indicated in Figure 5 [13].

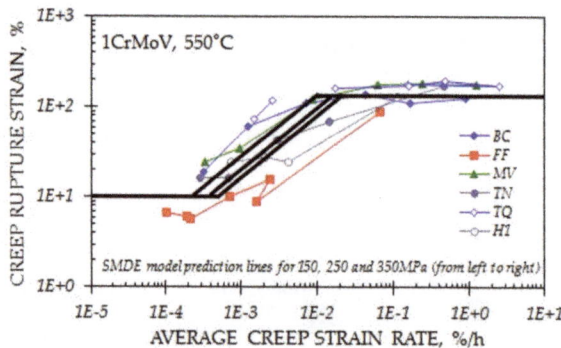

Figure 5. Comparisons of observed variations in creep rupture strain as a function of average creep strain rate for a number of heats of 1CrMoV rotor steel at 550 °C with Binda modified SMDE predicted behaviour (for $f(MP) = 1$, with the identifiers in the legend referring to different heats of the steel [13]).

In the original study, Binda indicated a more than 50% reduction in ductility prediction variance by the adoption of an $f(MP)$ material pedigree parameter [23].

3. Intrinsic Ductility

3.1. General

It has already been acknowledged that intrinsic ductility, as defined by ε'_{cu} in Figure 1, is regarded by many (e.g., [16,17]) as being the most practical indicator of creep ductility, with $\lambda =$

A_u/ε'_{cu}, (or $\varepsilon_{cu}/\varepsilon'_{cu}$) reflecting the capacity of a steel to redistribute stresses in structural components. As defined, the determination of ε'_{cu} requires the measurement of creep strain development during the course of test. However, it has been postulated that uniform elongation at rupture could provide an indication of intrinsic ductility. If this can be shown to be true, there would be a means of defining ε'_{cu} from tests without continuous strain measurement, i.e., through the measurement of $A_{u,u}$.

By assuming that $\lambda \sim A_u/A_{u,u}$, indicative upper and lower-bound loci represent λ for various turbine steels at 550–625 °C and have been included in Figure 2a. These compare with actual λ values of ~5–7 for $\frac{1}{2}$CrMoV steels at 565 °C [24] and for Monkman-Grant based λ values of ~3–9 for 18Cr8Ni steels at 600 °C (and ~13–30 for 18Cr8Ni steels at 650 °C) [25].

In the following sections, other quantities representing intrinsic ductility are examined, i.e., those due to Monkman-Grant and to Woodward.

3.2. Monkman-Grant

The sensitivity of the tertiary strain component of A_u (and Z_u) to specimen geometry has at least in part been responsible for the consideration of other creep ductility parameters, such as the Monkman-Grant ductility, $\varepsilon_{cu,mg}$ [26]. The Monkman-Grant ductility is not universally accepted, not least because it can ignore a significant component of primary creep strain (Figure 1). Primary creep can be as high as ~0.5% in some steels. Nevertheless, $\varepsilon_{cu,mg}$ is importantly not dominated by a specimen dependent tertiary creep strain component, and the ratio $\lambda = A_u/\varepsilon_{cu,mg}$ (or $\varepsilon_{cu}/\varepsilon_{cu,mg}$) is regarded as being a reasonable approximation of $\lambda = A_u/\varepsilon'_{cu}$, (or $\varepsilon_{cu}/\varepsilon'_{cu}$) [25].

While there is currently no known evidence to directly relate $\varepsilon_{cu,mg}$ to $\varepsilon_{cu,u}$, it is postulated that the results of future investigations will confirm the two quantities to be similar, in particular in circumstances for which instantaneous plastic strain accumulation is negligible and primary creep strain is small.

In terms of the Monkman-Grant ductility, ε'_{cu} (as defined in Figure 1) comprises two components, i.e.,

$$\varepsilon'_{cu} = \varepsilon'_{cP} + \varepsilon_{cu,mg} \tag{7}$$

where $\varepsilon_{cu,mg}$ is defined by a relationship proposed by Monkman-Grant [26], i.e.,

$$t_u \cdot \left(\dot{\varepsilon}_{c,min}\right)^{v} = D_{mg} \tag{8}$$

from which $\varepsilon_{cu,mg}$ may be derived by the substitution of $\varepsilon_{cu,mg} = \dot{\varepsilon}_{c,min} t_u$, i.e.,

$$\varepsilon_{cu,mg} = D_{mg} \cdot \left(\dot{\varepsilon}_{c,min}\right)^{1-v} \tag{9}$$

with v often regarded as being equal to unity, although this has been shown to be not always the case in practice [25].

It is not uncommon for ε'_{cP} in Equation (7) to be ignored (either for simplicity, or when the extent of primary creep is relatively low). In such circumstances, ε'_{cu} is simply defined by Equation (9).

3.3. Woodford

Woodford has developed an alternative concept of intrinsic ductility which is the strain rate sensitivity determined by differentiating log stress versus log strain rate curves determined from the results of stress relaxation tests (e.g., Reference [27]). Originally, the concept was based on a microstructurally independent experimental correlation between strain rate sensitivity and elongation at creep rupture for a wide range of metals and alloys [28], but is more recently regarded by Woodford as being verified by a theoretical treatment [29].

The advantage of the concept is said to be that intrinsic ductility minima for a range of engineering materials can be reliably identified in short duration stress relaxation tests [27,30]. Inherently, this claim is regarded as dubious since ductility troughs are entered not only because of mechanism changes associated with reducing stress (and strain rate), but also as a consequence of long-time thermally induced metallurgical changes that cannot possibly be reproduced in short duration stress relaxation tests (<~100 h) at the same temperature.

4. Multiaxial Ductility

In practice, the generation of creep damage typically occurs at stress concentrations, due to the local exhaustion of multiaxial ductility. Various models are available to predict the magnitude of multiaxial ductility relative to uniaxial ductility, and are invariably a function of the hydrostatic to equivalent stress ratio (σ_m/σ_{eq}). A number of these are reviewed below in terms of $\varepsilon_{cu}^*/\varepsilon_{cu}$, while recognizing that the multiaxial ductility ratio should ideally be expressed in terms of intrinsic ductility, i.e., $\varepsilon'^*_{cu}/\varepsilon'_{cu}$.

The model due to Rice & Tracey is based on their observations relating to the exponential amplification of plastic void growth rate as a consequence of stress triaxiality [7], i.e.,

$$\frac{\varepsilon_{cu}^*}{\varepsilon_{cu}} = \exp\left(\frac{1}{2} - \frac{3\sigma_m}{2\sigma_{eq}}\right) \tag{10}$$

An alternative physically based multiaxial ductility model is that proposed by Cocks & Ashby, which assumes that grain boundary cavities develop due to power law creep of the surrounding matrix [32], i.e.,

$$\frac{\varepsilon_{cu}^*}{\varepsilon_{cu}} = \sinh\left[\frac{2}{3} \cdot \left(\frac{n-0.5}{n+0.5}\right)\right] / \sinh\left[2 \cdot \left(\frac{n-0.5}{n+0.5}\right) \cdot \frac{\sigma_m}{\sigma_{eq}}\right] \tag{11}$$

It has been demonstrated that Equation (11) can strongly overestimate ε^*_{cu} in some circumstances, and has been modified with the intention of overcoming this problem [33], i.e., by:

$$\frac{\varepsilon_{cu}^*}{\varepsilon_{cu}} = \exp\left[\frac{2}{3} \cdot \left(\frac{n-0.5}{n+0.5}\right)\right] / \exp\left[2 \cdot \left(\frac{n-0.5}{n+0.5}\right) \cdot \frac{\sigma_m}{\sigma_{eq}}\right] \tag{12}$$

These physically based relationships compare with empirical models of the type proposed by Manjoine [34], the latest of which is given by:

$$\frac{\varepsilon_{cu}^*}{\varepsilon_{cu}} = 2^{\left(1-\frac{3\sigma_m}{\sigma_{eq}}\right)} \tag{13}$$

Based on the Rice & Tracey model, Spindler proposed [31]

$$\frac{\varepsilon_{cu}^*}{\varepsilon_{cu}} = \exp\left[p \cdot \left(1 - \frac{\sigma_1}{\sigma_{eq}}\right) + q \cdot \left(\frac{1}{2} - \frac{3\sigma_m}{2\sigma_{eq}}\right)\right] \tag{14}$$

with p and q as fitting parameters, being respectively 0.15 and 1.25 for TP316, and 2.38 and 1.04 for TP304.

The effectiveness of the cited multiaxial ductility models to represent available experimental observations for TP304 and TP316 stainless steels is given in Figure 6.

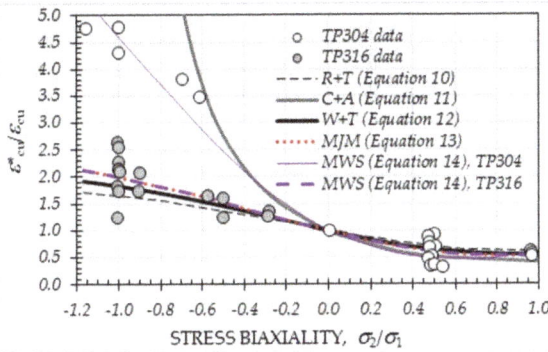

Figure 6. Comparison of multiaxial ductility model predictions with experimental observations for TP304 and TP316 stainless steels (data from [31]).

5. Practical Implications

5.1. Notch Sensitivity

Creep notch sensitivity is typically defined in terms of a ratio of the strength determined using a standard notched specimen relative to that measured in an un-notched specimen in the same time (usually in a time $<<\sim$10 kh). When this ratio is less than unity, the material is regarded as notch sensitive (or notch weakening). The duration at which a material becomes notch sensitive according to this definition relates to the time to the start of the ductility trough (as depicted in Figure 3).

Most engineering steels exhibit a creep-rupture strength scatterband reflecting the range of chemical compositions and heat treatment conditions permitted by the alloy specification (e.g., [13]). High strength materials in such scatterbands tend to be more susceptible to notch weakening, often this being because the time to entering the ductility trough is relatively short ($\leq\sim$1 kh). Bottom of the strength scatterband heats (casts) are often regarded as being notch insensitive because the time to enter the ductility trough for these materials is much greater than $\sim$10 kh (e.g., Figure 7).

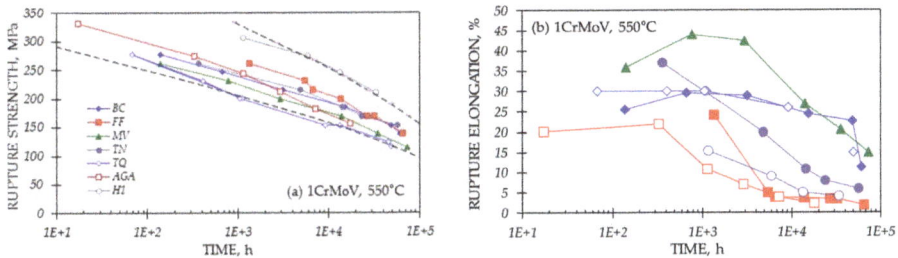

Figure 7. Variation of (a) rupture strength, and (b) rupture ductility, for different heats (casts) of 1CrMoV at 550 °C (only a small number of heats (casts) are shown from a much greater database to illustrate the variability trends [35–37]).

High temperature plant designers should be careful. Critical component lifetimes are typically well in excess of $\sim$100 kh, and the assessment of notch sensitivity should be based on the time to the start of the ductility trough with respect to such design lifetimes.

5.2. Creep Crack Incubation and Growth

Low creep ductility can influence creep crack incubation and creep crack growth times in power plant components containing pre-existing defects at high temperatures [38]. Typically, there is an incubation period prior to the onset of creep crack extension, during which time damage accumulates at the crack tip of the pre-existing defect, and the crack tip opens until a critical condition is exceeded in terms of damage accumulated (local ductility exhaustion of ε^*_{cu}) and the opening displacement ($\delta_{i,x}$), Figure 8. Creep crack incubation models reflect this through the inclusion of ε^*_{cu} (or better ε'^*_{cu}) in Equation (15) [5].

$$t_{i,x} = \left[\frac{\varepsilon^*_{cu}}{B \cdot \sigma^n}\right] \cdot \left[\frac{B \cdot I_n \cdot x}{C*}\right]$$ (15)

where $\dot{\varepsilon}_{min} = B \cdot \sigma^n$ and x is typically 0.2–0.5 mm (for engineering calculations).

There is also long standing experience of the influence of creep-ductility on creep crack growth rate (i.e., Equation (16)) [39], i.e.,

$$t_g = \int \frac{\varepsilon^*_{cu}}{\alpha' \cdot (C*)^\beta} \cdot da$$ (16)

It therefore appears that the overall creep lives of high temperature components containing pre-existing defects are directly influenced by the magnitude of creep ductility, i.e., $t_u = t_{i,x} + t_g$. In particular, low creep ductility is responsible for reductions in both creep crack incubation and creep crack growth times.

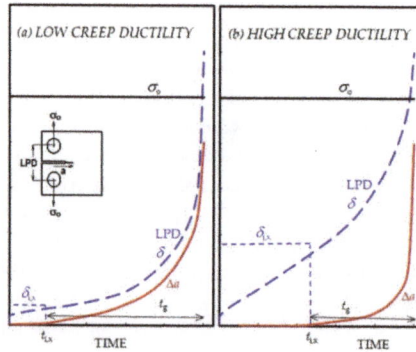

Figure 8. Crack incubation and crack growth contributions to lifetime of components with pre-existing defect subject to steady loading (σ_o) at high temperature, for steels with: (**a**) Low creep ductility, and (**b**) high creep ductility [38,40].

5.3. Creep-Fatigue

Increasingly, high temperature plant components are expected to operate at higher temperatures (to improve efficiency) and with greater operational flexibility. In such circumstances, the integrity of high temperature power plant components is strongly influenced by their resistance to creep-fatigue loading. The creep damage at critical locations, which are largely subjected to secondary self-equilibrating (thermal) loading, typically accumulates at the operating temperature later during steady running periods when creep strain rates and creep ductility are low (e.g., Figure 5). Here, the determination of accumulated creep damage using a ductility exhaustion model in conjunction with Equation (17) is a logical option [41,42].

$$D_c = \sum \left\{ N_i \left[\int_{.}^{t_h} \frac{\dot{\varepsilon}_c}{\varepsilon^*_{cu}(\dot{\varepsilon}_c, \sigma)} \cdot dt \right] \right\}$$ (17)

where $\dot{\varepsilon}_c = \frac{1}{E} \cdot \frac{d\sigma}{dt}$ and where the summed ductility is represented by the creep elements consumed during the respective stress relaxation phase of each cycle.

There is strong experimental evidence to indicate that creep-fatigue crack initiation endurances, for example for 1CrMoV rotor steel, are significantly influenced by creep ductility at 550 °C (Figure 9).

Creep ductility is strain rate sensitive for a number of engineering materials, and tends to the minimum value for the material heat (cast) at the lower relaxation rates experienced towards the end of longer hold (operating) times [43,44]. Creep ductility is exhausted earlier in lower ductility alloy heats (casts), these therefore being more susceptible to creep-fatigue damage accumulation and consequent crack initiation.

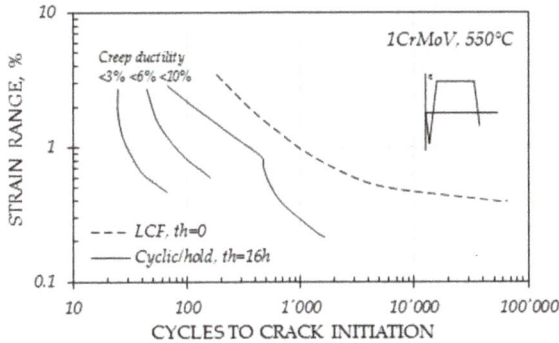

Figure 9. Influence of creep ductility on cyclic/hold creep-fatigue crack initiation endurance.

6. Concluding Remarks

Creep ductility is an important outcome of creep-rupture tests on high temperature steels, although there are a number of possible quantities which could be adopted and there is no consensus about which is the most appropriate parameter to be used to quantify this property. The candidates are reviewed. It is concluded that intrinsic ductility (as defined by Ainsworth and Goodall) is probably the most useful indicator of creep ductility, but acknowledged that, currently, this can only be determined from the results of tests with strain measurement. Unfortunately, there are many existing creep-rupture test results that do not include intermediate strain values (i.e., the results of stress-rupture tests).

Other indicators of intrinsic ductility are examined.

The way in which creep ductility varies with temperature and stress (and the underlying mechanisms) is relatively well known for many engineering alloys. The effective modelling of this parameter from the results of multi-temperature, multi-heat (multi-cast) data collations, in a comparable way to that of creep strength modelling, is currently a topic of on-going R&D. The complexity of large alloy creep ductility data collations is far greater than that of the equivalent creep strength data collations used to determine long time creep strength values for Design and Product Standards, and modelling therefore provides a challenge.

It is recognized that, in service, the exhaustion of creep ductility is most likely to occur at stress concentrations. This being the case, the important material property is not the creep ductility determined from uniaxial specimens, but multiaxial creep ductility. Models available to convert uniaxial ductility to multiaxial ductility are considered.

The assessment of creep notch sensitivity, creep crack incubation and growth, and creep damage accumulation during creep-fatigue transients, are practical examples requiring a knowledge of creep-ductility.

Funding: This research received no external funding.

Conflicts of Interest: The author declares no conflict of interest.

Nomenclature

a	Crack depth
$a_1 \ldots e_1$	Material constants (Equation (4))
A_u	Creep-rupture elongation ($\varepsilon_o + \varepsilon_{cu}$)
$A_{u,loc}$	Local creep rupture strain in region of necking ($\varepsilon_o + \varepsilon_{cu,loc}$)
$A_{u,u}$	Uniform creep rupture elongation ($\varepsilon_o + \varepsilon_{u,u}$)
B	Constant in steady state creep law
C_1	Constant in SMDE model
C^*	Parameter characterising crack tip stress and strain rate fields
d_o, d_{u1}	Initial net section diameter of gauge section, Net section diameter at rupture adjacent to necking
d_{u2}	Net section diameter at rupture at end of parallel length (Figure 2a)
D_c	Creep damage fraction
D_{mg}	Monkman-Grant constant
E	Elastic modulus
$f(mp)$	Individual condition (independent or simple interaction) material pedigree function
$f(MP)$	Overall material pedigree parameter (included in Equation (6))
I_n	Tabulated geometrical function [5] (Equation (15))
k	Original specimen dimension proportionality ($L_{r,o}/\sqrt{S_o}$) factor
$L_r, L_{r,o}, L_{r,u}$	Reference length, Initial reference length, Reference length at rupture
LPD	Load point displacement
m_1	Stress exponent in SMDE model (Equations (5) and (6))
n	Stress exponent for steady state creep
n_1	Strain rate exponent in SMDE model (Equations (5) and (6))
p	Material constant in Spindler multiaxial ductility model (Equation (14))
P, S, T	Primary, Secondary, Tertiary (creep)
q	Material constant in Spindler multiaxial ductility model (Equation (14))
Q^*	Constant relating to activation energy for diffusion creep
S_o, S_u	Initial cross sectional area of gauge section, Cross sectional area at rupture
SMDE	Stress modified ductility exhaustion
t, t_u	Time, Time to rupture
$t_{i,x}, t_g$	Creep crack incubation time (to crack initiation criterion, x), Creep crack growth time
T, T_K	Temperature (in °C), Temperature (in K)
T_c	Insignificant creep temperature
x	Crack initiation criterion
Z_u	Reduction of area at creep-rupture
α'	Creep crack growth constant in Equation (16)
β	C^* exponent in Equation (16)
$\delta, \delta_{i,x}$	Crack tip opening displacement, Critical crack tip opening displacement (at the onset of creep crack extension, $\Delta a = x$)
$\varepsilon, \varepsilon_c$	Strain, Creep strain
ε_{cu}	Creep rupture strain (equivalent to A_u, when there is insignificant instantaneous plastic strain on loading)
ε_{cn}^*	Multiaxial creep rupture strain
ε'_{cu}	Intrinsic creep-rupture strain (Figure 1)
$\varepsilon_{cu,L}, \varepsilon_{cu,U}$	Lower shelf creep rupture strain, Upper shelf creep rupture strain
$\varepsilon_{cu,loc}$	Local creep-rupture strain in region of necking (equivalent to $A_{u,loc}$, when there is insignificant instantaneous plastic strain on loading)
$\varepsilon_{cu,mg}$	Monkman-Grant ductility
$\varepsilon_{cu,u}$	Uniform creep rupture strain (equivalent to $A_{u,u}$, when there is insignificant instantaneous plastic strain on loading)
$\dot{\varepsilon}_{c,min}$	Minimum (steady secondary) creep rate, also referred to as $d\varepsilon_{c,min}/dt$ (in Figure 1)
ε'_{cP}	Fraction of primary creep not accommodated by Monkman-Grant ductility (see Figure 1)
ε_o	Instantaneous plastic strain

λ	Creep ductility ratio (A_u/ε'_{cu} or $\varepsilon_{cu}/\varepsilon'_{cu}$)
v	Monkman-Grant strain rate exponent
σ, σ_o	Stress, Initial stress in constant load creep test
σ_{eq}, σ_m	Equivalent stress, Hydrostatic stress
σ_{ref}, σ_1	Reference stress, Maximum principal stress

References

1. Bailey, R.W. Note on the softening of strain hardened metals and its relation to creep. *J. Inst. Met.* **1926**, *35*, 27–40.
2. Orowan, E. The creep of metals. *J. West Scotl. Iron Steel Inst.* **1946**, *54*, 45–53.
3. Norton, F.H. *Creep of Steel at High Temperatures*; McGraw-Hill: New York, NY, USA, 1929.
4. Weertman, J. Theory of steady-state creep based on dislocation climb. *J. Appl. Phys.* **1955**, *26*, 1213–1217. [CrossRef]
5. Riedel, H. *Fracture at High Temperatures*; Springer: Berlin/Heidelberg, Germany, 1987.
6. Pickering, F.B. Some aspects of creep deformation and fracture in steels. In Proceedings of the IOM Conference on Rupture Ductility of Creep Resistant Steels, York, UK, December 1990; The Institute of Metals: London, UK, 1990; pp. 17–48.
7. Rice, J.R.; Tracey, D.M. On the ductile enlargement of voids in triaxial stress fields. *J. Mech. Phys. Solids* **1969**, *17*, 201–217. [CrossRef]
8. Hull, D.; Rimmer, D.E. The growth of grain boundary voids under stress. *Philos. Mag.* **1959**, *4*, 673–687. [CrossRef]
9. Beere, W. Mechanism maps. In *Cavities and Crack in Creep and Fatigue*; Gittus, J., Ed.; Applied Science: London, UK, 1981; pp. 29–57.
10. Dyson, B.F. Constrained cavity growth, its use in quantifying recent creep fracture results. *Can. Metall. Q.* **1979**, *18*, 31–38. [CrossRef]
11. Ashby, M.F.; Gandhi, C.; Taplin, D.M.R. Fracture mechanism maps and their construction for FCC metals and alloys. *Acta Met.* **1979**, *27*, 699–729. [CrossRef]
12. ISO 204. *Metallic Materials—Uniaxial Creep Testing in Tension—Method of Test*; International Organisation for Standardisation: Geneva, Switzerland, 2018.
13. Holdsworth, S. Creep ductility of 1CrMoV rotor steel. *Mater. High Temp.* **2017**, *34*, 99–108. [CrossRef]
14. NIMS Creep Data Sheets. Japan National Institute for Materials Science. Available online: https://smds. nims.go.jp/creep/index_en.html (accessed on 31 January 2019).
15. Ewald, J.; Kern, T.-U. The role of the material parameter uniform elongation. In Proceedings of the 4th International ECCC Conference, Düsseldorf, Germany, 10–14 September 2017.
16. Goodall, I.V.; Ainsworth, R.A. Failure of structures by creep. In Proceedings of the 3rd International Conference on Pressure Vessel Technology, Tokyo, Japan, 19–22 April 1977; Volume II, pp. 871–885.
17. Hayhurst, D.R.; Webster, G.A. An overview on studies of stress state effects during creep of circumferentially notched bars. In Proceedings of the Symposium on Techniques for Multiaxial Creep Testing, Leatherhead, UK, 25–26 September 1985; pp. 137–175.
18. Holdsworth, S.R. The ECCC approach to creep data assessment. In Proceedings of the CREEP8 8th International Conference on Creep and Fatigue at Elevated Temperatures, San Antonio, TX, USA, 22–26 July 2007; pp. 661–667.
19. Holdsworth, S.R.; Beech, S.M. Microstructural factors affecting notch creep-rupture behaviour in high temperature power plant steels. In Proceedings of the IOM Conference on Rupture Ductility of Creep Resistant Steels, York, UK, December 1990; pp. 320–333.
20. Holdsworth, S.R.; Merckling, G. ECCC developments in the assessment of creep-rupture properties. In Proceedings of the 6th International Charles Parsons Turbine Conference on Engineering Issues in Turbine Machinery, Power Plant and Renewables, Trinity College, Dublin, Ireland, 16–18 September 2003; pp. 411–426.
21. Trunin, I.I.; Golobova, N.G.; Loginov, E.A. New method of extrapolation of creep test and long time strength results. In Proceedings of the 4th International Symposium on Heat Resistant Metallic Materials, Mala Fatra, Slovakia, 1971; p. 168.

22. Spindler, M.W. The multiaxial and uniaxial creep rupture ductility of Type 304 steel as a function of stress and strain rate. *Mater. High Temp.* **2004**, *21*, 47–52. [CrossRef]
23. Binda, L. Advanced Creep Damage and Deformation Assessment of Materials Subject to Steady and Cyclic Loading Conditions at High Temperatures. Ph.D. Thesis, ETH Zürich, Zürich, Switzerland, 2010.
24. Cane, B.J.; Browne, R.J. Representative stresses for creep deformation and failure of presurised tubes and pipes. *Int. J. Press. Vessel. Pip.* **1982**, *10*, 119–128. [CrossRef]
25. Skelton, R.P. Deformation, diffusion and ductility during creep—Continuous void nucleation and creep-fatigue damage. *Mater. High Temp.* **2017**, *34*, 121–133. [CrossRef]
26. Monkman, F.C.; Grant, N.J. An empirical relationship between rupture life and minimum creep rate in creep-rupture tests. *Proc. Am. Soc. Test. Mater.* **1956**, *56*, 593–620.
27. Woodford, D.A. Intrinsic ductility for structural materials as a function of stress and temperature. *Mater. High Temp.* **2017**, *34*, 134–139. [CrossRef]
28. Woodford, D.A. Strain rate sensitivity as a measure of ductility. *Trans. ASM* **1969**, *69*, 291–293.
29. Nichols, F.A. Plastic instabilities and uniaxial tensile ductilities. *Acta Metall.* **1980**, *28*, 663–673. [CrossRef]
30. Woodford, D.A. Performance-based creep strength and intrinsic ductility for a cast nickel-based superalloy. *Mater. High Temp.* **2018**, *35*, 399–409. [CrossRef]
31. Spindler, M.W. The multiaxial creep ductility of austenitic stainless steels. *Fatigue Fract. Eng. Mater. Struct.* **2004**, *27*, 273–281. [CrossRef]
32. Cocks, A.F.; Ashby, M.F. Intergranular fracture during power law creep under multiaxial stress. *Met. Sci.* **1980**, *8*, 395–402. [CrossRef]
33. Wen, J.-F.; Tu, S.-T.; Xuan, F.Z.; Zhang, X.W.; Gao, X.L. Effects of stress level and stress state on creep ductility: Evaluation of different models. *J. Mater. Sci. Technol.* **2016**, *32*, 695–704. [CrossRef]
34. Manjoine, M.J. Creep-rupture behaviour of weldments. *Weld. J. Res. Suppl.* **1982**, *61*, 50s–57s.
35. Branch, G.D.; Marriot, J.B.; Murphy, M.C. The creep and creep-rupture properties of six CrMoV rotor forgings for high temperature steam turbines. In Proceedings of the International Conference on Properties of Creep Resistant Steels, Dusseldorf, Germany, 1972.
36. Gooch, D.J.; Holdsworth, S.R.; McCarthy, P.R. The influence of net section area on the notched bar creep rupture lives of three power plant steels. In Proceedings of the Conference on Creep and Fracture of Engineering Materials and Structures, Swansea, UK, 5–10 April 1987; pp. 441–457.
37. Holdsworth, S.R.; Mazza, E. Using the results of crack incubation tests on 1CrMoV steel for predicting long-time creep-rupture properties. *Int. J. Press. Vessel. Pip.* **2009**, *86*, 838–844. [CrossRef]
38. Holdsworth, S.R. Initiation and early growth of creep cracks from pre-existing defects. *Mater. High Temp.* **1992**, *10*, 127–137. [CrossRef]
39. Neate, G.J. Creep crack growth behaviour in 0.5CrMoV steel at 838K. *Mater. Sci. Eng.* **1986**, *82*. Part I: Behaviour at a constant load, 59–76; Part II: Behaviour under displacement controlled loading, 77–84.
40. Holdsworth, S.R. Creep crack growth in low alloy steel weldments. *Mater. High Temp.* **1998**, *15*, 203–209. [CrossRef]
41. R5. *Assessment Procedure for the High Temperature Response of Structures*; EDF Energy Nuclear Generation Ltd.: Barnwood, UK, 2003.
42. Priest, R.H.; Ellison, E.G. A combined deformation map-ductility exhaustion approach to creep-fatigue analysis. *Mater. Sci. Eng.* **1981**, *49*, 7–17. [CrossRef]
43. Hales, R. A method of creep damage summation based on accumulated strain for the assessment of creep-fatigue analysis. *Fatigue Eng. Mater. Struct.* **1983**, *6*, 121–135. [CrossRef]
44. Holdsworth, S.R. Prediction of creep-fatigue behaviour at stress concentrations in 1CrMoV rotor steel. In Proceedings of the Conference on Life Assessment and Life Extension of Engineering Plant, Structures and Components, Churchill College, Cambridge, UK, 1996; pp. 137–146.

metals

MDPI

Article

Influence of Different Annealing Atmospheres on the Mechanical Properties of Freestanding MCrAlY Bond Coats Investigated by Micro-Tensile Creep Tests

Sven Giese [1,*], Steffen Neumeier [1], Jan Bergholz [2], Dmitry Naumenko [3], Willem J. Quadakkers [3], Robert Vaßen [2] and Mathias Göken [1]

[1] Materials Science & Engineering, Institute I, Friedrich-Alexander-Universität Erlangen-Nürnberg (FAU), Martensstr. 5, 91058 Erlangen, Germany; steffen.neumeier@fau.de (S.N.); mathias.goeken@fau.de (M.G.)
[2] IEK-1: Materials Synthesis and Processing, Institute of Energy and Climate Research, Forschungszentrum Jülich GmbH, Jülich 52425, Germany; Jan.Bergholz@de.rheinmetall.com (J.B.); r.vassen@fz-juelich.de (R.V.)
[3] IEK-2, Institute of Energy and Climate Research, Forschungszentrum Jülich GmbH, Jülich 52425, Germany; d.naumenko@fz-juelich.de (D.N.); j.quadakkers@fz-juelich.de (W.J.Q.)
* Correspondence: sven.giese@fau.de; Tel.: +49-09131-85-27475

Received: 22 May 2019; Accepted: 17 June 2019; Published: 19 June 2019

Abstract: The mechanical properties of low-pressure plasma sprayed (LPPS) MCrAlY (M = Ni, Co) bond coats, Amdry 386, Amdry 9954 and oxide dispersion strengthened (ODS) Amdry 9954 (named Amdry 9954 + ODS) were investigated after annealing in three atmospheres: $Ar–O_2$, $Ar–H_2O$, and $Ar–H_2–H_2O$. Freestanding bond coats were investigated to avoid any influence from the substrate. Miniaturized cylindrical tensile specimens were produced by a special grinding process and then tested in a thermomechanical analyzer (TMA) within a temperature range of 900–950 °C. Grain size and phase fraction of all bond coats were investigated by EBSD before testing and no difference in microstructure was revealed due to annealing in various atmospheres. The influence of annealing in different atmospheres on the creep strength was not very pronounced for the Co-based bond coats Amdry 9954 and Amdry 9954 + ODS in the tested conditions. The ODS bond coats revealed significantly higher creep strength but a lower strain to failure than the ODS-free Amdry 9954. The Ni-based bond coat Amdry 386 showed higher creep strength than Amdry 9954 due to the higher fraction of the β-NiAl phase. Additionally, its creep properties at 900 °C were much more affected by annealing in different atmospheres. The bond coat Amdry 386 annealed in an $Ar–H_2O$ atmosphere showed a significantly lower creep rate than the bond coat annealed in $Ar–O_2$ and $Ar–H_2–H_2O$ atmospheres.

Keywords: MCrAlY; TMA; creep; bond coat; hydrogen; water vapor

1. Introduction

Many scientific studies have been carried out to investigate the oxidation behavior of MCrAlY bond coats and intermetallic coatings in varying environments. However, far fewer investigations have been performed for bond coats used in environments bearing hydrogen or water vapor. One strategy to reduce the CO_2 emissions of power plants is to use a H_2-enriched syngas in gas turbines [1]. By using syngas with hydrogen or water vapor instead of natural gas, some problems may arise. The negative effect of hydrogen embrittlement in iron and steels has been well-known for decades [2]. Some studies investigated the influence of water vapor and especially hydrogen on the formation and agglomeration of nanovoids and vacancies in various materials [3,4]. Vacancy clusters were found in Ni alloys by measuring the variation in positron lifetime and the intensity of hydrogen-charged samples. These investigations revealed that grain boundaries are the preferred regions in which hydrogen can facilitate and stabilize vacancy clusters [3]. In addition, the diffusion of vacancies from the surface to the interior of the sample, which led to the formation of 20–200 nm holes, was observed in Ni samples after high temperature heat treatment [4]. The same behavior was found in Ni-base superalloys in which strain-induced agglomeration of vacancies during plastic deformation was observed by Takai et al., promoting fracture in Inconel 625 due to the creation of microvoids [5]. Furthermore, charging samples with hydrogen led to a remarkable reduction of the ductility of the Ni-base superalloy Inconel 718 during tensile tests by the coalescence of voids promoting crack initiation and propagation [6]. Localized deformation and stress concentration at grain boundaries and the growth and coalescence of voids led to the final failure of hydrogen-charged samples [7]. The formation of hydrogen bubbles and the agglomeration of voids was also observed, leading to blistering of the material and a general decrease of its mechanical properties [4,8]. Another problem caused by the addition of water vapor was the partially observed earlier spallation of thermal barrier coatings (TBCs) during cyclic tests [9].

The influence of H_2 and H_2O on the oxidation behavior of MCrAlY bond coats was investigated before by Zhou et al. as well as by Sullivan and Mumm [10,11]. The addition of water vapor led to an increase of $(Ni,Co)(Al,Cr)_2O_4$ spinel formation in bond coats, which influences the bond coat stability, the thermally grown oxide, and the TBC lifetime until spallation in syngas-fired turbines [11]. It was found that a higher P_{H_2O} led to stronger spinel growth for Amdry 386 at a higher P_{O_2}. The highest amount of spinel formation was observed at high P_{H_2O} and low P_{O_2}. Leyens et al. investigated the influence of environments containing water vapor and hydrogen on the oxidation resistance of Ni-based coatings and noticed that the addition of water vapor led to a 25% increase of the total mass gain after a heat treatment at 1100 °C for 28 h [12]. Subanovic et al. [8] found blistering of 2 mm thick freestanding MCrAlY bond coats exposed in a H_2/H_2O atmosphere. The effect was attributed to hydrogen release into the coating from the reaction between yttrium and water vapor and the subsequent formation of hydrogen gas at the internal coating defects (pores). Few experiments on freestanding bond coats have compared the influence of atmospheres containing H_2O and/or H_2 on creep properties. Micro-tensile creep testing of miniaturized samples under uniaxial loading has been used to evaluate the mechanical properties at the sub-micrometer scale [13–16]. A new testing methodology has recently been suggested in [17] where cylindrical samples were tested in a thermomechanical analyzer (TMA). The aim of this work is to investigate the effect of the heat treatment of freestanding Ni- and Co-bond coats under three different atmospheres containing oxygen, water vapor and hydrogen on creep resistance.

2. Experimental Methods

2.1. Material

Three MCrAlY bond coats were investigated in this work: A Ni-based bond coat, Amdry 386, and the two Co-based bond coats Amdry 9954, and oxide dispersion strengthened (ODS) Amdry 9954—the latter containing 2 wt.-% of Al_2O_3 for oxide dispersion strengthening. The chemical compositions are listed in Table 1.

Table 1. Atmospheres and nominal composition of Amdry 386 and oxide dispersion strengthened (ODS) Amdry 9954 in at.-% and O in ppm.

Material	Ni	Co	Al	Cr	Si	Y	Hf	O
Amdry 386	40.35	18.97	23.04	16.49	0.79	0.36	0.08	<500
Amdry 9954	28.68	34.18	15.57	21.25	0.13	0.32	-	<500
Amdry 9954 + ODS	28.82	34.01	17.14	20.14	0.05	0.20	-	9650

The coatings were produced by low-pressure plasma spraying (LPPS) on a steel substrate. To obtain freestanding bond coats, the steel substrate was spark eroded afterwards. A first homogenization annealing under an argon atmosphere for 2 h at 1100 °C was done. A second heat treatment was carried out for each bond coat in argon with additions of 20% O_2, 2% H_2O or 4% H_2 + 2% H_2O for 72 h at 1100 °C.

2.2. Sample Preparation

The exposed freestanding bond coat specimens were sliced with a precision wet abrasive cutting machine Brillant 220 (ATM GmbH, Mammelzen, Germany) from atm into a rectangular shape with a width of about 600 µm. For small-scale tensile testing, a circular sample cross section is recommended, resulting in a more reproducible manufacturing of the samples and equal testing conditions. These are huge advantages compared to the production of miniaturized samples with common fabrication methods like electrical discharge machining or mechanical milling [18,19]. To produce a cylindrical tensile specimen without notches and with high accuracy, the abrasive sliced samples were sent to Microsample GmbH in Austria for further preparation. The production of the miniaturized samples is described in detail in the work of Rathmayr et al. [20,21]. The cross sections of the samples were determined to be 450 ± 2 µm in diameter over a gauge length of 2500 µm. Microstructural analysis was performed with a Zeiss Crossbeam 540 (Zeiss GmbH, Oberkochen, Germany) with backscattered electron images (BSE). The grain sizes and individual phase fractions of all the bond coats were determined by electron backscatter diffraction (EBSD) (Oxford Instruments, Abingdon, UK) with the software ATZEC from Oxford Instruments. Polished sections of the tested specimens after creep experiments along the testing direction were characterized by ImageJ (version 1.46r, Wayne Rasband, Bethesda, Rockville, MD, USA).

2.3. Experimental Setup

Micro-tensile creep tests were carried out with a thermomechanical analyzer (TMA) type 402 F3 Hyperion from Netzsch (Netzsch, Selb, Germany)—see also [17] for more details. The experimental setup consists of a SiC furnace with a constant flow of 400 mL/min of Ar during testing. The sample holder, the pushrod and the clamping jaws for mounting the samples are made of alumina, which is thermally stable up to 1550 °C. The measuring system is thermally stabilized by constant water cooling during the experiments. Creep tests were performed at temperatures of 900–950 °C. A constant load of 50 mN was applied for fixing the samples during heating with 5 K/min until the testing temperature was achieved. A holding period of 90 min was used for reaching thermal equilibrium before testing. The force of the load-controlled experiment was increased from 0.05 N to 2.37 N in 1 min, resulting in a stress of 15 MPa in the 450 µm diameter at the beginning of the experiments. The TMA setup and a micro tensile specimen before testing can be seen in Figure 1.

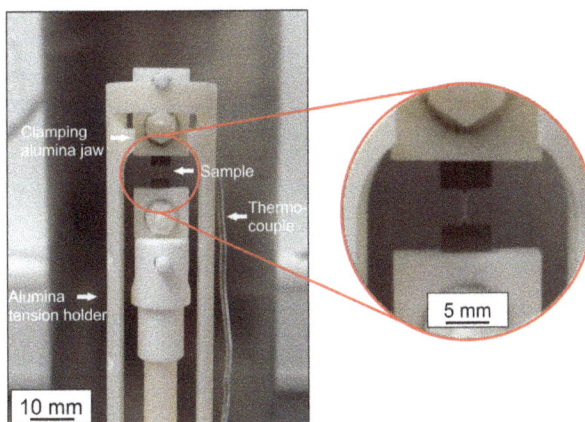

Figure 1. Thermomechanical analyzer (TMA) setup: alumina tension holder, pushrod and clamping jaw for mounting the sample. The polished tensile specimen has a diameter of 450 µm over the entire gauge length of 2500 µm before creep testing.

3. Results

3.1. Microstructure at the Initial State

The microstructures of Amdry 386, Amdry 9954 and Amdry 9954 + ODS after heat treatment at 1100 °C for 72 h in Ar–O_2, Ar–H_2O and Ar–H_2–H_2O are shown in Figure 2. No detrimental effect of oxygen, water vapor or hydrogen for all bond coats can be seen by microstructural analysis before the creep experiments. Blistering of the bond coats due to internal recombination of hydrogen, as described by Subanovic et al., was not found [8], probably due to a smaller Y-reservoir in the creep specimens associated with a smaller specimen thickness.

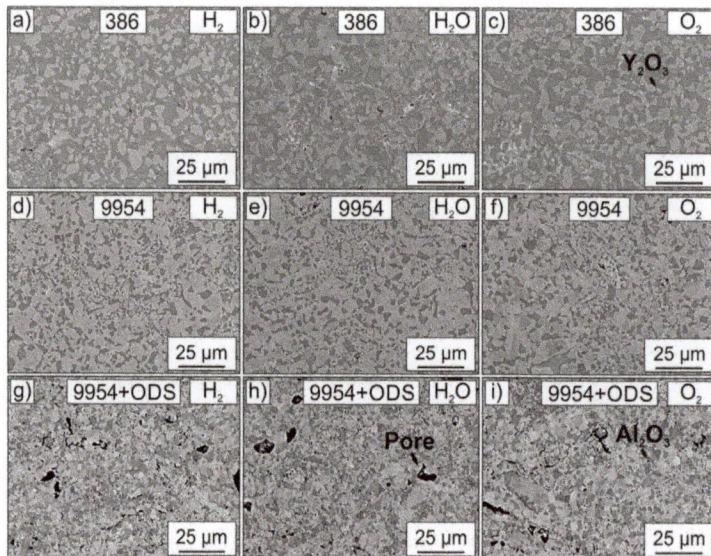

Figure 2. BSE (backscattered electron) images of the microstructures of Amdry 386 (**a–c**), Amdry 9954 (**d–f**) and Amdry 9954 + ODS (**g–i**) after a heat treatment of 72 h in Argon with the addition of 4% H_2 + 2% H_2O, 2% H_2O and 20% O_2 before creep testing.

Differences regarding the microstructure can only be attributed to the varying content of the β-NiAl phase and the γ-solid solution phase between Amdry 386 and Amdry 9954, as described in a previous study [17]. The formation of internal Y-rich oxides could be observed in all bond coats, similarly to the study by Huang et al. [22]. In addition to the two-phase β/γ-microstructure of Amdry 386 and Amdry 9954, the ODS-containing bond coats showed small, finely distributed alumina particles as well as accumulation of alumina.

Due to insufficient melting during the plasma spraying process, large, not completely molten particles were found in all specimens, especially in Amdry 9954 and 9954 + ODS. The inhomogeneous microstructure may have a negative effect on the mechanical properties of these alloys [23,24].

Due to the manufacturing process, a higher porosity of 2.1 ± 0.48% was found in the ODS-containing bond coats (Figure 2g–i) as opposed to 0.6 ± 0.17% in Amdry 386 and Amdry 9954. Figure 3 shows the grain size in all different conditions, analyzed by EBSD. As can be seen, the average grain size of 0.8–1.1 μm was not influenced by the varying atmospheres during the heat treatment. The ODS-containing bond coats revealed smaller areas of particles that were not completely molten and show a maximum grain size of 5–6 μm, significantly smaller than Amdry 386 and Amdry 9954. The sizes of all phases, as well as the phase fractions of each bond coat, were analyzed as well. Neither the individual size of the β-NiAl phase, γ-solid solution and the alumina particles, nor the phase fraction was significantly influenced by heating in oxygen-, water vapor- or hydrogen-containing atmospheres.

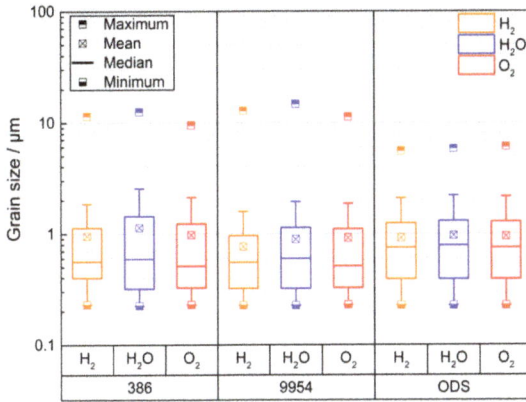

Figure 3. Distribution of the grain sizes of Amdry 386, Amdry 9954 and Amdry 9954 + ODS after a heat treatment at 1100 °C for 72 h in argon with the addition of 4% H_2 + 2% H_2O, 2% H_2O and 20% O_2.

3.2. Micro-Tensile Creep Experiments

The bond coats Amdry 386, Amdry 9954 and Amdry 9954 + ODS, which were heat treated in Ar–O_2, Ar–H_2O, and Ar–H_2–H_2O atmospheres at 1100 °C for 72 h were crept at testing temperatures of 900–950 °C. The results of the micro-tensile creep experiments are shown in Figure 4. Amdry 386 shows generally lower creep minima compared to Amdry 9954. Furthermore, it is obvious that the addition of 2 wt.-% of Al_2O_3 particles can significantly increase the creep resistance of MCrAlY bond coats.

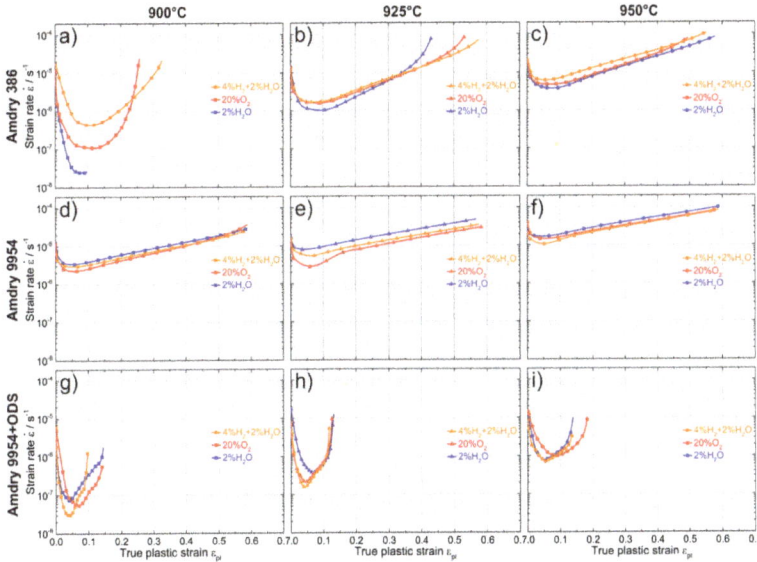

Figure 4. Strain rate versus true plastic strain plots at testing temperatures of 900–950 °C with an applied stress of 15 MPa. Before creep testing, Amdry 386 (**a–c**) Amdry 9954 (**d–f**) and Amdry 9954+ODS (**g–i**) were heat treated at 1100 °C for 72 h in Argon with the addition of 4% H_2 + 2% H_2O, 20% O_2 and 2% H_2O.

The most significant effect of heat treatment atmosphere on the creep behavior can be seen in Amdry 386 tested at 900 °C. The sample heat treated in Ar–H_2O showed the lowest creep minima, whereas the sample heat treated in Ar–H_2–H_2O showed a creep rate of more than one order of magnitude faster. The sample heat treated in Ar–H_2O was stopped after passing the creep minima at 627 h. With increasing temperature, the discrepancy diminished and the influence of the different atmospheres seemed to decline. The creep minima at all testing temperatures were reached at 0.06–0.1 plastic strain, though it was reached earlier at higher temperatures. Among the experiments with Amdry 386, the samples annealed in Ar–H_2O revealed lower creep minima than those annealed in Ar–H_2–H_2O and Ar–O_2, as indicated by the time to failure T_f in Table 2.

Table 2. Time to failure T_f for Amdry 386, Amdry 9954 and Amdry 9954 + ODS annealed in different atmosphere after creep experiments at 900–950 °C.

Material	Atmosphere	T_f/h 900 °C	T_f/h 925 °C	T_f/h 950 °C
Amdry 386	4% H_2 + 2% H_2O	94	37	11
	2% H_2O	627	52	17
	20% O_2	362	40	14
Amdry 9954	4% H_2 + 2% H_2O	27	14	7
	2% H_2O	23	10	5
	20% O_2	32	21	6
Amdry 9954 + ODS	4% H_2 + 2% H_2O	438	102	34
	2% H_2O	235	53	29
	20% O_2	355	84	31

For Amdry 9954, much higher creep rates than the Ni-based bond coat were observed in all experimental conditions. With increasing temperature, the creep rates of all samples annealed in different atmospheres increased slightly. Although the creep curves show almost the same progression, the samples heated in Ar–O_2 and Ar–H_2–H_2O had a tendency to have slightly lower creep minima than those heat treated in Ar–H_2O, at all temperatures. The high creep rate made it difficult to determine the possible influence of the individual atmospheres. Furthermore, the trend of the creep resistance of Amdry 9954 seemed to follow the opposite way to the sequence of Amdry 386. The highest creep rate and the longest time to failure could be related to the samples annealed in Ar–H_2O, whereas those annealed in Ar–H_2–H_2O and Ar–O_2 showed slightly lower creep rates. In the temperature range of 900–950 °C, all heat treatment variations showed the same creep curve progressions and strain to failure.

A remarkable decrease of the strain to failure from 0.6 without ODS particles to 0.1–0.2 with ODS particles was observed. The creep curves of Amdry 9954 + ODS were characterized by distinct creep minima, followed by a subsequent steep rise of the creep curves. The influence of oxygen, water vapor and hydrogen is difficult to interpret. However, the creep curves showed a similar tendency as in the oxide dispersion free alloy Amdry 9954. Samples annealed in Ar–H_2–H_2O and Ar–O_2 showed lower creep minima than those annealed in Ar–H_2O, especially at lower testing temperatures of about 900 °C and 925 °C. At 950 °C, the creep minima were virtually independent of the annealing atmosphere.

In Figure 5, the comparison of the creep minima for all nine experimental conditions and the corresponding activation energies are summarized.

Amdry 386 showed higher activation energies than Amdry 9954 due to the higher amount of β-NiAl. Associated with the lower creep minima of Amdry 386 annealed in Ar–H_2O, especially at 900 °C, it revealed the highest activation energy. The ODS particles significantly improved the creep resistance. However, the difference in activation energies is moderate due to the uniform changes of the creep minima with the increasing temperature.

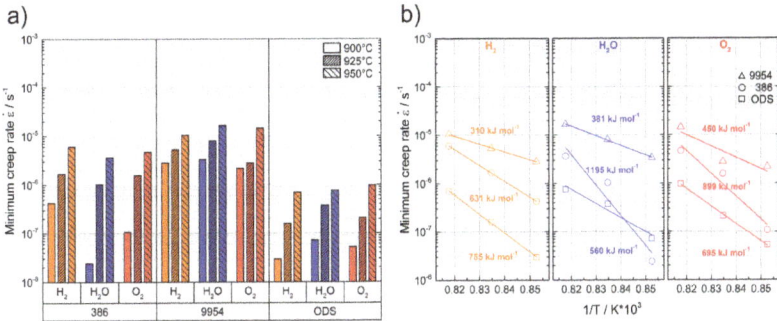

Figure 5. Minima creep rates (**a**) and activation energies (**b**) of testing temperatures of 900–950 °C. The samples of Amdry 386, Amdry 9954 and Amdry 9954 + ODS were heat treated before creep testing in Argon with the additions of 4% H_2 + 2% H_2O, 2% H_2O and 20% O_2.

3.3. Microstructure after Creep Experiments

Figure 6 exemplifies an overview of all three bond coats tested in small-scale creep experiments at 900 °C with a stress of 15 MPa. The microstructures of the crept samples revealed significant differences between the three kinds of MCrAlY bond coats. These differences, which are due to the varying amounts of the high temperature stable β-NiAl phase and the oxide dispersion particles, were already investigated in a previous study [17]. The most substantial influence of the different atmospheres during annealing was observed in Amdry 386. The experiment of the sample annealed in Ar–H_2O was stopped after 627 h. Afterwards, it was cut at the thinnest position. No significant inner failure in the form of internal cracking could be noticed, as shown in Figure 6a. This was more pronounced in the Ar–H_2–H_2O and Ar–O_2 samples since they were crept until failure, as shown in Figure 6b,c. It can be assumed that the cracks and pores were generated in the tertiary creep regime during tensile testing because almost no porosity was found in the regions of the grip sections of the Amdry 386 and Amdry 9954 specimens.

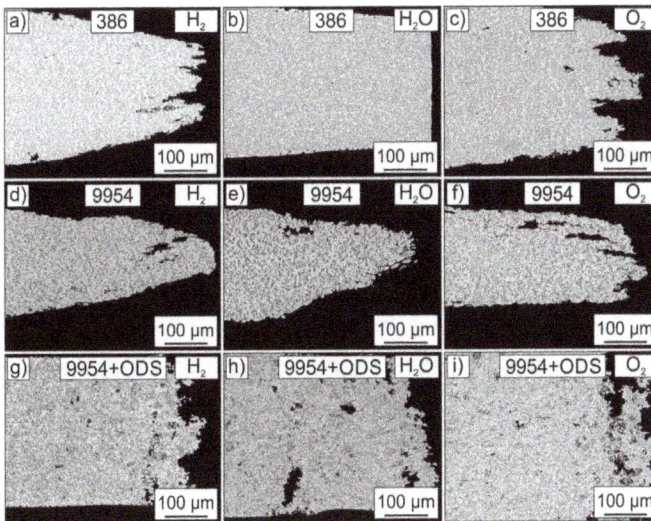

Figure 6. BSE images of the microstructures of Amdry 386 (**a–c**), Amdry 9954 (**d–f**) and Amdry 9954 + ODS (**g–i**) after tensile creep tests at 900 °C. Before creep testing, the samples were heat treated in Argon with the additions of 4% H_2 + 2% H_2O, 2% H_2O and 20% O_2.

Based on the strong ODS effect of the Al_2O_3 particles, the ODS-containing bond coats failed much earlier than Amdry 9954, resulting in a more furrowed fracture surface.

4. Discussion

Amdry 386 showed superior creep properties compared to the Co-based bond coat Amdry 9954. The higher creep strength of β-NiAl, in contrast with the Ni- and Co-containing γ-solid solution phase, at temperatures above 900 °C, is responsible for the better creep resistance [24–26]. The higher yttrium content in Amdry 386 compared to Amdry 9954 led to the formation of smaller Y_2O_3 particles. These nanoscale oxide particles provide an additional hindering effect for deformation and therefore improve the mechanical properties. The beneficial influence of small oxide particles is also shown by the lower creep minima of the ODS-containing bond coat. Investigations by Arzt et al. of yttria-oxide particle-strengthened NiAl with a similar grain size as the bond coats investigated in this study showed that the additional hindering effect can be attributed to the interaction between dislocations and dispersoids in the material [27–29]. It is assumed that the nanoscale Al_2O_3 particles in Amdry 9954 + ODS have a similar dispersion strengthening effect and hinder the dislocation motion during creep testing. The better creep properties of Amdry 9954 + ODS come along with a lower ductility than Amdry 9954. In contrast, the observed differences in the creep behavior of the bond coats annealed in different atmospheres are more difficult to interpret. At low testing temperatures of 900 °C, significant differences were observed for Amdry 386, while testing at higher temperatures, as well as for the two Co-based bond coats, revealed only a slightly different creep behavior due to the annealing in different atmospheres. The reason for this difference will be discussed in the following section.

The influence of external oxidation during heat treatment at 1100 °C for 72 h in different atmospheres on the mechanical properties can be ignored in this study because the micro tensile samples were taken from the center of the bulk material. However, according to the literature, the addition of water vapor and hydrogen can promote internal oxidation [30]. Water vapor can lead to an increase of the oxygen vacancy concentration and an enhanced outward diffusion of aluminum [30–32]. Furthermore, water vapor extends the early initial stage of oxidation of Al-rich bond coats, hence stabilizing the transient θ-alumina stage [33]. Therefore, the less dense θ-alumina allows more oxygen to pass through the oxide layer, promoting the formation of external and internal α-Al_2O_3 [11,27]. Huang et al. showed that samples annealed in Ar–O_2 revealed an increased grain boundary diffusion for oxygen, resulting in a more pronounced internal oxidation than in Ar–H_2–H_2O [22]. However, according to microstructural analysis, no internal oxidation could be observed in all bond coats and atmosphere compositions in the present study.

Further, neither the grain size, phase fraction, nor the phase size varied in the bond coats annealed in different atmospheres. The high creep rate of Amdry 386 annealed in Ar–H_2–H_2O could be due to the interaction of hydrogen inside the bulk material with grain boundaries and interfaces. Investigations of tensile testing of Ni-based superalloys under hydrogen charging have revealed that hydrogen can enhance localized plasticity, which is known as the hydrogen-enhanced localized plasticity (HELP) mechanism. Fracture was mainly due to this localized plasticity and transgranular cracking was observed [7]. It was suggested that hydrogen promotes the coalescence and widening of voids during deformation, resulting in an easier crack propagation [6]. Takahashi et al. found that the hydrogen-enhanced decohesion (HEDE) mechanism, which was originally developed to explain the weakening of metallic bonds in bulk materials, can also be applied to small-scale specimens [34]. Therefore, the poorer creep behavior of the Amdry 386 sample annealed in Ar–H_2–H_2O might be due to a combination of the interaction of hydrogen with grain boundaries and a decrease of the energy barrier for dislocation motion, which leads to an enhanced plasticity and earlier crack propagation.

5. Conclusions

One Ni-based and two Co-based freestanding bond coats with and without ODS particles annealed at 1100 °C for 72 h in argon atmospheres, containing additions of 20% O_2, 2% H_2O and 4% H_2 + 2%

H$_2$O were investigated by micro-tensile creep tests. The force-controlled creep experiments were performed in a TMA at testing temperatures of 900–950 °C under constant Ar-flow and a constant load of 2.37 N. The higher amount of β-NiAl in the NiCoCrAlY—bond coat Amdry 386–led to superior creep properties than in the CoNiCrAlY—bond coat Amdry 9954. The addition of ODS particles to Amdry 9954 led to the highest creep resistance due to particle strengthening. For the Ni-based bond coat Amdry 386, the best creep properties were observed for samples annealed in Ar–H$_2$O. The influence of the atmospheres diminished at higher testing temperatures. For Amdry 9954, the influence of the different atmospheres was difficult to distinguish given the high creep rates at 900–950 °C with an applied stress of 15 MPa. However, both Co-based bond coats with and without ODS particles showed the lowest creep resistance after the addition of water vapor.

Author Contributions: Conceptualization, W.J.Q., R.V. and M.G.; methodology, M.G.; software, S.G.; validation, S.G. and S.N.; formal analysis, S.G.; investigation, S.G.; resources, J.B. and D.N.; data curation, S.G., S.N. and M.G.; writing—original draft preparation, S.G.; writing—review and editing, S.N. and M.G.; visualization, S.G.; supervision, S.N. and M.G.; project administration, S.N. and M.G.; funding acquisition, M.G.

Funding: This research was funded by the Deutsche Forschungsgemeinschaft (DFG) through projects A6 and B6 of the collaborative research centre SFB/TR 103 "From Atoms to Turbine Blades-a Scientific Approach for Developing the Next Generation of Single Crystal Superalloys".

Acknowledgments: The authors are grateful to Ralf Laufs at IEK-1, Forschungszentrum Jülich GmbH, for supporting the production of samples. The authors would like to acknowledge Georg Rathmayr from Microsample GmbH for manufacturing the micro tensile specimen.

Conflicts of Interest: The authors declare no conflict of interest.

References

1. Chiesa, P.; Lozza, G.; Mazzocchi, L. Using Hydrogen as Gas Turbine Fuel. *J. Eng. Gas Turbines Power* **2005**, *127*, 73–80. [CrossRef]

2. Geng, W.T.; Wan, L.; Du, J.-P.; Ishii, A.; Ishikawa, N.; Kimizuka, H.; Ogata, S. Hydrogen bubble nucleation in α-iron. *Scr. Mater.* **2017**, *134*, 105–109. [CrossRef]

3. Lawrence, S.K.; Yagodzinskyy, Y.; Hänninen, H.; Korhonen, E.; Tuomisto, F.; Harris, Z.D.; Somerday, B.P. Effects of grain size and deformation temperature on hydrogen-enhanced vacancy formation in Ni alloys. *Acta Mater.* **2017**, *128*, 218–226. [CrossRef]

4. Osono, H.; Kino, T.; Kurokawa, Y.; Fukai, Y. Agglomeration of hydrogen-induced vacancies in nickel. *J. Alloy. Compd.* **1995**, *231*, 41–45. [CrossRef]

5. Takai, K.; Shoda, H.; Suzuki, H.; Nagumo, M. Lattice defects dominating hydrogen-related failure of metals. *Acta Mater.* **2008**, *56*, 5158–5167. [CrossRef]

6. Zhang, Z.; Obasi, G.; Morana, R.; Preuss, M. Hydrogen assisted crack initiation and propagation in a nickel-based superalloy. *Acta Mater.* **2016**, *113*, 272–283. [CrossRef]

7. Tarzimoghadam, Z.; Ponge, D.; Klöwer, J.; Raabe, D. Hydrogen-assisted failure in Ni-based superalloy 718 studied under in situ hydrogen charging: The role of localized deformation in crack propagation. *Acta Mater.* **2017**, *128*, 365–374. [CrossRef]

8. Subanovic, M.; Naumenko, D.; Kamruddin, M.; Meier, G.; Singheiser, L.; Quadakkers, W.J. Blistering of MCrAlY-coatings in H$_2$/H$_2$O-atmospheres. *Corros. Sci.* **2009**, *51*, 446–450. [CrossRef]

9. Déneux, V.; Cadoret, Y.; Hervier, S.; Monceau, D. Effect of Water Vapor on the Spallation of Thermal Barrier Coating Systems During Laboratory Cyclic Oxidation Testing. *Oxid. Met.* **2010**, *73*, 83–93. [CrossRef]

10. Zhou, C.; Yu, J.; Gong, S.; Xu, H. Influence of water vapor on the isothermal oxidation behavior of low pressure plasma sprayed NiCrAlY coating at high temperature. *Surf. Coat. Technol.* **2002**, *161*, 86–91. [CrossRef]

11. Sullivan, M.H.; Mumm, D.R. Transient stage oxidation of MCrAlY bond coat alloys in high temperature, high water vapor content environments. *Surf. Coat. Technol.* **2014**, *258*, 963–972. [CrossRef]

12. Leyens, C.; Fritscher, K.; Gehrling, R.; Peters, M.; Kaysser, W.A. Oxide scale formation on an MCrAlY coating in various H$_2$-H$_2$O atmospheres. *Surf. Coat. Technol.* **1996**, *82*, 133–144. [CrossRef]

13. Alam, Z.; Eastman, D.; Jo, M.; Hemker, K. Development of a High-Temperature Tensile Tester for Micromechanical Characterization of Materials Supporting Meso-Scale ICME Models. *JOM* **2016**, *68*, 2754–2760. [CrossRef]
14. Hemker, K.J.; Mendis, B.G.; Eberl, C. Characterizing the microstructure and mechanical behavior of a two-phase NiCoCrAlY bond coat for thermal barrier systems. *Mater. Sci. Eng. A* **2008**, *483–484*, 727–730. [CrossRef]
15. Hebsur, M.G.; Miner, R.V. Stress rupture and creep behavior of a low pressure plasma-sprayed NiCoCrAlY coating alloy in air and vacuum. *Thin Solid Films* **1987**, *147*, 143–152. [CrossRef]
16. Jaya, N.B.; Alam, M.Z. Small-scale mechanical testing of materials. *Curr. Sci.* **2013**, *105*, 1073–1099.
17. Giese, S.; Neumeier, S.; Amberger-Matschkal, D.; Bergholz, J.; Vaßen, R.; Göken, M. Micro-tensile creep testing of freestanding MCrAlY bond coats. *J. Mater. Res.* **2019**. [CrossRef]
18. Kumar, S.; Singh, R.; Singh, T.P.; Sethi, B.L. Surface modification by electrical discharge machining: A review. *J. Mater. Process. Technol.* **2009**, *209*, 3675–3687. [CrossRef]
19. Lee, L.C.; Lim, L.C.; Narayanan, V.; Venkatesh, V.C. Quantification of surface damage of tool steels after EDM. *Int. J. Mach. Tools Manuf.* **1988**, *28*, 359–372. [CrossRef]
20. Rathmayr, G.B.; Bachmaier, A.; Pippan, R. Development of a New Testing Procedure for Performing Tensile Tests on Specimens with Sub-Millimetre Dimensions. *J. Test. Eval.* **2013**, *41*, 20120175. [CrossRef]
21. Rathmayr, G.B.; Hohenwarter, A.; Pippan, R. Influence of grain shape and orientation on the mechanical properties of high pressure torsion deformed nickel. *Mater. Sci. Eng. A* **2013**, *560*, 224–231. [CrossRef] [PubMed]
22. Huang, T.; Bergholz, J.; Mauer, G.; Vassen, R.; Naumenko, D.; Quadakkers, W.J. Effect of test atmosphere composition on high-temperature oxidation behaviour of CoNiCrAlY coatings produced from conventional and ODS powders. *Mater. High Temp.* **2018**, *35*, 97–107. [CrossRef]
23. Chen, H. Microstructure characterisation of un-melted particles in a plasma sprayed CoNiCrAlY coating. *Mater. Charact.* **2018**, *136*, 444–451. [CrossRef]
24. Chen, H.; Si, Y.Q.; McCartney, D.G. An analytical approach to the β-phase coarsening behaviour in a thermally sprayed CoNiCrAlY bond coat alloy. *J. Alloy. Compd.* **2017**, *704*, 359–365. [CrossRef]
25. Raj, S.V. Tensile creep of polycrystalline near-stoichiometric NiAl. *Mater. Sci. Eng. A* **2003**, *356*, 283–297. [CrossRef]
26. Karashima, S.; Oikawa, H.; Motomiya, T. Steady-state creep characteristics of polycrystalline nickel in the temperature range 500° to 1000 °C. *Trans. Jpn. Inst. Met.* **1969**, *10*, 205–209. [CrossRef]
27. Arzt, E.; Behr, R.; Göhring, E.; Grahle, P.; Mason, R.P. Dispersion strengthening of intermetallics. *Mater. Sci. Eng. A* **1997**, *234*, 22–29. [CrossRef]
28. Arzt, E.; Grahle, P. High temperature creep behavior of oxide dispersion strengthened NiAl intermetallics. *Acta Mater.* **1998**, *46*, 2717–2727. [CrossRef]
29. Grahle, P.; Arzt, E. Microstructural development in dispersion strengthened NiAl produced by mechanical alloying and secondary recrystallization. *Acta Mater.* **1997**, *45*, 201–211. [CrossRef]
30. Maris-Sida, M.C.; MEIER, G.H.; Pettit, F.S. Some Water Vapor Effects during the Oxidation of Alloys that are α-Al$_2$O$_3$ Formers. *Metall. Mater. Trans. A* **2008**, *34*, 2609–2619. [CrossRef]
31. Xing, L.; Zheng, Y.; Cui, L.; Sun, M.; Shao, M.; Lu, G. Influence of water vapor on the oxidation behavior of aluminized coatings under low oxygen partial pressure. *Corros. Sci.* **2011**, *53*, 3978–3982. [CrossRef]
32. Kaplin, C.; Brochu, M. Effects of water vapor on high temperature oxidation of cryomilled NiCoCrAlY coatings in air and low-SO$_2$ environments. *Surf. Coat. Technol.* **2011**, *205*, 4221–4227. [CrossRef]
33. Saunders, S.R.J.; Monteiro, M.; Rizzo, F. The oxidation behaviour of metals and alloys at high temperatures in atmospheres containing water vapour: A review. *Prog. Mater. Sci.* **2008**, *53*, 775–837. [CrossRef]
34. Takahashi, Y.; Kondo, H.; Asano, R.; Arai, S.; Higuchi, K.; Yamamoto, Y.; Muto, S.; Tanaka, N. Direct evaluation of grain boundary hydrogen embrittlement: A micro-mechanical approach. *Mater. Sci. Eng. A* **2016**, *661*, 211–216. [CrossRef]

metals MDPI

Article

Nanoindentation Investigation on the Size-Dependent Creep Behavior in a Zr-Cu-Ag-Al Bulk Metallic Glass

Z. Y. Ding [1,2], Y. X. Song [1,2], Y. Ma [1,*], X. W. Huang [1] and T. H. Zhang [3,*]

[1]   College of Mechanical Engineering, Zhejiang University of Technology, Hangzhou 310014, China;
      zyding@zjut.edu.cn (Z.Y.D.); songyux@zjut.edu.cn (Y.X.S.); huangxw@zjut.edu.cn (X.W.H.)
[2]   Institute of Process Equipment and Control Engineering, Zhejiang University of Technology,
      Hangzhou 310014, China
[3]   Institute of Solid Mechanics, Beihang University, Beijing 100191, China
*   Correspondence: may@zjut.edu.cn (Y.M.); zhangth66@buaa.edu.cn (T.H.Z.);
      Tel.: +86-571–88320132 (Y.M. & T.H.Z.)

Received: 1 May 2019; Accepted: 23 May 2019; Published: 27 May 2019

Abstract: Nanoindentation technology has been widely adopted to study creep behavior in small regions. However, nanoindentation creep behavior of metallic glass is still not well understood. In the present work, we investigated nanoindentation size effects on creep deformation in a Zr-based bulk metallic glass at room temperature. The total creep strain and strain rate of steady-state creep were gradually decreased with increasing holding depth under a Berkovich indenter, indicating a length-scale-dependent creep resistance. For a spherical indenter, creep deformations were insignificant in elastic regions and then greatly enhanced by increasing holding strain in plastic regions. Strain rate sensitivities (SRS) decreased with increasing holding depth and holding strain at first, and then stabilized as holding depth was beyond about 500 nm for both indenters. SRS values were 0.4–0.5 in elastic regions, in which atomic diffusion and free volume migration could be the creep mechanism. On the other hand, evolution of the shear transformation zone was suggested as a creep mechanism in plastic regions, and the corresponding SRS values were in the range of 0.05 to 0.3.

Keywords: metallic glass; nanoindentation; creep; size effect; strain rate sensitivity

---

1. Introduction

As a relatively new member of the glass family, metallic glasses have great potential to be an excellent candidate as engineering materials due to their attractive mechanical properties, such as super high yield strength, large elastic limit (~2%), and strong wear resistance [1–3]. Nevertheless, the brittleness inherited from the amorphous structure seriously hinders the commercial application of bulk metallic glasses [4]. Distinct to the dislocation move in crystalline alloys, plastic mechanism in metallic glass is still under debate and on the cutting edge of structure investigations [5,6]. In recent years, size effect on mechanical properties of metallic glass has attracted numerous attention [7–9]. By reducing physical dimensions, ductility could be greatly enhanced at the nanoscale without sacrificing high strength. Nanoindentation is the most powerful technology to use to study mechanical properties in small regions, which has been widely used to reveal the size effect in metals and alloys [10].

Creep is a time-dependent plastic deformation, which is an inevitable process and vital to engineering materials on service [11]. Relying on nanoindentation, creep behaviors can be studied at small region, ignoring the limitation of required standard size in conventional creep test [12–14]. For metallic glass, creep resistance is not fully studied and creep mechanism is far from being understood [15,16]. In order to avoid undesired influence by thermal drift, the nanoindentation creep test is commonly conducted at low temperatures. The sample size effect on creep deformation has also been examined by nanoindentation. Yoo et al. studied room-temperature creep behavior in metallic glass nanopillars with 250~2000 nm diameter by elastic holding [17]. As with most nanoindentation creep experiments, creep behaviors were detected in plastic regions by adopting a three-sided pyramidal, namely a Berkovich indenter. Wang et al. performed creep tests in Cu-Zr films with thicknesses from 1000 to 3000 nm [18], while Ma et al. investigated creep behaviors in 500–1500 nm Cu-Zr-Al films [19]. Creep features in Ni-Nb thin films and ribbons were also compared by Ma [20]. Under both elastic and plastic holdings, creep deformation was more pronounced in smaller samples. That is to say, the creep feature seems to be in conflict with the "smaller is stronger" principle in metallic glass.

On the other hand, the size effect on creep deformation could also be studied by changing the nanoindentation holding depth. In this scenario, the creep feature is linked to the deformation volume beneath an indenter whilst the inner structure state is unchanged. Superficially, in previous reports, the recorded creep displacements were increased with holding depth under both Berkovich and spherical indenters [21–24]. This phenomenon was commonly attributed to the more excess free volume generated in deep nanoindentation. In fact, it is conceivable that creep displacement is in proportion to the length scale of the sample and holding strain in conventional uniaxial holding. Therefore, true creep resistance needs to be carefully examined in metallic glass and linked to deformation volume and holding strain during nanoindentation holding. In the present work, we aim to reveal the intrinsic nanoindentation size effect on creep deformation in a Zr-Cu-Ag-Al metallic glass. By adopting a standard Berkovich indenter, creep deformation can be studied at various depths while the holding strain is constant. Using a spherical indenter, the holding strain effect on creep deformation could be investigated.

2. Materials and Methods

$Zr_{46}Cu_{37.6}Ag_{8.4}Al_8$ alloy ingots were prepared from high pure elements (99.99%) by arc mixing in a Ti-gettered argon atmosphere. Alloy sheets with a rectangular cross-section of 2 mm × 30 mm were obtained by injecting alloy melt into a copper mold. Prior to nanoindentation, the specimen surface was precisely polished to a mirror finish. The structure of the as-cast Zr-Cu-Ag-Al specimen was detected using X-ray diffraction (XRD) with Cu Kα radiation. By means of X-ray energy dispersive spectrometer (EDS) attached on a SEM, the chemical composition was detected, which was equal to the alloy ingot.

Nanoindentation creep tests were conducted at a constant temperature of 20 °C on Agilent Nano Indenter G200. The displacement and load resolutions of the machine are 0.01 nm and 50 nN, respectively. Constant-load holding method was adopted, during which displacement of indenter into the surface at a prescribed load could be continuously recorded. A standard Berkovich indenter and a spherical indenter with a nominal radius of 20 μm were used, respectively. Upon calibration on fused silica, the true contact radius of spherical indenter was obtained as 9.8 μm. The Berkovich indenter was held at maximum loads of 5, 10, 25, 50, 100, 200, and 400 mN. For spherical nanoindentation, the peak loads were 2, 5, 10, 30, 60, 120, 200, 300, 470, and 700 mN. The loading rate and holding time were fixed, equal to 2 mN/s and 500 s. More than twenty-five independent measurements were performed under each testing condition. All the nanoindentation tests were carried out until thermal drift reduced to below 0.02 nm/s, and drift correction was strictly performed at 10% of the maximum load during the unloading process. Upon drift correction, the thermal drift effect could be greatly alleviated, especially for creep displacement during the holding stage.

3. Results and Discussion

Figure S1 shows the typical X-ray diffraction pattern of the as-prepared Zr-Cu-Ag-Al alloy. It is clear that only a broad diffraction peak can be detected, which represents a crystal free structure. The representative creep load versus displacement (*P-h*) curves at various holding loads under a

Berkovich indenter are exhibited in Figure 1a. The *P-h* curves at shallow depths were enlarged, as shown in the insets. The permanent displacement, i.e., creep flow, could be observed in the holding stage, though it was subtle particularly under small loads. As exhibited in Figure 1b, the creep displacement during the holding stage was plotted as a function of holding time. For a clear view, the onsets of both *x*-axis (holding time) and *y*-axis (displacement) in graph were set to zero. The creep *P-h* curves and creep flow curves under spherical indenter were similar to those under Berkovich indenter, as exhibited in Figure S2 in the Supplementary Materials. For both indenters, creep displacement was more pronounced at larger holding loads and/or depths.

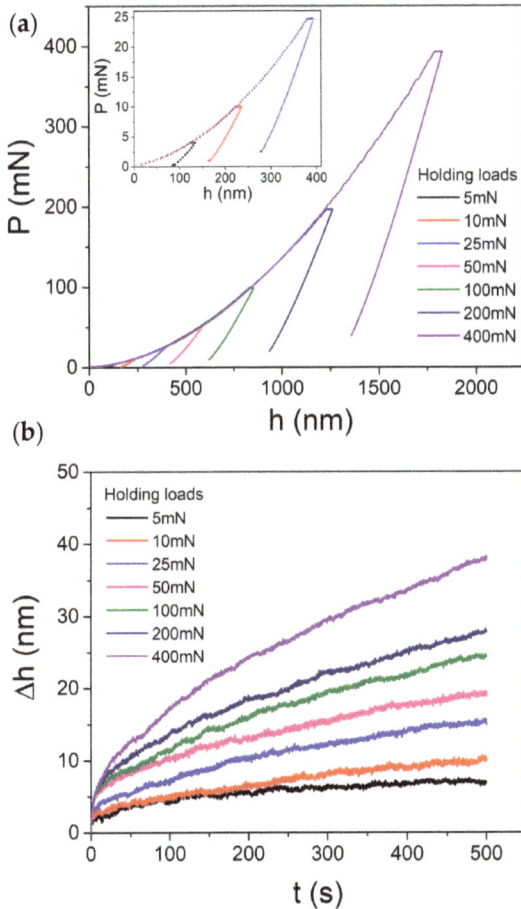

Figure 1. (a) The typical creep *P-h* curves at various holding depths under Berkovich indenter. *P-h* curves at shallow depths were enlarged in the inset. (b) Creep displacements at various holding loads were plotted with holding time.

The total creep displacements at the end of the holding stage were recorded, which were plotted with holding depths, as shown in Figure 2. The creep displacement in the Berkovich nanoindentation was increased with holding depth, while in spherical nanoindentation, creep displacement was almost independent of holding depth at first and then quickly increased. Creep deformations under Berkovich indenter were more pronounced than those under spherical indenter. The holding-depth-facilitated nanoindentation creep displacements were consistent with previous reports in metallic glasses.

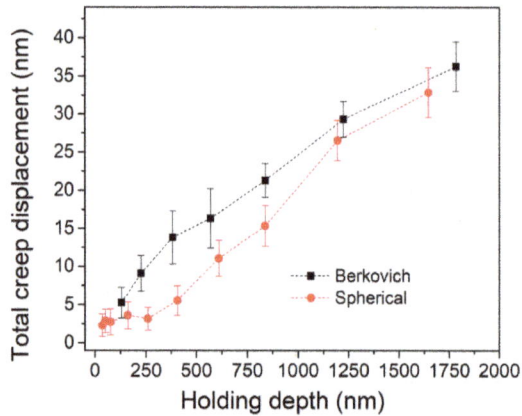

Figure 2. The total creep displacements in the end of holding stage were plotted with holding depth for both indenters.

For a standard Berkovich indenter (without tip bluntness), the imposed plastic volume and stress distribution during nanoindentation are self-similar at various pressed depths. Nanoindentation strains at various holding depths were constant, equal to 7.1% (0.2cot70.5°). Theoretically, creep displacement under a Berkovich indenter would be in proportion to the holding depth whilst creep strain would be invariable, regardless of structure agitation at various holding depth. Furthermore, the anticipation that more excess free volume could be generated at deep nanoindentation lacks strict verification, while for a spherical indenter, the deformation zone gradually evolved from elastic to elastoplastic with increasing pressed depth. The nanoindentation strain was continuously increased to the limit about 11% (the configuration of spherical indenter is a conical body with spherical tip, the maximum indentation strain would be approaching to 11% by 0.2cot61°). Under spherical nanoindentation, more severe structural agitation and better atomic mobility could be expected due to the increased plastic strain at deeper location. Thus, the situation under spherical indenter was more complicated than that under a Berkovich indenter. The increased creep displacement could be attributed to the combined effects of deformation volume, holding strain, and atomic mobility. The creep deformation under spherical indenter needs to be discussed separately at elastic and plastic holdings.

According to Bei's work [25], the first pop-in event on the loading sequence could be linked to incipient plasticity. Figure 3a shows the typical spherical *P-h* curve at 200 mN (holding time was 5 s) with loading rate of 2 mN/s. The pop-in events with length scales of 20~30 nm clear occurred, which represent the generation of shear bands. The initial loading sequence could be well fitted by the Hertzian elastic contact equation [26], given by

$$P = \frac{4}{3} E_r \sqrt{R} h^{1.5} \tag{1}$$

where E_r is the reduced elastic modulus which accounts for the elastic displacement occurred in both the tip and sample, given by

$$\frac{1}{E_r} = \frac{1 - v_i^2}{E_i} + \frac{1 - v_s^2}{E_s} \tag{2}$$

where E and v are the elastic modulus and Poisson's ratio, and the subscripts s and i represent the sample and the indenter, respectively. For commonly used diamond tip, $E_i = 1141$ GPa and $v_I = 0.07$. For the Zr-Cu-Ag-Al metallic glass, elastic modulus was measured as 110 GPa as shown in Figure S3 in the Supplementary Materials and the Poisson's ratio is 0.36 [27]. The Hertzian fitting line deviated from the *P-h* curve at the position of the first pop-in. This clearly indicates the transition from elastic deformation to plastic deformation once the first pop-in emerges. Shear banding nucleation in metallic glass is a stochastic process, which could result in a scatter distribution of the first pop-in event [28]. By conducting 81 measurements at a 9 × 9 matrix at intervals of 50 μm, the critical loads at first pop-in events were detected to be uniformly scattered in the range between 90 and 150 mN, as shown in the inset of Figure 3a. Thus, 2–60 mN holdings were at elastic regions and 200–700 mN holdings were at plastic regions under spherical indenter. The 120 mN holding was exactly around the yielding point of Zr-Cu-Ag-Al sample. We can regard approximately 120 mN holding as the elastic holding because plastic deformation was still negligible. The holding strains at various holding depths were estimated for spherical indenter by $\varepsilon_i = 0.2(a/R)$, a is the contact radius. As exhibited in Figure 3b, the holding strain increased from 1.2% to 10% as peak load increased from 2 to 700 mN. The elastic limit under 9.8 μm spherical indenter was about 4%, which was far beyond the typical ~2% for bulk metallic glasses. This could be explained from the complicated stress distribution beneath the indenter where incipient plasticity is unable to be triggered immediately when the maximum stress attains yield stress [29]. To form a shear band during nanoindentation, there needs a certain space along the shear path of which stress has been beyond the yield stress.

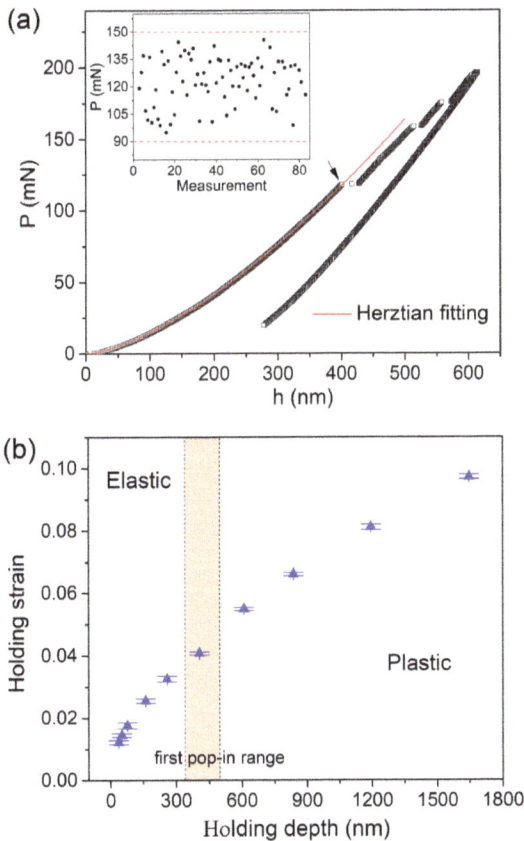

Figure 3. (**a**) The typical *P-h* curve under 200 mN spherical indenter with 5 s holding. The distribution range of the critical load at the first pop-in event is shown in the inset, which was plotted with measurement. (**b**) The corresponding holding strain for each holding load was calculated for spherical nanoindentation and was plotted with holding depth. The creep tests could be divided as elastic holding and plastic holding.

For creep deformation under Berkovich indenter, we defined creep strain as $\Delta h/h_c$, in which Δh is the total creep displacement and h_c is the contact displacement at the beginning of holding stage. Contact displacement was estimated by $h_c = h - \varepsilon \times P/S$, in which $\varepsilon = 0.72$, S is the stiffness deduced from the unloading curve. The creep strain rate of steady-state part was estimated by $\dot{\varepsilon} = \frac{1}{h_c}\frac{dh_c}{dt}$. The mean value of creep strain rate at the last 200 s segment of holding stage was adopted. Figure 4a exhibits creep strain and strain rate of steady-state creep under Berkovich indenter at various holding depths. The creep strain was continuously decreased from 0.04 to 0.02 with increasing holding depth from 130 to 1780 nm. Meanwhile, creep strain rate decreased from 1.1×10^{-4} to 5.4×10^{-5} s^{-1}. That is to say, creep deformation was actually depressed with increasing nanoindentation depth under a Berkovich indenter. This result confirms previous reports about sample-size-dependent creep flow [17–20]. From the perspective of structure agitation, the density of shear bands could be decreased at deep nanoindentation, i.e., lower density of excess free volume. On the other hand, size effect on plastic deformation has been largely reported in metallic glass that plastic flow is facilitated at the nanoscale [7–9], which suggests a better atomic mobility. Therefore, the enhanced creep deformation at shallow depth under a Berkovich indenter could be explained qualitatively.

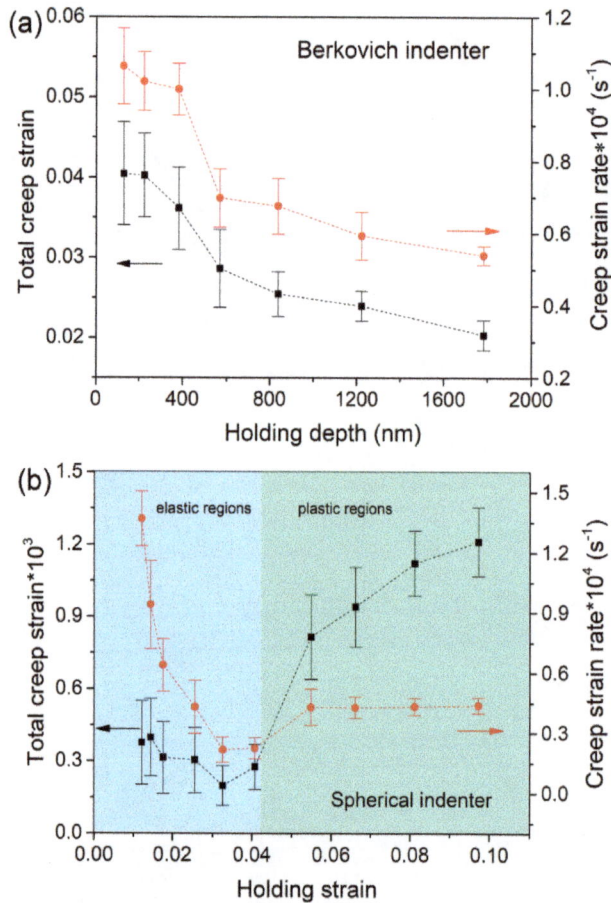

Figure 4. The total creep strain in the end of holding stage and strain rate of steady-state creep were estimated for (**a**) Berkovich and (**b**) spherical indenters, which were plotted with holding depth and holding strain.

Creep strain under spherical indenter was calculated by $0.2(a - a_0)/R$, where a and a_0 are the contact radii at the beginning and ending of holding stage, respectively. Creep strain rate was calculated by $\dot{\varepsilon} = \frac{1}{\sqrt{A}}\frac{d\sqrt{A}}{dt}$, where A is the contact area, equal to $2\pi Rh_c$ at plastic region and πRh at elastic region. Figure 4b depicts the correlations between creep strain, creep strain rate, and holding strain under spherical indenter. At elastic regions, creep strains were insignificant (lower than 0.04%) and slightly decreased as increasing holding strain. It is rational that creep flow hardly occurred under elastic holding at room temperature for the high-melting bulk metallic glass in such short duration (compared to conventional creep measurement). The present result indicates that creep deformation at elastic region is mainly dependent on structure state, rather than holding strain. Atomic diffusion on the contact surface and the migration of pre-existing free volume could be the creep mechanism at elastic holdings. As the holding strain turned to plastic, a sudden increase of creep strain was observed. At plastic regions, creep strain was greatly increased from about 0.04% to 0.1% as holding strain increased from 4% to 10%. For plastic holdings, the evolution of shear transformation zone (STZ) dominated during creep deformation. With increasing holding strain, plastic zone beneath indenter increased and more STZs were activated to carry creep flow. In this scene, the increased plastic strain and more severe structure agitation jointly stimulated creep deformation as increasing holding strain. On the other hand, creep strain rate precipitously decreased as increasing holding strain at elastic regions and was independent on holding strain at plastic regions. As holding strain increased from elastic to plastic, the enlargement of creep strain rate was apparently less dramatic than the increase of creep strain. The unexpected high creep strain rates at shallow depths below 100 nm (2–10 mN holdings) were mysterious, probably due to the thermal drift effect on such weak creep deformations. Furthermore, it was indicated that steady-state creep deformation was holding-depth-independent at plastic regions, which represented the true creep resistance.

The present creep feature under nanoindentation was similar to conventional testing. Hence, it has merits to estimate strain rate sensitivity (SRS) in order to reveal the creep mechanism and its correlation with nanoindentation length scale. Here, we selected 200 mN-holding testing by spherical indenter as an illustration to calculate SRS. As exhibited in Figure 5a, creep curve could be perfectly fitted ($R^2 > 0.99$) by an empirical law:

$$h(t) = h_0 + a(t - t_0)^b + kt \tag{3}$$

where h_0, t_0 are the displacement and time at the beginning of holding stage, a, b, k are the fitting constants. The value of SRS exponent m can be evaluated via

$$m = \frac{\partial \ln \sigma}{\partial \ln \dot{\varepsilon}} \tag{4}$$

The creep flow stress σ can be obtained from the mean pressure P_m beneath indenter via Tabor's mode, $P_m = 3\sigma$ [30] in the elastic region, $P_m = \frac{P}{\pi Rh}$. At plastic region, the mean pressure is also defined as hardness, which is $H = \frac{P}{2\pi Rh_c}$ for spherical tip and $H = \frac{P}{Ch_c^2}$ for standard Berkovich indenter. C is the tip area coefficient for Berkovich indenter and was rectified upon testing on standard fused silica, equal to 24.3 here. $\dot{\varepsilon}$ is the creep strain rate. Figure 5b,c show the changes of strain rate and hardness during holding stage, which were deduced from the fitting line of creep curve. Figure 5d shows the Logar–Logar correlation between hardness and strain rate during the holding stage. Then SRS can be obtained by linearly fitting on the part of steady-state creep.

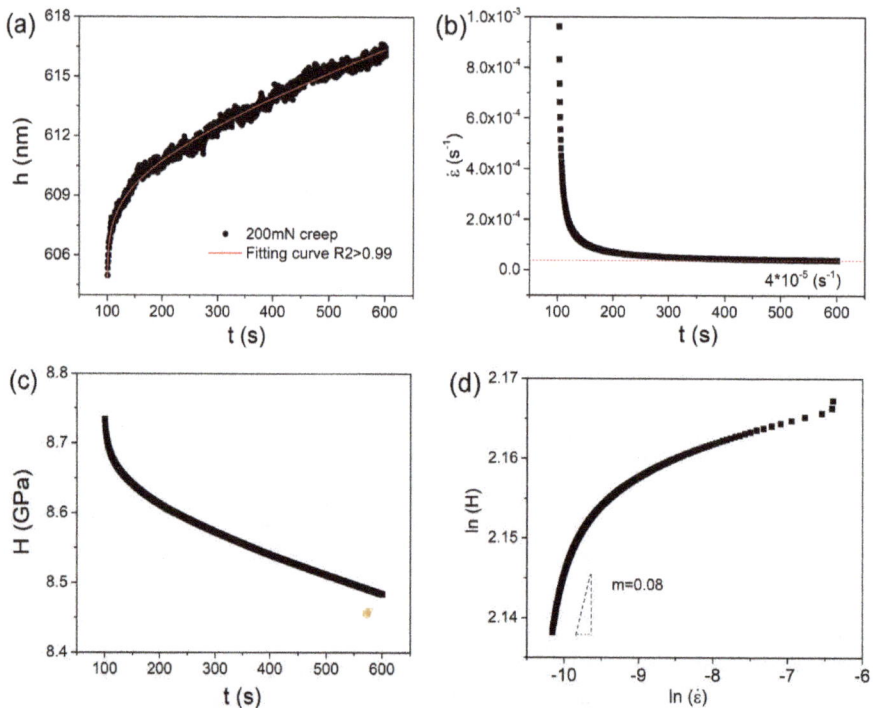

Figure 5. (**a**) The creep displacements versus holding time, which is perfectly fitted by an empirical law; (**b**) the creep strain rate versus holding time; (**c**) the hardness versus holding time; (**d**) the log–log correlation between hardness and strain rate obtained from the creep, strain rate sensitivity was estimated by linear fitting of the steady-state part.

Figure 6a shows the correlation between *m* and holding depth for Berkovich indenter. SRS was decreased from 0.17 to 0.075 as holding depth increased from 130 to 570 nm, and then tended to stabilize around 0.07. In the present work, the tip bluntness effect on creep behavior at shallow depth could be excluded since the minimum pressed depth was beyond 100 nm. Figure 6b shows the correlation between *m* and holding strain for spherical indenter. At elastic regions, *m* greatly decreased from 0.53 to 0.16, as holding strain increased from 1.2% to 4%. At plastic regions, *m* slightly decreased, from 0.09 to 0.055, as holding strain increased from 5.5% to 10%. Apparently, SRS decreased with increasing holding depth and settled as the pressed depth was larger than about 500 nm for both indenters.

The value of strain rate sensitivity *m* or stress exponent *n* ($n = 1/m$) is widely used as an indication to creep mechanism in crystalline alloy or metals [31]. For example, dislocation move is dominating in creep flow as *m* falls in the range between 0.1 and 0.3. In metallic glasses, free volume generation and annihilation, shear transformation zone (STZ) evolution and atomic diffusion (under elastic contact) are thought to be the possible creep mechanisms [23], while the relationship between *m* and creep mechanism in metallic glass is inconclusive. For creep flow under a Berkovich indenter, the STZ creep flow could be mainly actuated by STZ evolution. The gentle change of *m* was probably due to that STZ size and density were also changed with nanoindentation depth. For spherical nanoindentation creep, it is worth mentioning that *m* was in between 0.25 and 0.055 within holding strain range from 2.6% to 10%. It is reasonable that *m* at plastic region under spherical indenter was much comparable to Berkovich nanoindentation, due to the same creep mechanism, while for 30–120 mN elastic holdings (2.6–4% holding strains), the maximum stress beneath indenter had already exceeded yield stress as it was aforementioned. Though it could not meet the requirement of shear banding generation, the stress level and atomic surrounding were satisfied for STZ activation [23]. Creep deformation beneath indenter was prone to occur at the region that suffered high stress. Thus, STZ evolution might also be the creep mechanism under the nominal elastic holding at 30–120 mN. The 2–10 mN holdings, of which holding strains were below 2%, could be regarded as purely elastic under nanoindentation. In this scenario, STZs were unable to be activated. Atomic diffusion and free volume migration could be suggested as creep mechanism. Accordingly, the high values of SRS (0.4~0.5) at shallow depths under spherical indenter could be explained by the transition of creep mechanism. In the current work, we investigated creep behavior and its correlation with nanoindentation length scale and holding strain. Relying on the suggested modes for plastic deformation in metallic glass, we bridged the connection between creep mechanisms and SRS values at elastic holding and plastic holding.

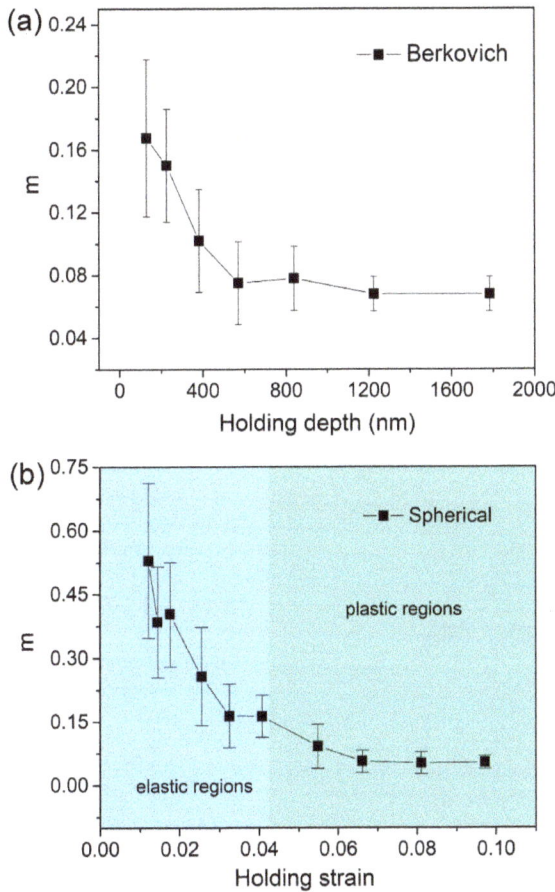

Figure 6. Strain rate sensitivities were plotted with holding depth for (**a**) a Berkovich indenter and plotted with holding strain for (**b**) a spherical indenter.

4. Conclusions

In summary, we systematically studied nanoindentation size effect on the creep deformation of a Zr-based metallic glass upon a standard a Berkovich indenter and a spherical indenter. At a given holding strain ~7.1%, creep deformation decreased with increasing holding depth, which suggested a size-dependent creep resistance. In the elastic regions, creep deformation was insignificant and weakly decreased with holding depth. In the plastic regions, creep deformation was evidently enlarged and increased with holding strain. The estimated strain rate sensitivities (SRS) were decreased at first and then tended to stabilize with increasing holding depth and holding strain. The evolution of the shear transformation zone was presumed as a creep flow carrier in those tests with a holding strain larger than 2%. The corresponding values of SRS for STZ evolution were approximately between 0.05 and 0.3. Atomic diffusion and free volume migration were thought to be the creep mechanism at purely elastic holdings, and the characteristic values of SRS were 0.4~0.5.

Supplementary Materials: The following are available online at http://www.mdpi.com/2075-4701/9/5/613/s1, Figure S1: The typical XRD pattern of as-cast Zr-Cu-Ag-Al alloy, Figure S2: (**a**) The typical creep *P-h* curves at various holding depths under spherical indenter. *P-h* curves at shallow depths are enlarged in the inset. (**b**) Creep displacements at various holding loads were plotted with holding time, Figure S3: Elastic modulus was measured by continuous stiffness module (CSM) and was depicted as a function of displacement. The mean elastic modulus was 110 GPa.

Author Contributions: Data curation, Y.X.S.; formal analysis, Y.X.S.; investigation, Y.M.; software, X.W.H.; supervision, T.H.Z.; writing—original draft, Y.M.; funding acquisition, Z.Y.D.; writing—review and editing, Z.Y.D.

Funding: This research was funded by the National Natural Science Foundation of China (11727803, 11672356, 51705457) and Zhejiang Provincial Natural Science Foundation of China (LY18E010006).

Conflicts of Interest: The authors declare no conflict of interest.

References

1. Inoue, A.; Shen, B.; Koshiba, H.; Kato, H.; Yavari, A.R. Cobalt-based bulk glassy alloy with ultrahigh strength and soft magnetic properties. *Nat. Mater.* **2003**, *2*, 661–663. [CrossRef]
2. Jun, W.K.; Willens, R.H.; Duwez, P.O.L. Non-crystalline structure in solidified gold–silicon alloys. *Nature* **1960**, *187*, 869–870.
3. Schuh, C.A.; Hufnagel, T.C.; Ramamurty, U. Mechanical behavior of amorphous alloys. *Acta Mater.* **2007**, *55*, 4067–4109. [CrossRef]
4. Loffler, J.F. Bulk metallic glasses. *Intermetallics* **2003**, *11*, 529–540. [CrossRef]
5. Argon, A.S. Plastic deformation in metallic glasses. *Acta Metall.* **1979**, *27*, 47–58. [CrossRef]
6. Mayr, S.G. Activation energy of shear transformation zones: A key for understanding rheology of glasses and liquids. *Phys. Rev. Lett.* **2006**, *97*, 195501. [CrossRef] [PubMed]
7. Jang, D.; Greer, J.R. Transition from a strong-yet-brittle to a stronger-and-ductile state by size reduction of metallic glasses. *Nat. Mater.* **2010**, *3*, 215–219. [CrossRef]
8. Guo, H.; Yan, P.F.; Wang, Y.B.; Tan, J.; Zhang, Z.F.; Sui, M.L.; Ma, E. Tensile Ductility and Necking of Metallic Glass. *Nat. Mater.* **2007**, *10*, 735–739. [CrossRef]
9. Ma, Y.; Cao, Q.P.; Qu, S.X.; Jiang, J.Z. Stress-state-dependent deformation behavior in Ni–Nb metallic glassy film. *Acta Mater.* **2012**, *60*, 4136–4143. [CrossRef]
10. Schuh, C.A. Nanoindentation studies of materials. *Mater. Today* **2006**, *9*, 32–40. [CrossRef]
11. Nabarro, F.R. Creep in commercially pure metals. *Acta Mater.* **2006**, *54*, 263–295. [CrossRef]
12. Ma, Y.; Peng, G.J.; Wen, D.H.; Zhang, T.H. Nanoindentation creep behavior in a CoCrFeCuNi high-entropy alloy film with two different structure states. *Mat. Sci. Eng. A* **2015**, *621*, 111–117. [CrossRef]
13. Ma, Y.; Feng, Y.H.; Debela, T.T.; Zhang, T.H. Nanoindentation study on the creep characteristics of high-entropy alloy films: Fcc versus bcc structures. *Int. J. Refract. Met. H* **2016**, *54*, 395–400. [CrossRef]
14. Chen, H.; Zhang, T.H.; Ma, Y. Effect of Applied Stress on the Mechanical Properties of a Zr-Cu-Ag-Al Bulk Metallic Glass with Two Different Structure States. *Materials* **2017**, *10*, 711. [CrossRef] [PubMed]
15. Zhang, T.H.; Ye, J.H.; Feng, Y.H.; Ma, Y. On the spherical nanoindentation creep of metallic glassy thin films at room temperature. *Mat. Sci. Eng. A* **2017**, *685*, 294–299. [CrossRef]
16. Ma, Y.; Ye, J.H.; Peng, G.J.; Zhang, T.H. Loading rate effect on the creep behavior of metallic glassy films and its correlation with the shear transformation zone. *Mat. Sci. Eng. A* **2015**, *622*, 76–81. [CrossRef]
17. Yoo, B.G.; Kim, J.Y.; Kim, Y.J.; Choi, I.C.; Shim, S.; Tsui, T.Y.; Jang, J.I. Increased time-dependent room temperature plasticity in metallic glass nanopillars and its size-dependency. *Int. J. Plasticity* **2012**, *37*, 108–118. [CrossRef]

18. Wang, Y.; Zhang, J.; Wu, K.; Liu, G.; Kiener, D.; Sun, J. Nanoindentation creep behavior of Cu–Zr metallic glass films. *Mater. Res. Lett* **2018**, *6*, 22–28. [CrossRef]

19. Ma, Y.; Peng, G.J.; Jiang, W.F.; Chen, H.; Zhang, T.H. Nanoindentation study on shear transformation zone in a Cu-Zr-Al metallic glassy film with different thickness. *J. Non-Cryst. Solids* **2016**, *442*, 67–72. [CrossRef]

20. Ma, Y.; Ye, J.H.; Peng, G.J.; Zhang, T.H. Nanoindentation study of size effect on shear transformation zone size in a Ni–Nb metallic glass. *Mat. Sci. Eng. A* **2015**, *627*, 153–160. [CrossRef]

21. Yoo, B.G.; Oh, J.H.; Kim, Y.J.; Park, K.W.; Lee, J.C.; Jang, J.I. Nanoindentation analysis of time-dependent deformation in as-cast and annealed Cu–Zr bulk metallic glass. *Intermetallics* **2010**, *18*, 1898–1901. [CrossRef]

22. Wang, F.; Li, J.M.; Huang, P.; Wang, W.L.; Lu, T.J.; Xu, K.W. Nanoscale creep deformation in Zr-based metallic glass. *Intermetallics* **2013**, *38*, 156–160. [CrossRef]

23. Ma, Y.; Peng, G.J.; Feng, Y.H.; Zhang, T.H. Nanoindentation investigation on the creep mechanism in metallic glassy films. *Mat. Sci. Eng. A* **2016**, *651*, 548–555. [CrossRef]

24. Yoo, B.G.; Kim, K.S.; Oh, J.H.; Ramamurty, U.; Jang, J.I. Room temperature creep in amorphous alloys: Influence of initial strain and free volume. *Scr. Mater.* **2010**, *63*, 1205–1208. [CrossRef]

25. Bei, H.; Lu, Z.P.; George, E.P. Theoretical strength and the onset of plasticity in bulk metallic glasses investigated by nanoindentation with a spherical indenter. *Phys. Rev. Lett.* **2004**, *93*, 125504. [CrossRef] [PubMed]

26. Johnson, K.L. *Contact Mechanics*; Cambridge University Press: Cambridge, UK, 1987.

27. Jiang, Q.K.; Wang, X.D.; Nie, X.P.; Zhang, G.Q.; Ma, H.; Fecht, H.-J.; Bendnarcik, J.; Franz, H.; Liu, Y.G.; Cao, Q.P.; et al. Zr–(Cu,Ag)–Al bulk metallic glasses. *Acta Mater.* **2008**, *56*, 1785–1796. [CrossRef]

28. Packard, C.E.; Homer, E.R.; Al-Aqeeli, N.; Schuh, C.A. Cyclic hardening of metallic glasses under Hertzian contacts: Experiments and STZ dynamics simulations. *Philos. Mag.* **2010**, *90*, 1373–1390. [CrossRef]

29. Packard, C.E.; Schuh, C.A. Initiation of shear bands near a stress concentration in metallic glass. *Acta Mater.* **2007**, *55*, 5348–5358. [CrossRef]

30. Tabor, D. The hardness of solids. *Rev. Phys. Technol.* **1970**, *1*, 145–179. [CrossRef]

31. Su, C.; Herbert, E.G.; Sohn, S.; LaManna, J.A.; Oliver, W.C.; Pharr, G.M. Measurement of power-law creep parameters by instrumented indentation methods. *J. Mech. Phys. Solids* **2012**, *61*, 517–536. [CrossRef]

metals

Article

Creep Buckling of 304 Stainless-Steel Tubes Subjected to External Pressure for Nuclear Power Plant Applications

Byeongnam Jo [1,2,*], Koji Okamoto [2] and Naoto Kasahara [2]

[1] Department of Mechanical Engineering, Ajou University, Suwon 16499, Korea
[2] Department of Nuclear Engineering and Management, The University of Tokyo, Hongo, Tokyo 113-8656, Japan; okamoto@n.t.u-tokyo.ac.jp (K.O.); kasahara@n.t.u-tokyo.ac.jp (N.K.)
* Correspondence: jo798@ajou.ac.kr; Tel.: +82-31-219-2684

Received: 19 April 2019; Accepted: 7 May 2019; Published: 9 May 2019

Abstract: The creep-buckling behaviors of cylindrical stainless-steel tubes subjected to radial external pressure load at elevated temperatures—800, 900, and 1000 °C—were experimentally investigated. Prior to the creep-buckling tests, the buckling pressure was measured under each temperature condition. Then, in creep-buckling experiments, the creep-buckling failure time was measured by reducing the external pressure load for two different tube specimens—representing the first and second buckling modes—to examine the relationship between the external pressure and the creep-buckling failure time. The measured failure time ranged from <1 min to <4 h under 99–41% loading of the buckling pressure. Additionally, an empirical correlation was developed using the Larson–Miller parameter model to predict the long-term buckling time of the stainless-steel tube column according to the experimental results. Moreover, the creep-buckling processes were recorded by two high-speed cameras. Finally, the characteristics of the creep buckling under radial loading were discussed with regard to the geometrical imperfections of the tubes and the material properties of the stainless steel at the high temperatures.

Keywords: creep buckling; external pressure; Larson–Miller parameter; elevated temperature; visualization

---

1. Introduction

Creep is an important deformation mechanism at high temperatures, which unfavorably affects structural integrity. Because the creep rate increases with the temperature under a certain stress condition, the creep behaviors of various components used in high-temperature environments, such as power plants, have been studied to ensure the structural integrity of nuclear components [1–3]. However, studies into the mechanical behaviors of components under beyond-design-basis-accident (BDBA), such as the 2011 Fukushima accident, have to be carried out to handle the accidents. A BDBA causes an extremely high temperature and pressure. Therefore, the creep behaviors of the components used in nuclear power plants are studied under extreme conditions to prevent catastrophic disasters and to design BDBA-controllable nuclear power plants.

Stainless steel has been widely used in construction applications, including nuclear power plants, because of its outstanding and desirable characteristics [2,4]. In previous studies, the creep-related behaviors of various stainless-steel specimens at elevated temperatures have been investigated. Brinkman et al. investigated the effects of irradiation on the creep-fatigue properties of type 316 stainless steel at 593 °C (1100 °F). They found that aging is beneficial in improving the creep-fatigue properties [5]. Wareing developed a crack-propagation model to estimate the fatigue life at an elevated temperature between 538 and 760 °C and determined that creep damage influences the growth rate of surface fatigue cracks [6]. Furumura et al. experimentally and numerically investigated the creep-buckling behaviors (lateral deflection and stress–strain relationship) for steel columns with an H cross-section in the range of 475–550 °C. It was obtained that creep buckling behaviors were influenced by the temperature and eccentricity [7]. A numerical study of the creep behaviors of steel columns under axial loads was conducted to examine the effects of the heating rate and boundary restraint [8]. Kobayashi et al. performed multiaxial (uniaxial, biaxial, and triaxial) tension tests to investigate the positive mean stress effect on the creep damage and rupture lifetimes of 304 stainless steel (SS-304). Using electron microscopy, they observed void formation at the center of the specimen under equi-triaxial loading [9]. Additionally, after the Fukushima accidents, Jo et al., performed buckling and creep-buckling tests on slender SS-304 plates with rectangular cross sections under axial compression [10–12]. The buckling loads were experimentally measured at varying temperatures from room temperature (25 °C) to 1200 °C. It was found that geometrical imperfections such as initial bending reduced the buckling load; additionally, creep affected the buckling behaviors at temperatures above 800 °C [10,11]. They also performed creep-buckling experiments at 800, 900, and 1000 °C and developed empirical models of the creep-buckling failure time based on the Larson–Miller parameter (LMP) and the lateral deflection of the specimen [12]. Je et al., performed buckling experiments on thin cylindrical tube columns subjected to an external pressure in the radial direction from room temperature to 1200 °C. The effects of the tube dimensions (radius-to-thickness ratio) and temperature on the buckling pressure were examined for two different buckling modes [13]. As mentioned previously, knowledge regarding creep-buckling behaviors is necessary to clearly understand the failure of the tube column under BDBA conditions, because the creep effects unfavorably influence the metallic components at extremely high temperatures from the viewpoint of structural integrity.

Thus, the objective of the present study was to investigate the creep-buckling behaviors of SS-304 cylindrical tubes under external pressures in the radial direction. The creep-buckling failure time was measured at 800, 900, and 1000 °C with the variation of the external pressure (164 ~ 957 kPa). The relationship between the creep-buckling failure time and the external pressure was obtained for two different tube dimensions (different radius-to-thickness ratios), which yielded different buckling failure modes (different numbers of circumferential waves). Moreover, empirical correlations based on the creep-buckling failure time and the external pressure were developed for each temperature. The tube columns were visually recorded during the creep-buckling tests to elucidate the creep deformation.

2. Experiments

Figure 1 shows the experimental configuration for the creep-buckling tests performed in this study. A cylindrical pressure chamber (SOWA, Hitachi, Japan) was used to apply the external pressure loading to the tube specimen. The chamber was designed to be pressurized up to 1 MPa by supplying argon gas. For safe experiments, a safety valve designed to open at the design pressure (1 MPa) was installed at the top of the chamber head. The pressure of the chamber was measured by a transducer (Sensez, HLV-001MP, Tokyo, Japan) and recorded by a data logger (Keyence Wave Logger, NR500, Osaka, Japan) every second. The tube column was heated via Joule heating using a direct-current (DC) power supply (Yamabishi, YTR-8-500NX, Tokyo, Japan); therefore, the specimen was vertically placed between two copper electrodes that were connected to the power supply. The temperature of the tube specimen was measured by K-type thermocouples (SAKAGUCHI E.H VOC Corp., T35051, Tokyo, Japan), which were attached to the tube surface and covered by 0.1-mm-thick stainless-steel (same material as the tube) plates. These thin cover-plates ensured accurate measurement by reducing the heat loss. As shown in Figure 1, three thermocouples were used for the creep-buckling experiment, which were located 0, +20, and −20 mm from the center of the tube column in the vertical direction. Temperature data were recorded by the data logger at a frequency of 1 Hz. To examine the time evolution of the radial deformation of the tube specimen during the test, the tube column was recorded by two high-speed cameras: One for the front view (Photron, Fastcam SA-X, Tokyo, Japan) and the other for the side view (Photron, Fastcam APX-RS, Tokyo, Japan). The recording speed of the cameras was only 1 frame per second (fps), because of the long duration of the creep buckling and the limit of the recordable time due to the memory of the cameras.

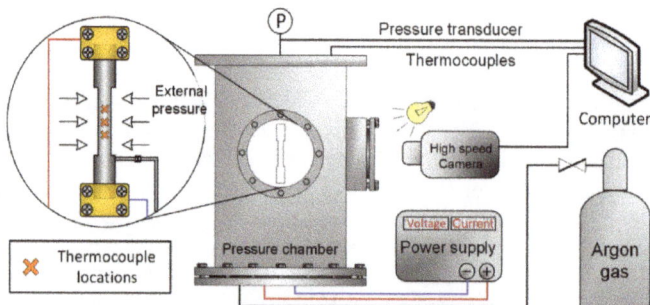

Figure 1. Schematic of the experimental setup.

Figure 2 presents photographs of two cylindrical tube columns, along with schematics showing their dimensions. In this study, 304 stainless-steel tubes provided by Nilaco Corp. (Tokyo, Japan) were used. The chemical compositions of the SS-304 tubes are presented in Table 1. As shown in Figure 2, the tube columns consisted of two sections: A testing part and holder parts. The testing part was the middle of the columns and buckled under the external pressure load. The holder parts were the upper and lower parts of the testing part, which were connected to the electrodes via special holders. The two tube columns had the same dimensions, except for the diameter. The tube columns in Figure 2A,B were denoted as Tube 1-03 and Tube3-03, respectively. Tube1-03 had an inner diameter of 13 mm and a thickness of 0.3 mm in the testing part. The inner diameter and the thickness of the holder parts for Tube1-03 were 13 and 1 mm, respectively. Tube 3-03 had an inner diameter of 22.1 mm and a 0.3-mm thickness of the testing part. The outer diameter of the holder parts was 25.4 mm, and the inner diameter was equal to that of the testing part. The length of the testing part was 100 mm for both the tubes. Table 2 presents the dimensions of the tube columns tested in this study. As shown in Figure 2, a small stainless-steel pipe 3.2 mm in diameter was inserted into the lower holder part and welded to connect the inside of the tube to the outside of the pressure chamber (atmosphere). Thus, the atmospheric pressure (1 atm) was maintained inside the tube during the experiment.

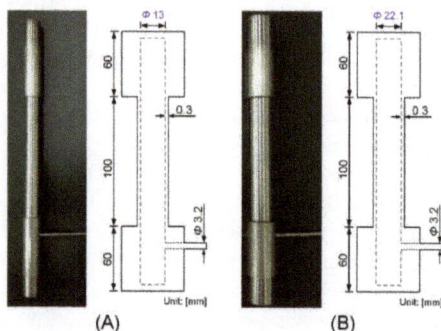

Figure 2. Photographs and dimensions of two tube specimens: (**A**) Tube1-03 and (**B**) Tube3-03.

Table 1. Chemical composition of SS-304 (mass fraction, %).

C	Si	Mn	P	S	Ni	Cr
0.003	< 1.0	< 2.0	0.045	< 0.03	8–11	18–20

Table 2. Dimensions of the two tube columns tested in this study.

Specimen Name	Inner Diameter [mm]	Thickness [a] [mm]	Length [a] [mm]	Radius-to-Thickness Ratio (R/t) [b]	Length-to-Radius Ratio (L/R) [b]
Tube 1-03	13.0	0.3	100	22.2	15.0
Tube 3-03	22.1	0.3	100	37.3	8.9

[a] Testing part of tube columns; [b] Mean radius was used to determine R/t and L/R values.

According to a previous study [13], the buckling mode induced by a radial pressure load is determined by the dimensions of the tube columns, i.e., the radius-to-thickness ratio (R/t) and length-to-radius ratio (L/R). As shown in Table 2, the two tube columns (Tube 1-03 and Tube 3-03) had different R/t and L/R values. The R/t values of Tube 1-03 and Tube 3-03 were 22.2 and 37.3, respectively. The L/R values of Tube 1-03 and Tube 3-03 were 15.0 and 8.9, respectively. Accordingly, Tube 1-03 buckled in the first mode, where two circumferential waves were formed; thus, the tube column was compressed in two opposite directions. In contrast, Tube 3-03 buckled in the second mode, where three circumferential waves were generated; thus, the tube column was deformed in three different directions. The two buckling modes were compared using the commercial finite element software ANSYS (mechanical APDL 14.0: element type: Solid185, Tokyo, Japan) as shown in Figure 3.

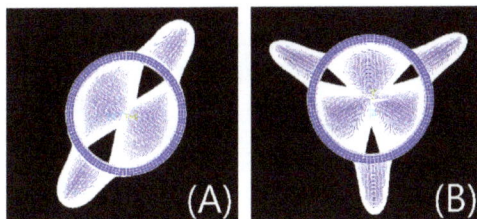

Figure 3. Comparison of the buckling modes. Top views of the specimens: (**A**) Tube 1-03 and (**B**) Tube 3-03.

The simple procedures for the creep-buckling experiment are summarized as follows.

- A tube column was vertically mounted between the copper electrodes in the pressure chamber.
- The chamber was pressurized to a target value to apply the external pressure to the tube radially.
- The temperature of the tube column was quickly increased to an experimental condition (800, 900, and 1000 °C).
- The temperature and pressure were maintained during the test.
- The time taken for the tube to exhibit creep buckling (creep-buckling failure time) was measured.

The temperature was increased by manually controlled voltages of the DC power supply. Because the buckling occurred in a short time, the temperature measured at the center deviated from the temperatures of the upper and lower (±20 mm from the center) thermocouples. However, in our previous study, the vertical temperature profile was uniform in creep-buckling experiments because of the long experimental duration. Additionally, in this study, the maximum deviation of the average temperature measured by three thermocouples for all of the tests was 2.0%. Figure 4 shows the typical temperature and pressure profiles obtained from the creep-buckling experiment of the present study. The time periods for the transient regions of the temperature (temperature-increasing regions) were not identical among the experiments, because the temperature was manually controlled. Nevertheless, the effect of the transient period was minimized by increasing the temperature. The transient heating time varied from approximately 1.3 to 8.2 min (average value: 3.7 min). These times are shorter than the creep-buckling failure time of the tube columns. The time period when the temperature of the tube column exceeded 700 °C was not significant for estimating the creep-buckling time. Hence, in the present study, the transient heating time was neglected in the estimation of the creep-buckling failure time.

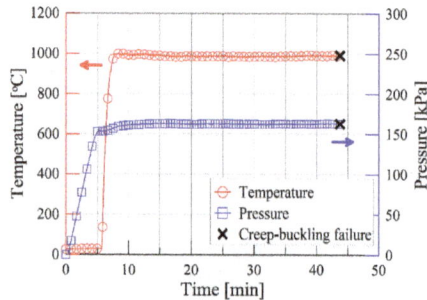

Figure 4. Typical temperature and pressure behaviors during the creep-buckling experiment.

3. Measurement Uncertainty

There were several measurement uncertainties for the creep-buckling failure time, temperature, and pressure in this study. The measurement uncertainty for the creep-buckling failure time was easily quantified, because the buckling failure of the tube column induced by the creep progressed quickly, and the deformation process was recorded by two cameras. According to the recording speed of 1 fps, the maximum measurement uncertainty for the creep-buckling failure time was obtained as 2.9%. The uncertainty of the pressure measurement was 0.5% according to the specifications of the pressure transducer. According to the 0.5 °C error of the K-type thermocouple, the uncertainty for the temperature measurement was negligible, because the minimum temperature was approximately 800 °C. However, as mentioned previously, the vertical temperature profile of the tube column was not uniform, because of the conduction heat loss in both the upper and lower copper electrodes. Nevertheless, temperature uniformity in the vertical direction of the tube column was achieved for most of the tested part of the specimen. Figure 5 compares the vertical temperature profiles for Tube1-03 between the experimental measurements and numerical predictions made using ANSYS. As the temperature increased, temperature uniformity was achieved for a greater length of the tube specimen. This is attributed to the longer time needed for the creep deformation to induce the failure (buckling) of the tube specimen.

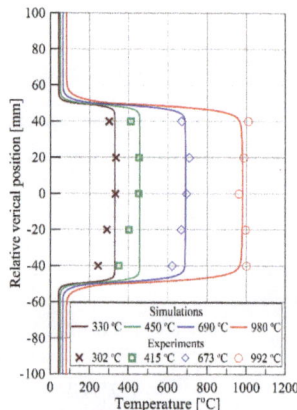

Figure 5. Comparison of the vertical temperature profiles of Tube 1-03 between the experimental measurements and numerical predictions.

4. Results and Discussions

4.1. Visualization of Creep-Buckling Deformation

Figures 6 and 7 show the time evolution of the creep-buckling deformation of Tube 1-03 at 800 °C and Tube 3-03 at 1000 °C, respectively. As mentioned above, it was experimentally confirmed that Tube 1 and Tube 3 buckle in the first and second mode of buckling. In other words, Tube 1 was compressed from two opposite directions as generating two circumferential waves. Additionally, in spite of the creep-buckling, buckling occurred very quickly; less than in 1 s. However, gradual deformation was also captured as shown in Figure 6. A white vertical line (reflection of the lamp) on the tube column was bent slowly, but it was clearly observed. Eventually, the deformation initially began at the center of the column and then, spread in both vertical directions. Similar features were observed for Tube 3-03 which buckled in the second mode. At 300 s, a white vertical light was observed on the left side of the column, but it was faded out as time goes by. Finally, the initial deformation of the creep-buckling was formed right at the white line. Thus, it means that the creep deformation proceeded slowly but obviously.

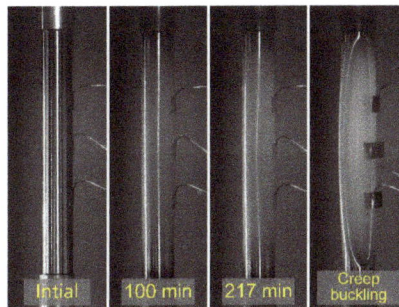

Figure 6. Time evolution images of the creep-buckling deformation (Tube 1-03 at 800 °C).

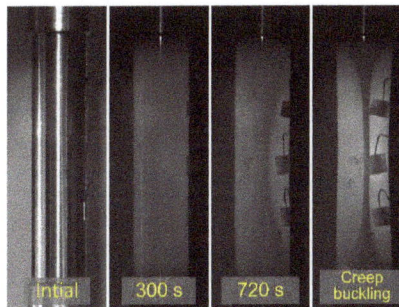

Figure 7. Time evolution images of the creep-buckling deformation (Tube 3-03 at 1000 °C).

4.2. Creep-Buckling Failure Time

Figures 8 and 9 show experimental measurements of the creep-buckling failure time, which was the time required for the tube column to buckle owing to the creep effect under radial external pressure loading lower than the buckling pressure at a given temperature. Figure 8 shows the results for Tube 1-03. As indicated by the plot with linear axes in Figure 8A, the failure time exponentially increased with the decrease of the external pressure; thus, the curve fit appeared to follow a decreasing log function (*external pressure* $= -C \log(failure\ time)$, where C is a constant). At 800 °C, the tube column was sustained for approximately 50 min under a load of 88% of the buckling pressure. When the external pressure was reduced to 80% of the buckling pressure, the specimen buckled at approximately

224 min. A similar but distinguishable trend of the creep-buckling failure time was observed in the cases of 900 and 1000 °C. At these temperatures, the failure time did not rapidly increase with the large decrease of the external pressure. For example, at 900 °C, the failure time for a load of approximately 61% of the buckling pressure was approximately 27 min. However, the failure time of the tube column subjected to approximately half of the buckling pressure was 59 min, which is more than double the failure time for the 61% loading case. Furthermore, the failure time was measured to be 133 min for a load of 43% of the buckling pressure at 900 °C. Therefore, the curve fits for the creep-buckling failure time at 900 and 1000 °C appeared to follow the decreasing log function more rapidly than those for 800 °C. This feature is clearly indicated by the logarithmic plot in Figure 8B. Linear behaviors of the creep-buckling failure time were observed, as shown in Figure 8B, and the slopes of the curve fits for 900 and 1000 °C were larger than that for 800 °C.

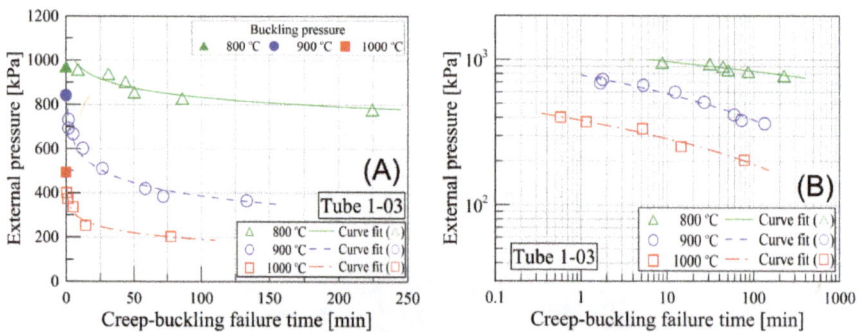

Figure 8. Measurements of the creep-buckling failure for Tube 1-03: (**A**) linear horizontal axis and (**B**) logarithmic horizontal axis.

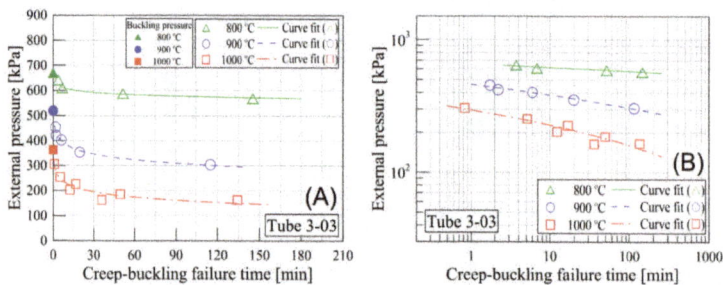

Figure 9. Measurements of the creep-buckling failure for Tube 3-03: (**A**) Linear horizontal axis and (**B**) logarithmic horizontal axis.

The creep-buckling failure time for Tube 3-03 is plotted in Figure 9. Because Tube 3-03 had a larger radius-to-thickness ratio than Tube 1-03, it deformed in a different buckling mode (second mode). However, the relationship between the external pressure and the failure time was similar to that for Tube 1-03. The failure-time increment with the decrease of the external pressure was more drastic at 800 °C than at 900 and 1000 °C. As indicated by the plots with logarithmic axes in Figure 9B, the slopes of the curve fits obtained at 900 and 1000 °C were larger than that for 800 °C.

The temperature-dependent creep-buckling features can be explained by the variations of the creep-buckling failure time with respect to the circumferential stress (hoop stress) normalized by the Young's modulus at each temperature. Figure 10 shows the variations of the creep-buckling failure time with respect to the circumferential stress (S) normalized by the Young's modulus (E) for Tube 1-03

(Figure 10A) and Tube 3-03 (Figure 10B). The circumferential stress values were estimated at the inner radius using Equation (1), according to the external pressures.

$$S = \left(\frac{P_i r_i^2 - P_o r_o^2}{r_o^2 - r_i^2} \right) - \left(\frac{r_i^2 \cdot r_o^2 (P_o - P_i)}{r^2 \left(r_o^2 - r_i^2 \right)} \right) \tag{1}$$

Here, P_i and P_o represent the pressures inside and outside the tube, respectively. r_i and r_o represent the inner and outer radii of the tube, respectively. The circumferential stress was normalized by the Young's modulus at each temperature [14]. The failure times in Figure 10 are classified into two groups—800 °C and 900/1000 °C —regardless of the buckling mode (tube dimension). For both tubes, the 800 °C group exhibits larger S/E values than the 900/1000 °C group, because of the high external pressures. The failure time for 800 °C drastically increased with the decrease of the S/E value. In contrast, the failure time for the 900/1000 °C group exhibited a relatively moderate increase with the decrease of the S/E value. This characteristic is clearly indicated by the graphs in the insets (logarithmic axes) of Figure 10.

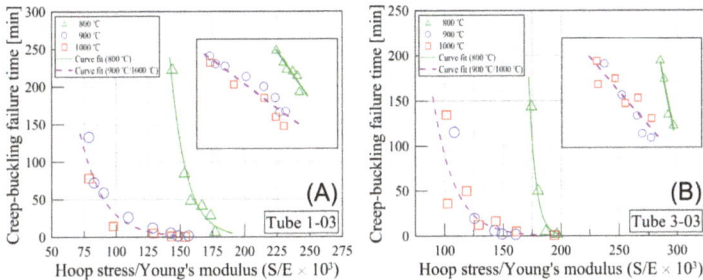

Figure 10. Variation of the creep-buckling failure time with respect to the circumferential stress (hoop stress) normalized by the Young's modulus (small window: Logarithmic axes): (**A**) Tube 1-03 and (**B**) Tube 3-03.

The two groups of the creep-buckling failure time in Figure 10 are attributed to change of the microstructures of the SS-304 specimens. A recrystallization process may have occurred in the specimens during the creep-buckling tests at 900 and 1000 °C. Thus, the microstructures of the austenitic stainless steel changed, and the grain size of the SS-304 tubes increased [15,16]. Since the failure time (duration of the tests) widely varied in this study from less than 1 min to more than 3 hours, the changes in the grain size might not be identical in all specimens. Nevertheless, the distinct relation between the failure time and the S/E value was obtained for the different buckling modes of the SS-304 tube specimens, as shown in Figure 10. Thus, it is postulated that the microstructures of the SS-304 were changed in the tests at 900 and 1000 °C. It is well known that the grain size of austenitic stainless steel is proportional to the heat-treatment temperature. Moreover, the strength of the stainless steel increases with the decrease of the grain size [16–18]. Bregliozzi et al. experimentally determined that the grain size of SS-304 varied according to the annealing temperature [15]. The average grain sizes were 2.5 and 40 μm for annealing temperatures of 780 and 1100 °C, respectively. Additionally, Schino et al. reported that the grain size of SS-304 was determined by both the annealing temperature and duration [16]. That is, the different microstructures based on the different grain sizes of SS-304 divided the creep-buckling behaviors of the cylindrical tubes under the external pressure into two groups. Similarly, the lattice self-diffusion coefficient of the SS-304 explains the effect of the temperature on the behaviors of the creep-buckling failure time. The diffusion coefficient is expressed by Equation (2), considering the temperature.

$$D_L = D_0 \exp(-kT_m/T) \tag{2}$$

$$\dot{\varepsilon} = A\left(D_L/b^2\right)(S/E)^5 \tag{3}$$

Here, D_L, D_0, k, T_m, and T represent the lattice self-diffusion coefficient, the experimental diffusion prefactor, a constant for a given crystal structure, the melting temperature, and the temperature of the specimen. The lattice self-diffusion coefficient decreases with the increase of the temperature. The strain rate ($\dot{\varepsilon}$) is proportional to the lattice self-diffusion coefficient, as shown in Equation (3). A and b are a constant for a given material and the magnitude of the Burgers vector, respectively. SS-304 has a high diffusion coefficient under high-temperature conditions; accordingly, the strain rate increases with the temperature, and the failure time decreases for specimens at high temperatures. Ultimately, this may result in a nonlinearly decreasing Young's modulus of SS-304 above 800 °C [13].

4.3. Empirical Correlation: LMP

Several parametric models have been proposed for predicting the long-term failure time based on short-term measurement data [19–21]. The LMP model and the Monkman–Grant model are widely used for the estimation of the creep-buckling failure time [12,22–24]. These models are both based on experimental measurements, but the Monkman–Grant model needs the minimum strain rate of the specimen. Hence, the LMP model was employed to propose an empirical correlation for the prediction of the creep-buckling failure time. In the LMP method, first, the LMP is calculated using experimental measurements (temperature and failure time), as shown in Equation (4), where T is the temperature, t_c is the creep-buckling failure time, and K is the Larson–Miller constant for a given material. In this study, K was assumed to be 20 for the SS-304 [12,25].

$$\mathrm{LMP} = T{\cdot}(\log(t_c) + K) \tag{4}$$

$$\log(t_c) = \left(\frac{4534 - P}{0.1736 \times T} - 20\right) \text{ for Tube1} - 03 \tag{5}$$

$$\log(t_c) = \left(\frac{2438 - P}{0.08979 \times T} - 20\right) \text{ for Tube3} - 03 \tag{6}$$

Figures 9B and 11A show the LMP variations with respect to the external pressure for Tube 1-03 and Tube 3-03, respectively. As noted in each figure, the mathematical relationships between the LMP and the external pressure were obtained from the curve fits (P in the curve-fit equations represents the external pressure). The creep-buckling failure time for each tube column was obtained using Equation (4) and the curve fit. Equations (5) and (6) give the empirical correlations for predicting the creep-buckling failure time at a given temperature and external pressure for Tube 1-03 and Tube 3-03, respectively. According to the empirical correlations (Equations (5) and (6)), the creep-buckling failure time was estimated for the two tube columns at 800, 900, and 1000 °C with the variation of the external pressure and was compared with the experimental measurements in Figure 12. The prediction of the failure time by the LMP models was sufficiently accurate to elucidate the long-term behaviors of the tube columns subjected to the external pressure. However, in the case of Tube 1-03 at 1000 °C, the difference between the prediction and the measurement drastically increased with the decrease of the external pressure. The tendency of the prediction curve is strongly dependent on the mathematical model of the curve fit chosen for the LMP–external pressure relationship in Figure 11. Three parallel prediction lines can be produced by using a different curve-fit equation for the LMP–external pressure relationship. Therefore, additional measurements are necessary for more accurate prediction of the creep-buckling failure time using the LMP-based empirical correlation.

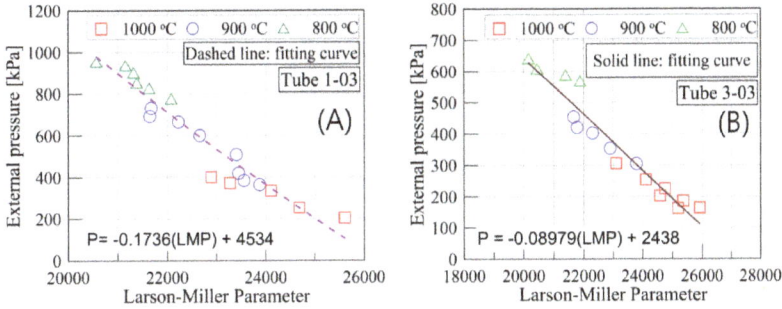

Figure 11. LMP variations with respect to the external pressure for (**A**) Tube 1-03 and (**B**) Tube 3-03.

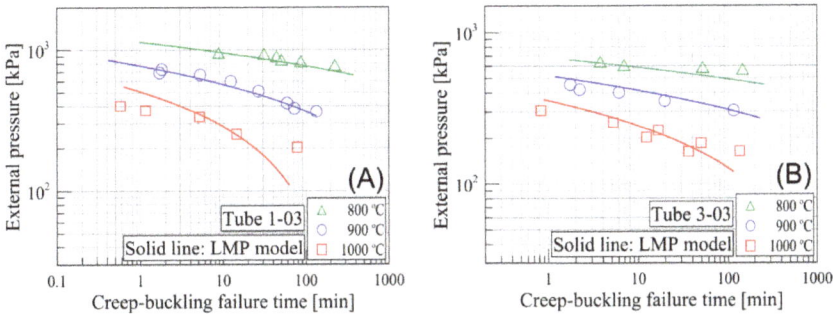

Figure 12. Prediction of the creep-buckling failure time using the LMP models (symbols: Experimental measurements) for (**A**) Tube 1-03 and (**B**) Tube 3-03.

5. Conclusions

The creep buckling of thin cylindrical tube columns made of 304 stainless steel subjected to an external pressure at 800, 900, and 1000 °C was experimentally investigated for two different buckling modes. The interesting temperature dependent creep-buckling characteristic (failure time) was obtained in this study. The empirical correlations for predicting the failure time of the tube column were developed for each tube dimension based on the experimental measurements. According to the results of the experimental study and the analysis, the following conclusions are drawn.

(a) The creep-buckling failure time for varying the external pressure is significantly affected by the temperature. For Tube 1-03, the failure time was measured to be >145 min for loading of 85% of the buckling pressure at 800 °C. However, the failure time under a similar external pressure (approximately 84% of the buckling pressure) was >1 min for Tube 1-03 at 1000 °C.

(b) The relationships between the failure time and the circumferential stress normalized by the Young's modulus (S/E) differed significantly between 800 and 900 °C. The experimentally-measured failure time was divided into two groups (800 °C and 900/1000 °C) as a function of the ratio of the circumferential stress to the Young's modulus at each temperature. The increase of the failure time caused by the decrease of the S/E value was more drastic at 800 °C than 900/1000 °C.

(c) LMP-based empirical correlations were developed for the two tube columns, and the prediction of the failure time agreed with the measurements. According to the mathematical model of the curve fit for the LMP–external pressure relationship, the temperature sensitivity of the failure-time predictions using the empirical correlations increased with the temperature.

Author Contributions: Conceptualization, K.O. and N.K.; methodology, B.J.; formal analysis, B.J., K.O., N.K.; investigation, B.J., K.O. and N.K.; writing—original draft preparation, B.J.; writing—review and editing, B.J.

Metals **2019**, *9*, 536

Acknowledgments: This study was conducted as part of the "Study on failure mechanism of nuclear components under ultimate loadings and prevention of catastrophic failure modes" entrusted to the University of Tokyo by the Ministry of Education, Culture, Sports, Science and Technology of Japan (MEXT).

Conflicts of Interest: The authors declare no conflict of interest.

References

1. Majumdar, S. Prediction of structural integrity of steam generator tubes under severe accident conditions. *Nucl. Eng. Des.* **1999**, *194*, 31–55. [CrossRef]
2. Zinkle, S.J.; Was, G.S. Materials challenges in nuclear energy. *Acta Mater.* **2013**, *61*, 735–758. [CrossRef]
3. Takahashi, Y.; Shibmoto, H.; Inoue, K. Study on creep-fatigue life prediction methods for low-carbon nitrogen-controlled 316 stainless steel (316FR). *Nucl. Eng. Des.* **2008**, *238*, 322–335. [CrossRef]
4. Baddoo, N.R. Stainless steel in contruction: A review of research, applications, challenges and opportunities. *J. Constr. Steel Res.* **2008**, *64*, 1199–1206. [CrossRef]
5. Brinkman, C.R.; Korth, G.E.; Hobbins, R.R. Estimates of creep-fatigue interaction in irradiated and unirradiated austenitic stainless steels. *Nucl. Technol.* **1972**, *16*, 297–307. [CrossRef]
6. Wareing, J. Creep-fatigue interaction in austenitic stainless steels. *Metall. Mater. Trans. A* **1977**, *8*, 711–721. [CrossRef]
7. Furumura, F.; Ave, T.; Kim, W.J. Creep buckling of steel columns at high temperatures Part II Creep buckling tests and numerical analysis. *J. Str. Constr. Eng.* **1986**, *361*, 142–151.
8. Huang, Z.F.; Tan, K.H.; Ting, S.K. Heating rate and boundary restraint effects on fire resistance of steel columns with creep. *Eng. Str.* **2006**, *28*, 805–817. [CrossRef]
9. Kobayashi, H.; Ohki, R.; Takamoto, I.; Masao, S. Multiaxial creep damage and lifetime evaluation under biaxial and triaxial stresses for type 304 stainless steel. *Eng. Fract. Mech.* **2017**, *174*, 30–43. [CrossRef]
10. Jo, B.; Sagawa, W.; Okamoto, K. Buckling behaviors of metallic columns under compressive load at extremely high temperatures, In Proceedings of the ASME 2014 Pressure Vessels and Piping Conference, Anaheim, CA, USA, July 20–24, 2014.
11. Jo, B.; Sagawa, W.; Okamoto, K. Measurement of buckling load for metallic plate columns in severe accident conditions. *Nucl. Eng. Des.* **2014**, *274*, 118–128. [CrossRef]
12. Jo, B.; Okamoto, K. Experimental investigation into creep buckling of a stainless steel plate column under axial compression at extremely high temperatures. *J. Press. Vessel Technol.* **2017**, *139*, 011406. [CrossRef]
13. Jo, B.; Kasahara, N.; Okamoto, K. Buckling of cylindrical stainless-steel tubes subjected to external pressure at extremely high temperatures. *Eng. Fail. Anal.* **2018**, *92*, 61–70. [CrossRef]
14. Kasahara, N. *Report for the MEXT Nuclear System Research Project, Study on Failure Mechanism of Nuclear Components Under Ultimate Loadings and Prevention of Catastrophic Failure Modes*; Ministry of Education, Culture, Sports, Science and Technology: Tokyo, Japan, 2016. (In Japanese)
15. Bregliozzi, G.; Ahmed, S.I.U.; Schino, A.D.; Kenny, J.M.; Haefke, H. Friction and wear behavior of austenitic stainless steel: influence of atmospheric humidity, load range, and grain size. *Tribol. Lett.* **2004**, *17*, 697–704. [CrossRef]
16. Schino, A.D.; Barteri, M.; Kenny, J.M. Effects of grain size on the properties of a low nickel austenitic stainless steel. *J. Mater. Sci.* **2003**, *38*, 4725–4733. [CrossRef]
17. Wang, N.; Wang, Z.; Aust, K.T.; Erb, U. Effect of grain size on mechanical properties of nanocrystalline materials. *Acta Metall. Mater.* **1995**, *43*, 519–528. [CrossRef]
18. Lee, Y.S.; Kim, D.W.; Lee, D.Y.; Ryu, W.S. Effect of grain size on creep properties of type 316LN stainless steel. *Met. Mater. Int.* **2001**, *7*, 107–114. [CrossRef]
19. Larson, F.R.; Miller, J. A time-temperature relationship for rupture and creep stresses. *Trans. ASME* **1952**, *74*, 765–771.
20. Orr, R.L.; Sherby, O.D.; Dorn, J.E. Correlation of rupture data for metals at elevated temperatures. *Trans. ASME* **1954**, *46*, 113–128.
21. Monkman, F.C.; Grant, N.J. An empirical relationship between rupture life and minimum creep rate in creep rupture test. *Proc. ASTM* **1956**, *56*, 593–620.
22. Sundararajan, G. The Monkman-Grant relationship. *Mat. Sci. Eng. A* **1989**, *112*, 205–214. [CrossRef]

23. Fedoseeva, A.; Dudova, N.; Kaibyshev, R.; Belyakov, A. Effect of tungsten on creep behavior of 9%Cr—3%Co martensitic steels. *Metals* **2017**, *7*, 573. [CrossRef]
24. Dewa, R.T.; Park, J.H.; Kim, S.J.; Lee, S.Y. High-temperature creep-fatigue behavior of alloy 617. *Metals* **2018**, *8*, 103. [CrossRef]
25. Masuyama, F. Creep rupture life and design factors for high-strength ferritic steels. *Int. J. Pres. Vessel. Pip.* **2007**, *84*, 53–61. [CrossRef]

metals

MDPI

Article

Low Cycle Fatigue and Relaxation Performance of Ferritic–Martensitic Grade P92 Steel

Maria Jürgens, Jürgen Olbricht *, Bernard Fedelich and Birgit Skrotzki

Bundesanstalt für Materialforschung und -prüfung (BAM), Division 5.2: Experimental and Model Based Mechanical Behaviour of Materials, 12205 Berlin, Germany; maria.juergens@bam.de (M.J.); bernard.fedelich@bam.de (B.F.); birgit.skrotzki@bam.de (B.S.)
* Correspondence: juergen.olbricht@bam.de, Tel.: +49-30-8104-3137

Received: 12 December 2018; Accepted: 10 January 2019; Published: 18 January 2019

Abstract: Due to their excellent creep resistance and good oxidation resistance, 9–12% Cr ferritic–martensitic stainless steels are widely used as high temperature construction materials in power plants. However, the mutual combination of different loadings (e.g., creep and fatigue), due to a "flexible" operation of power plants, may seriously reduce the lifetimes of the respective components. In the present study, low cycle fatigue (LCF) and relaxation fatigue (RF) tests performed on grade P92 helped to understand the behavior of ferritic–martensitic steels under a combined loading. The softening and lifetime behavior strongly depend on the temperature and total strain range. Especially at small strain amplitudes, the lifetime is seriously reduced when adding a hold time which indicates the importance of considering technically relevant small strains.

Keywords: ferritic–martensitic steel; P92; low cycle fatigue; relaxation fatigue; cyclic softening

---

1. Introduction

The growing share of renewable energy sources in the electricity markets has forced many power plants into a more "flexible" operation with frequent load shifts and shutdowns. The associated temperature and stress gradients lead to complex loading scenarios in plant components, which may result in superimposed creep deformation, creep-fatigue, and thermo-mechanical fatigue [1]. A fundamental understanding of the mechanical behavior and damage evolution in 9–12% Cr ferritic–martensitic steels under a combined static and cyclic loading, at a high temperature is, therefore, required.

Comprehensive analyses of the mechanical behavior of 9–12% Cr steels have been carried out during the market introduction of these steels [2,3], but they focused on creep and creep-rupture testing, in line with the constant operation modes of the power plants at these times. With the introduction of the so-called "second generation" Cr steels, like P91 and HCM12 [4], more attention was paid to the material behavior under combined loading scenarios involving creep and low-cycle fatigue. For P91, the seminal work of Kim and Weertman [5] identified a number of key effects of fatigue loading on subsequent mechanical behavior and microstructure evolution. As a most obvious feature of fatigue on P91, substantial initial softening of the specimens was observed in strain-controlled testing. It continued throughout the fatigue experiments, though with decreasing rates, leading to reductions of the stress range exceeding 250 MPa [5]. Hold times at the peak tensile or compressive strain were found to reduce the number of cycles to fracture by, nearly, a factor of two. While stress relaxation during these holds was pronounced, it hardly affected the peak stress in the subsequent cycles. Aging of the specimens for 5000 h, at 593 °C, prior to testing, showed little effect on time to fracture in fatigue testing [5]. Transmission electron microscopy (TEM) revealed that aging indeed had no obvious effect on the ferritic–martensitic microstructure, while a rapid loss of the original lath type microstructure, by a formation of equiaxed dislocation cells and carbide coarsening, was observed, after only few hours of fatigue loading [5].

In the following decades, the fatigue behavior and microstructure evolution have been investigated in more detail in P91 [6,7], as well as in other 9–12% Cr-steels [8–10]. The topic of combined loading (creep-fatigue) has been extensively studied for P91 [11]. Generally, these studies confirmed the above-mentioned trends and indicated qualitatively similar behaviors for all ferritic–martensitic steel grades. However, the number of studies dealing with fatigue and creep-fatigue of these materials is still quite limited, compared to the great efforts spent on characterizing the creep behavior [12], especially when considering the large parameter fields which would need to be covered for a thorough analysis of the material behavior, under the different combined loadings. Consequently, the amount of available fatigue data is quite restricted, as is reflected, for example, in the number of creep data and fatigue data sheets provided by NIMS in Japan [13].

Recently, combined loading involving creep and cyclic fatigue loads has received increased attention due to the more flexible operation of power plants in many countries. Consequently, reliable data on the fatigue and creep-fatigue behavior are required for current construction materials. Since cyclic operation may especially affect thick-walled components (due to the build-up of thermal stresses [1]), the common material grades for live steam piping, headers, etc., need to be comprehensively tested. One current candidate material is P92, an optimized version of the 9% Cr grades with an increased tungsten content. Due to its enhanced creep resistance, compared to earlier grades, P92 can be used up to maximum operation temperatures of 620 °C [14]. A number of studies on the fatigue behavior of P92 have been published recently [15–18]. It was found that P92 exhibits a considerable cyclic softening, as it is known from P91. No quantitative comparison of the effect was given but it was confirmed that softening depends on a number of parameters, like the amount of plastic strain and temperature. The lifetime data were found to well follow the Coffin-Manson relationship. The effect of hold times has been previously investigated [19]; it was demonstrated that hold times may result in significantly reduced lifetime, as is known from P91.

However, it turns out that existing studies are quite restricted in terms of the investigated parameter fields. Considering the actual operation modes of power plants, which involve frequent warm starts or load variations for fast reaction to changing demands, data are required for (a) intermediate temperatures in the 300 °C range, (b) low-strain amplitudes [20], and (c) hold times at these conditions which resemble periods of constant operation [21]. In the present contribution, the fatigue behavior of a 9% Cr steel (P92) was investigated, with a focus on these loading types and ranges of parameters. Standard low-cycle fatigue (LCF) tests were carried out to create a database of fatigue data and to establish the Coffin-Manson coefficients, at different temperature levels, in addition to those that have been tested so far. In a second step, hold times under strain control were included into the LCF test procedures to study the influence of combined fatigue/relaxation processes on lifetime. Symmetric load and hold profiles with strain ratios of $R_\varepsilon = -1$ and low-peak strains were applied in the current work to simulate the constrained thermal expansion and contraction of plant components, which may occur during fast load shifts due to inhomogeneous temperature distributions in the plant structure. In contrast to classical creep deformation, due to live steam pressure with continuously rising strain, the load scenario is then characterized by almost constant strains and related stress relaxation.

2. Materials and Methods

The material used for the mechanical tests is the ferritic–martensitic steel P92 (X12CrMoWVNbN10-1), according to DIN-EN 10216-2 [22]. The chemical composition of the investigated batch, as measured by spark spectrometry, is given in Table 1. All specimens were extracted from fully annealed sections of steam pipes with a wall thickness of 47 mm. The microstructure consists of tempered martensite, which results from a typical heat treatment, including normalizing (1040–1080 °C) and tempering (730–800 °C). During normalizing, austenite is formed and carbonitrides are dissolved. During subsequent air-cooling, the austenite transforms to martensite with a high density of free dislocations. To soften the material and to precipitate fine carbonitrides, a tempering treatment follows, leading to partial rearrangement of dislocations and formation of subgrains. The as-received microstructure is shown in Figure 1a,b; a schematic view of the typical substructure of a prior austenite grain is shown based on [23]; corresponding schemes have also been suggested elsewhere [6,24]. The prior austenite grains, which are transformed to a ferritic condition during tempering, still exhibit signs of the previous hierarchical martensitic structure with packets, blocks, and elongated cells that are often denominated as "laths". $M_{23}C_6$ carbides are formed at lath, packet, and at prior austenite grain boundaries (PAGB). They are elongated in shape, with a length of 60–150 nm. Very fine MX-type carbonitrides (20–80 nm) are homogeneously distributed in the matrix within the laths [25,26].

Table 1. Chemical composition of P92 (mass percent, %) measured by spark spectrometry.

C	Si	Mn	P	S	Cr	Ni	Mo	Co	Nb	V	W
0.126	0.114	0.446	0.012	0.005	8.93	0.167	0.5	0.019	0.092	0.169	1.95

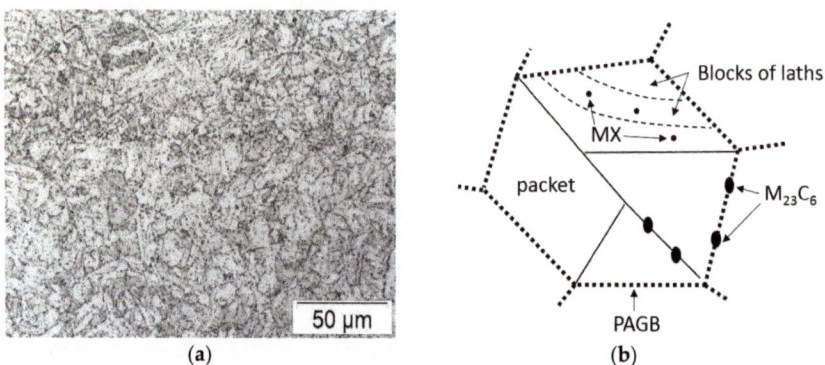

(a) (b)

Figure 1. Microstructure of P92, (**a**) an optical micrograph and (**b**) schematic view of the internal substructure within prior austenite grains of ferritic–martensitic steels (adapted by permission from Springer Nature Customer Service Centre GmbH: MDPI AG, Journal of Materials Engineering and Performance, Low-Cycle Fatigue Properties of P92 Ferritic–Martensitic Steel at Elevated Temperature, Z. Zhang et al., Copyright license 4432511224994, 2016).

For mechanical testing, cylindrical specimens of 18 mm gauge length and 6 mm gauge diameter (Figure 2) were machined from the pipe in a tangential direction. After testing, some specimens were prepared for metallographic inspection. For this purpose, the tested specimens were sectioned parallel to the loading direction, using a laboratory cutting machine. The relevant pieces were mounted in resin, and then ground and mechanically polished.

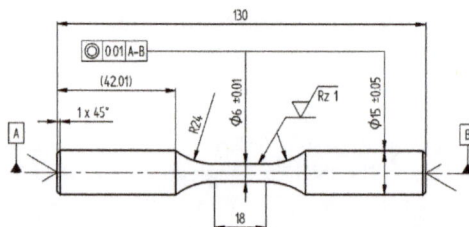

Figure 2. Specimen dimensions (unit: mm).

All tests were conducted on servo-hydraulic testing machines of type MTS Landmark with 100 kN force transducers. The machines were equipped with induction heating systems, which allowed fast heating of the test pieces. Initial heating and soaking was completed in less than 10 minutes at all test temperature levels. Strain was measured with high-temperature extensometers (water-cooled, MTS-632.51F.04), with a gauge length of 12 mm. The temperature variation in the gauge length section of the specimens was max. ±5 °C. Four thermocouples of type S were spot-welded to the specimens, to measure the temperature. The control thermocouple was placed in the center of the gauge section, a second thermocouple was placed also in the middle, but at a 90° rotated position. The two other thermocouples were placed 5 mm above and below the control thermocouple. In preceding tests, it was verified that the thermocouples did not have a systematic influence on the crack initiation of the specimens.

In a first campaign, standard low-cycle fatigue (LCF) tests according to ISO 12106 [27] were carried out. The tests were conducted in air, under strain control, with constant strain rates of 1.0×10^{-3} 1/s. Mechanical strains of ±0.2% to ±0.6% were applied in the test series. A triangular wave form with a mechanical strain ratio of $R_\varepsilon = -1$ was employed (Figure 3a). The failure criterion was a 10% drop in the peak tensile stress, following the procedure discussed in Reference [15]. LCF-tests were conducted at three temperature levels: 300 °C, 500 °C, and 620 °C. In a second step, hold periods t_h under strain control (at the maximum and minimum strain) were included in the LCF-test procedure (Figure 3b). They will be referred to as the relaxation-fatigue (RF) tests in the following, since the hold period allows for stress relaxation while the strain is kept constant. The hold time was 3 min, unless otherwise noted.

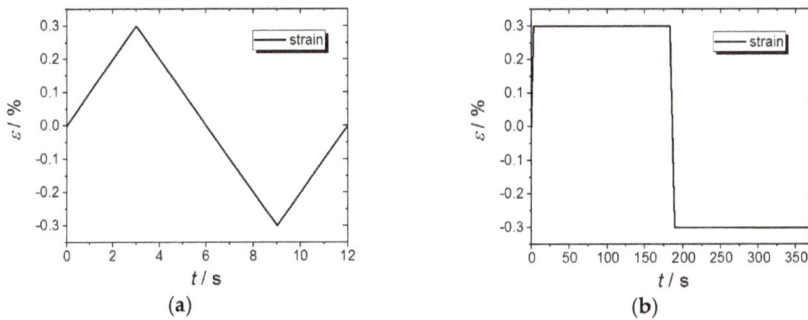

Figure 3. Schematic representation of loading waveform for one cycle of (**a**) a low-cycle fatigue (LCF)-test and (**b**) a LCF-test with a hold time of 3 min in tension and compression.

In Figure 4, all important mechanical parameters have been defined with the help of an example hysteresis loop. The stress range $\Delta\sigma$ is defined by the maximum and minimum stress (Equation (1)), as well as the mean stress σ_m (Equation (2)). The total strain range, $\Delta\varepsilon_t$, is equal to twice the strain amplitude, $\varepsilon_{a,t}$, and consists of an elastic ($\Delta\varepsilon_e$), and a plastic part ($\Delta\varepsilon_p$) (Equations (3) and (4)). Assuming that the elastic deformation of the material follows Hooke's law, the elastic and plastic contributions can be separated with the help of a straight line representing the unloading modulus (Equation (5)). Therefore, it is very important to determine the modulus of elasticity correctly. In this work, it was measured by the sonic resonance method in accordance to ASTM E 1875:2013 [28]. The values obtained at the different test temperatures are given in Table 2. In this study, the plastic strain contribution was determined at a stabilized cycle, which was at half-lifetime. At high temperatures, time-dependent processes take place. The total strain then consists of an elastic strain contribution, an instantaneous plastic strain, and a time-dependent creep strain. The sum of plastic strain and creep strain is the inelastic strain (Equation (6)). The energy dissipated during a cycle represents the plastic work, W_p, and is the area within the hysteresis loop (Equation (7)).

$$\Delta\sigma = \sigma_{max} - \sigma_{min} = 2\sigma_a \tag{1}$$

$$\sigma_m = \frac{\sigma_{max} + \sigma_{min}}{2} \tag{2}$$

$$\Delta\varepsilon_t = \Delta\varepsilon_e + \Delta\varepsilon_p = 2\varepsilon_{a,t} \tag{3}$$

$$\varepsilon_{a,t} = \varepsilon_{a,e} + \varepsilon_{a,p} \tag{4}$$

$$E = \frac{\sigma_a}{\varepsilon_{a,e}} \tag{5}$$

$$\Delta\varepsilon_t = \Delta\varepsilon_e + \underbrace{\Delta\varepsilon_p + \Delta\varepsilon_{cr}}_{\Delta\varepsilon_{in}} \tag{6}$$

$$W_p = \int \sigma \cdot d\varepsilon \tag{7}$$

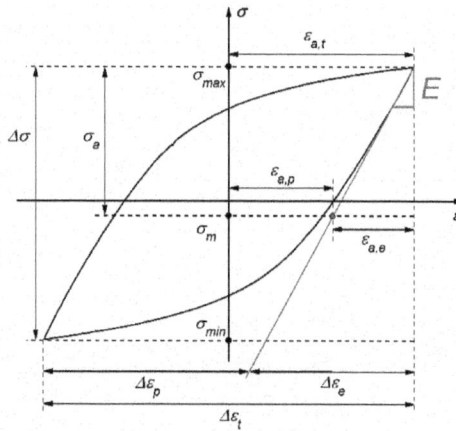

Figure 4. Schematic hysteresis loop for LCF-tests with characteristic quantities.

Table 2. Young's modulus, *E*, of P92 at different temperatures obtained by sonic resonance.

Temperature [°C]	E [GPa]
24	216
300	197
500	178
620	162

3. Results

An overview of all low-cycle fatigue (LCF) and relaxation fatigue (RF) tests performed in this work, including the test conditions and the lifetimes reached, is given in Table 3.

3.1. Cyclic Stress Response Behavior of the LCF-Tests

Ferritic–martensitic steels, such as P92, show a continuous cyclic softening behavior in LCF testing. In Figure 5a, the maximum and minimum stress of each cycle are plotted, exemplarily, for one LCF test. The curves can be divided into three phases. In the initial part, which lasted for about the first 250 cycles in the given example, the major part of the softening occurred. After this pronounced initial drop of stress, a stabilized part followed. It was characterized by a small but continuous further change in the stress values. Towards the end of the test, a final failure part, with rapidly falling stress values followed. The curves of the peak cyclic stress, at the different total strains and the three temperature levels investigated here, have been presented in Figure 5b–d. For reasons of clarity, only the maximum peak stress has been plotted, and the number of cycles has been given in a logarithmic scale to better display the initial material behavior. Due to the temperature dependency of the mechanical properties the maximum stresses are highest at 300 °C, Figure 5b. Additionally, the maximum stresses at 300 °C were spaced more closely between the different strain amplitudes than at higher temperatures and the softening was similar. At 500 °C and 620 °C, the range of initial maximum stresses was wider than that in Figure 5b, due to the extended range of the total strain ranges. Tests at total strain ranges below ±0.3% result in considerably reduced maximum stresses, especially at the highest test temperature (Figure 5d). For example, at 620 °C, the maximum stresses ranged between 340 MPa and 360 MPa for tests with $\Delta\varepsilon_t \geq \pm0.3\%$, whereas they ranged from 260 MPa to 340 MPa for strain ranges between ±0.2% and ±0.3%. Although the maximum stresses at $N = 1$, for the different total strain ranges differed by 35 MPa, at 300 °C and 100 MPa at 620 °C, the transition from the second phase (gradual lowering of the maximum stress) to the third phase (sudden decrease of stress towards the end of the test) occurred nearly at the same stress level, at each temperature.

Table 3. Test parameters and lifetime of all conducted tests.

Test Type	Temperature [°C]	Total Strain Range [%]	Lifetime [-]	Test Duration [h]
LCF	300	±0.3	11276	37.6
LCF	300	±0.4	3397	15.1
LCF	300	±0.5	2474	13.7
LCF	500	±0.2	41880	93.1
LCF	500	±0.23	18200	46.5
LCF	500	±0.25	5317	14.8
LCF	500	±0.3	3660	12.2
LCF	500	±0.4	1510	6.7
LCF	500	±0.5	1190	6.6
LCF	620	±0.2	8920	19.8
LCF	620	±0.23	5011	12.8
LCF	620	±0.3	1809	6.0
LCF	620	±0.4	1124	5.0
LCF	620	±0.5	725	4.0
LCF	620	±0.6	689	4.6
RF (3 min)	300	±0.5	2305	243.3
RF (3 min)	500	±0.23	5320	545.6
RF (3 min)	500	±0.3	2307	238.4
RF (3 min)	500	±0.5	955	100.8
RF (3 min)	620	±0.2	2050	209.6
RF (10 min)	620	±0.2	1582	530.8
RF (3 min)	620	±0.23	2230	224.4
RF (3 min)	620	±0.3	1324	136.8
RF (3 min)	620	±0.4	996	104.0
RF (10 min)	620	±0.4	883	298.3

The results given in Figure 5 indicate that the softening is more pronounced at higher temperatures. To quantify this observation the amount of softening was expressed according to the following equation:

$$softening\ ratio = \frac{\sigma_{max} - \sigma_{max,50\%\ Nf}}{\sigma_{max}} \cdot 100 \tag{8}$$

σ_{max} is the overall maximum stress of the complete test, whereas $\sigma_{max,50\%Nf}$ is the maximum stress of the stabilized cycle at half-lifetime. The equation was adapted from References [15,23], which used the maximum stress of the first cycle (see chapter 4. Discussion). In Table 4, the values for the softening ratio for all LCF tests have been summarized. For total strain ranges between ±0.3% and ±0.6%, the ratio appeared to be independent of the total strain for each temperature, but it decreased for ranges smaller than ±0.3%. It seems that a minimum amount of inelastic strain is required to cause significant softening. This can also be seen in Figure 5d, where the initial maximum stress of the cyclic softening curves is nearly independent of the applied strain for $\Delta\varepsilon_t \geq \pm0.3\%$.

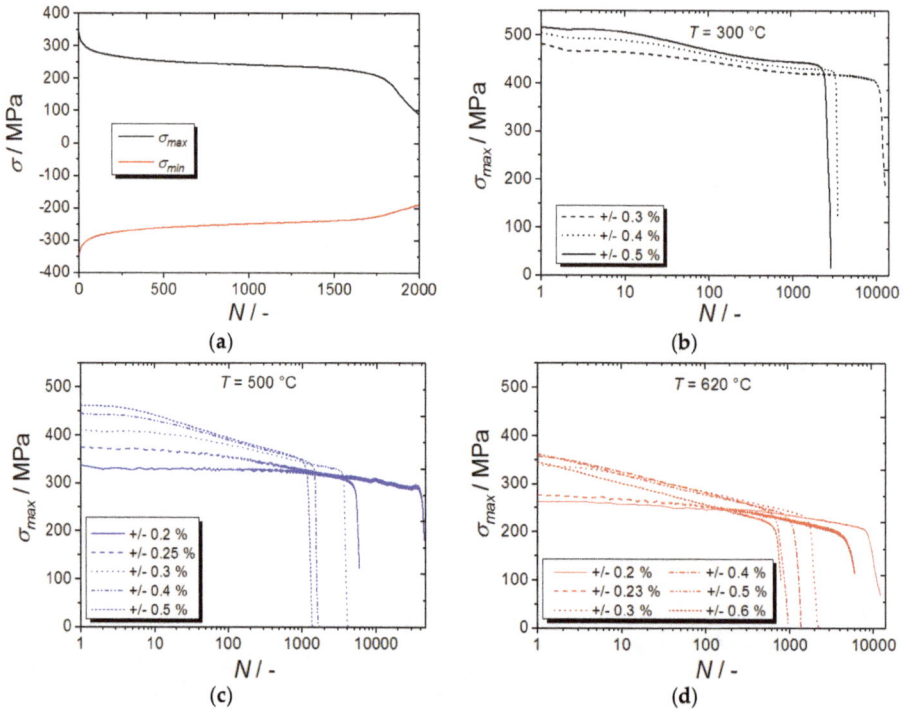

Figure 5. Cyclic stress response curves of one LCF test at 620 °C, ±0.3% (**a**), and of all LCF-tests at (**b**) 300 °C, (**c**) 500 °C and (**d**) 620 °C.

Table 4. Softening ratio as a function of total strain range and temperature for all conducted LCF-tests.

T [°C]	$\Delta\varepsilon_t$ [%]					
	±0.2	±0.23	±0.3	±0.4	±0.5	±0.6
300	-	-	14.2%	14.7%	14.1%	-
500	12.4%	16%	18%	20.2%	21.5%	-
620	17.9%	24.6%	27.8%	30.8%	29.5%	33.3%

To further analyze the cyclic stress response, the inelastic strain contribution calculated by Equations (5) and (6) is considered. As shown in Figure 6, linear correlations exist between the inelastic strain amplitude and the total strain amplitude of P92, at each test temperature. The constants of the fitted linear functions are also given in the diagram, for all temperatures. At each temperature level, the inelastic strain amplitude increased with the rising total strain amplitude, because the amount of elastic strain remained nearly the same. In turn, this led to rising relative contributions of inelastic strain. For example, at 500 °C and a total strain amplitude of 0.2% the inelastic strain amplitude was 0.03%, representing a relative share of 15%. At a total strain amplitude of 0.4%, the inelastic strain amplitude was 0.2%, which was a relative amount of 49.5%. Similarly, the inelastic contribution rose with temperature. This was obviously due to a reduced strength of the material at higher temperatures.

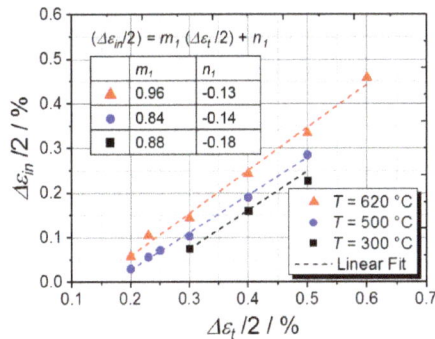

Figure 6. Relation between total strain amplitude and inelastic strain amplitude for the LCF-tests at $N = 50\% \, N_f$.

This increasing amount of plasticity could also be detected in the hysteresis loops. In Figure 7, the hysteresis loops for 500 °C and the different total strain ranges are plotted (first cycle and at half-lifetime). Qualitatively similar hysteresis loops were obtained at 300 °C and 620 °C and are, therefore, not shown. With increasing total strain range, the hysteresis loops became wider. The plot also shows how the stress–strain hysteresis loops developed, upon subsequent unloading and reloading. The maximum tension and compression stresses at half-lifetime were reduced, compared to the first cycle, while the width of the loops systematically increased upon cycling. This was a clear visual indication of the cyclic softening of P92.

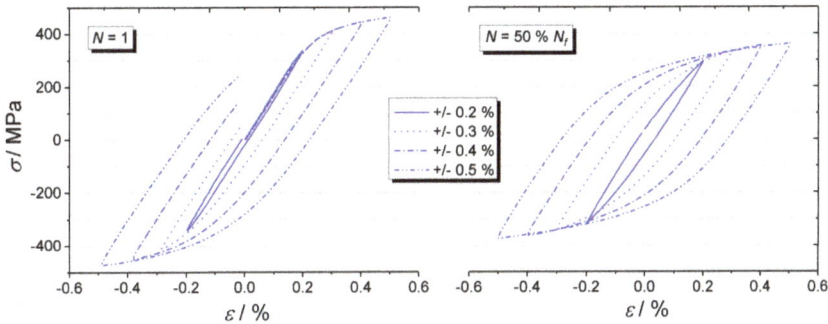

Figure 7. Evolution of the hysteresis loops at 500 °C and different total strain ranges in first cycle (**left**) and at half-lifetime (**right**).

The plastic work, W_p, at half-lifetime, as a function of the total strain amplitude is shown in Figure 8, for all three temperature levels. The data suggest that there was a linear correlation with the applied total strain. The parameters for the fitted linear functions are also listed in the plot. The fitted linear curves suggest that there was an intersection point around a strain amplitude of 0.4%. For smaller strain amplitudes, the plastic work at 620 °C exceeded the corresponding values for the 500 °C and 300 °C tests. For strain amplitudes higher than 0.4%, the relation was opposite. This phenomenon could be rationalized by an interplay of reduced mechanical strength at high temperature and material softening. At 300 °C and 500 °C, the mechanical strength was higher than at 620 °C, which meant that the main part of the applied total strain amplitude could be established via elastic deformation. As a result, the hysteresis loops below 0.4% were narrow for 300 °C and 500 °C. At $\Delta\varepsilon_t/2 = 0.5\%$, the plastic strain contributions at 300 °C and 500 °C were higher. Consequently, the hysteresis loops were wider in the x-direction and in the y-direction (due to the simultaneous increase of maximum stress). At 620 °C, the plastic strain contribution was again higher, and the hysteresis loops were broader, but the maximum stresses were now lower, due to the strong softening at this temperature level.

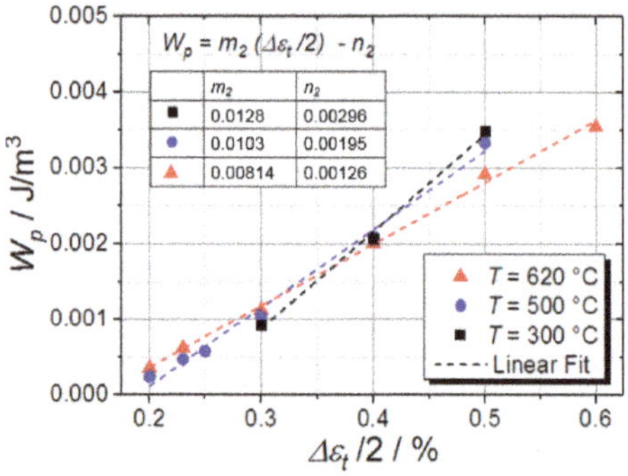

Figure 8. Relationship between the total strain amplitude and plastic work for the LCF-tests (cycles at 50% lifetime).

3.2. Lifetime Behavior of LCF-Tests

In Table 3, the lifetimes of the LCF-tests on P92 are listed, for each temperature level. As would be expected, the lifetime decreased with increasing temperatures, at the same total strain range. Within each temperature level, the lifetime decreased with a rising total strain range. The lifetime curves could be described by the Manson-Coffin Basquin relationship. It is given by the following equation [29]:

$$\frac{\Delta \varepsilon_t}{2} = \frac{\sigma'_f}{E} \cdot \left(2N_f\right)^b + \varepsilon'_f \cdot \left(2N_f\right)^c \tag{9}$$

The first part is known as the Basquin relationship and corresponds to the lifetime dependence on elastic strain. The second part, known as the Manson-Coffin relationship, represents the influence of the plastic strain. $2N_f$ is the number of strain reversals (1 cycle = 2 reversals), σ'_f is the fatigue ductility coefficient, b is the fatigue strength exponent, ε'_f is the fatigue strength coefficient, and c is the fatigue ductility exponent. The respective parameters, resulting from the present tests, are given in Table 5 for all three temperature levels. In Figure 9, the plots of both individual relationships are shown together with the total strain amplitude for all temperature levels. The values are in good agreement with the values recently reported by Zhang et al. [16]. Moreover, the values at 300 °C are in good accordance with the values at room temperature that were obtained by Zhang et al. This suggests that, at medium temperatures, no major changes in the mechanical behavior, compared to deformation at room temperature, occur.

Table 5. Parameters for the Basquin and Manson-Coffin relationship.

T [°C]	σ'_f/E	b	ε'_f	c
300	0.003573	−0.053	0.7712	−0.676
500	0.002984	−0.052	0.2158	−0.581
620	0.002672	−0.078	0.427	−0.698

3.3. Effect of Hold Times

Introducing periodic hold times turns a strain-controlled LCF-test into a relaxation fatigue (RF) test, which simulates the effects of combined cyclic and monotonous loads. The strain is kept constant at the maximum and minimum strain and, as exemplified in Figure 10 for the initial cycles of an RF-test, considerable stress relaxation both in tension and compression might be obtained under these conditions. In the present work, the hold time was 3 min, unless otherwise noted. The obtained (absolute) amounts of the stress relaxation are listed in Table 6. The results showed that, within tests at one temperature level, the amount of stress relaxation within the first cycle increased with the total applied strain range. With increasing cycle numbers, the stress relaxation decreased as is reflected in the smaller relaxation at $N = N_f/2$. Relaxation generally tended to be higher during the compressive holds than in tension. In the first cycle, the absolute differences between the amount of relaxation at ε_{min} and ε_{max} might reach considerable values of up to 63 MPa, in tests with low total strain range ($\pm 0.2\%$, 620 °C) and extended hold times of 10 min. However, this asymmetry in relaxation was only observed in the initial cycle. Already in the second cycle, the stress relaxation at ε_{max} was almost similar to the values at ε_{min}, for all tests. However, the values at ε_{max} were still up to 10 MPa lower than those at ε_{min}.

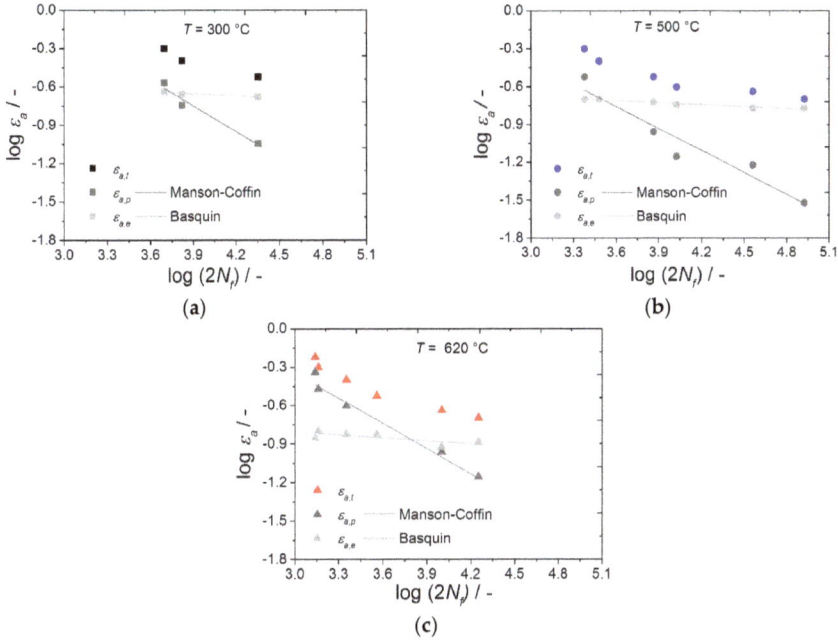

Figure 9. Manson-Coffin Basquin plots for the LCF-tests conducted at (**a**) 300 °C, (**b**) 500 °C, and (**c**) 620 °C.

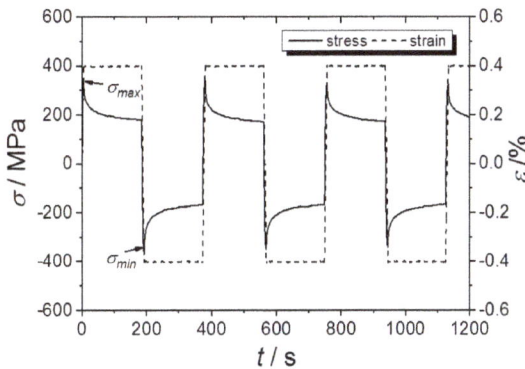

Figure 10. Stress and strain evolutions during the initial cycles of a tension–compression relaxation fatigue (RF) test with strain-controlled holds ($t_h = 3$ min).

Table 6. Amounts of stress relaxation in the different RF-tests, given in absolute values. For comparison purposes, the individual values obtained in tension and compression for $N = 1$ and $N = N_f/2$ are given.

T [°C]	$\Delta\varepsilon_t$ [%]	Stress Relaxation [MPa]			
		$N = 1$		$N = \frac{1}{2} N_f$	
		ε_{max}	ε_{min}	ε_{max}	ε_{min}
300	±0.3	31	32	33	33
500	±0.23	66	87	55	55
	±0.3	75	113	68	68
	±0.5	103	108	77	82
620	±0.2	107	156	80	81
	±0.2 (t_h = 10 min)	117	180	92	97
	±0.23	129	170	80	86
	±0.3	153	200	82	94
	±0.4	161	192	80	96
	±0.4 (t_h = 10 min)	184	210	105	116

In Figure 11, the maximum stress of each cycle has been plotted for the RF-tests. Compared to the corresponding LCF-tests, cf. Figure 5, all RF-tests exhibited a qualitatively similar cyclic softening behavior. But some parameters resulted in a more pronounced softening and strongly reduced lifetime, as has been outlined in the following section.

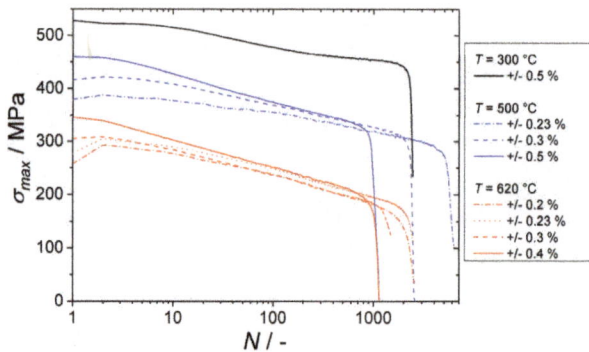

Figure 11. Cyclic stress response curves for the different RF-tests with tensile and compressive holds of 3 min.

3.3.1. Influence of Hold Times at Different Temperatures

In Figure 12a, the cyclic softening curves of three LCF-tests (one for each temperature level) and the corresponding RF-tests are plotted. At 300 °C, the LCF- and RF-test followed the same path. The corresponding values for softening ratio and lifetime are listed in Table 7. The softening ratio, as well as the lifetime (in terms of cycle number), did not significantly change as a result of the hold time at 300 °C. However, at 500 °C and 620 °C, the softening in the RF-tests was more pronounced, compared to the LCF-tests, leading to a continuously growing separation of the RF/LCF softening curves. The lifetime was not strongly affected, but the lifetime of the RF-tests was systematically lower than that of the LCF-tests. In Figure 12b, the corresponding hysteresis loops at half-lifetime are illustrated. The hysteresis loop at 300 °C showed only a very small stress relaxation (31–33 MPa), during both hold times. Apparently, the temperature was too low to cause significant effects of any time-dependent processes. In contrast, at 500 °C and 620 °C, a clearly detectable stress relaxation occurred during the hold time, due to stronger thermal activation. Relaxations of around 80 MPa (absolute) were obtained at the 50% lifetime, at both temperatures, cf. Table 6. This corresponded to relative relaxations of about 23% at 500 °C and 44% at 620 °C, since the maximum stress decreased with increasing temperature. The values for the relaxation in compression were similar.

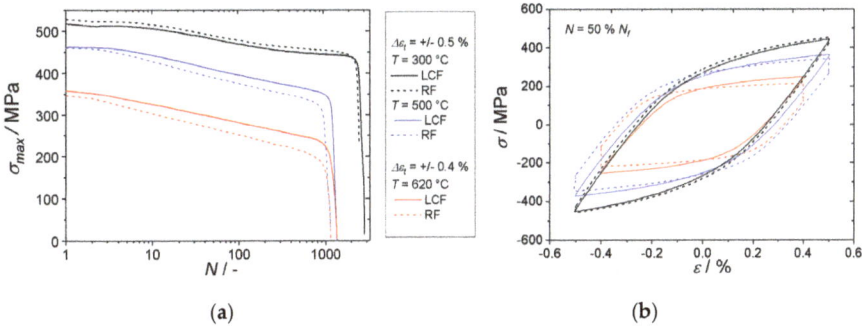

Figure 12. Cyclic stress response curves (**a**) and hysteresis loops at half-lifetime (**b**) of different LCF- and RF-tests.

Table 7. Softening ratio, lifetime, and lifetime ratio for the different LCF- and RF-tests.

T [°C]	Δε_t [%]	Softening Ratio [%]		Lifetime, N_f [-]		$\frac{N_{f,RF}}{N_{f,LCF}}$ [−]
		LCF	RF	LCF	RF	
300	±0.5%	14.4	14.8	2474	2305	0.93
500	±0.5%	21.5	24.6	1190	955	0.80
620	±0.4%	30.2	38.6	1124	996	0.89

3.3.2. Influence of Hold Times at Different Total Strain Ranges

In Table 8, the softening ratio and lifetime, as well as the lifetime ratio of all LCF- and RF-tests at 500 °C and 620 °C are listed for the different applied total strain ranges. Lifetimes (in terms of cycle number) of the RF-tests are systematically reduced, compared to the respective LCF-values. A clear dependence on imposed strain is obtained—the lifetime ratios show that, at smaller total strain ranges, the lifetime is decreased to a greater extent in the RF-tests than at higher total strain ranges. In Figure 13a, the cyclic softening curves of an LCF-test with a small total strain range and a large total strain range (solid lines), as well as the corresponding RF-test curves (dashed lines), at 620 °C, are plotted. In the RF-test with a total strain range of ±0.2% (red curves), an initial increase in maximum stress, between the first and the second cycle, occurred. It was marked by an arrow inside the figure. Consequently, the maximum stress of the RF-test remained higher than the stress of the LCF-test for the first 50 cycles, although the curve of the RF-test had a significantly higher (negative) slope than that of the LCF-test. This was also supported by the values for the softening ratio in Table 8, which were about two times higher for the RF-test. At a total strain range of ±0.4%, the maximum stress of the RF-test ran constantly below the stress of the LCF-test. Moreover, the transition to failure occurred at a lower stress in all RF-tests than in the LCF-tests. Figure 13b shows the hysteresis loops of the first cycle for all four tests of Figure 13a (presented is a full cycle, plus the loading to the second tensile strain maximum, following thereafter). Due to the stress relaxation, the hysteresis loops of the RF-tests showed an expansion in the x-direction of the diagram suggesting that the inelastic strain was larger. Again, in the diagram the initial increase in the maximum stress between the first and the second cycle, at the RF-test with ±0.2%, is marked by an arrow. The diagram also illustrates that this increase in peak stress already occurred in the preceding compression segment, where the stress minimum of the RF-test with ±0.2% was lower than the stress minimum of the related LCF-test. In conjunction with the strong relaxation, the subsequent tensile loading led to a higher maximum stress value, marked by an arrow in Figure 13b, than in the first tensile loading. An error in the strain control was excluded since the effect of the initial hardening between the first and the second cycle could be reproduced in a subsequent test with a 10 min hold time, at the same strain amplitude (Figure 14).

Table 8. Softening ratio, lifetime, and lifetime ratio for different LCF- and RF-tests.

T [°C]	Δε_t [%]	Softening Ratio [%]		Lifetime N_f [-]		$\frac{N_{f,RF}}{N_{f,LCF}}$ [−]
		LCF	RF	LCF	RF	
500	±0.23	16.4	22.4	18200	5320	0.29
	±0.3	18.0	23.0	3660	2307	0.63
	±0.5	21.5	25.1	1190	955	0.80
620	±0.2	17.5	38.2	8920	2050	0.23
	±0.3	27.8	37.6	1809	1324	0.73
	±0.4	30.2	38.6	1124	996	0.89

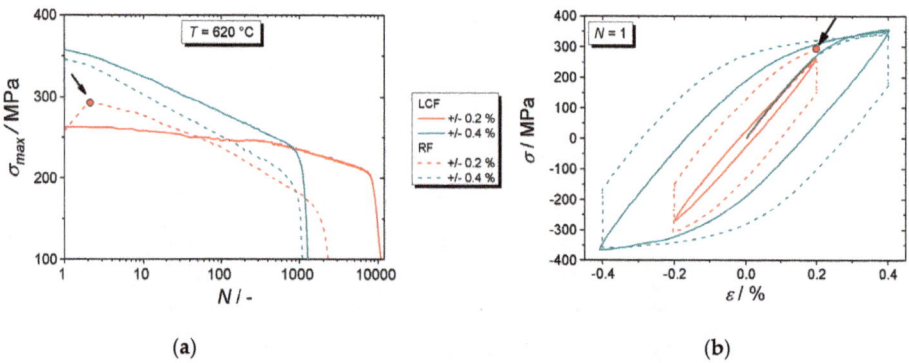

Figure 13. Cyclic softening curves (a) and hysteresis loops of the first cycle (b) for selected LCF- and RF-tests at 620 °C.

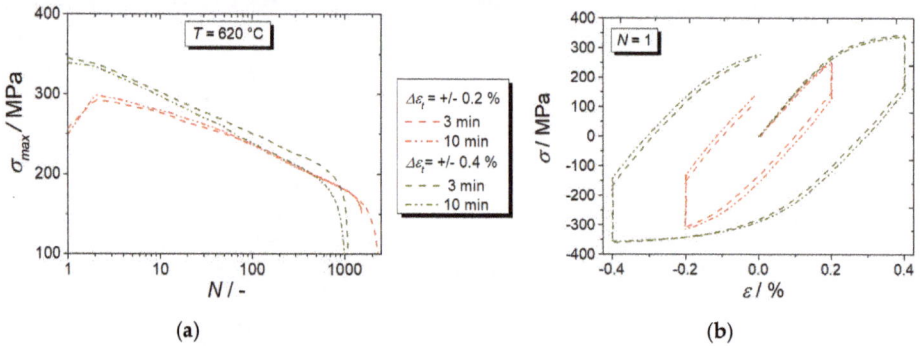

Figure 14. Cyclic softening curves (a) and hysteresis loops of the first cycle (b) for selected RF-tests with different hold times.

3.3.3. Influence of the Length of a Hold Time

In Figure 14a, the cyclic softening curves of RF-tests, at a similar temperature (620 °C) but different hold times of 3 min and 10 min, are plotted. Two total strain range levels of ±0.2% and ±0.4% were considered. The corresponding values for the softening ratio and lifetime are given in Table 9. Only little differences were obtained in the lifetime ratios for the extended hold time, yet with a clear tendency towards a stronger lifetime reduction (in terms of cycle number), at the extended hold time. In the hysteresis loops of the first cycles in Figure 14b, it can be seen that most of the stress relaxation occurred within the first 3 min of the hold time, resulting in pronounced stress drops at the extremes of the loops. Only a little further variation of the stress values was obtained at the extended hold times. For the tests with a total strain range of ±0.2%, the stress initially relaxed by 98 MPa, in the first 3 min and by another 12 MPa, between minute 3 and 10.

Table 9. Softening ratio and lifetime for the RF-tests at different hold times and total strain ranges.

$\Delta\varepsilon_t$ [%]	Softening Ratio [%]		Liftetime N_f [-]		$\frac{N_{f,RF}}{N_{f,LCF}}$ [−]	
	3 min	10 min	3 min	10 min	3 min	10 min
±0.2	38.2	37.6	2050	1582	0.23	0.18
±0.4	38.6	40.7	996	883	0.89	0.79

3.4. Fracture Characteristics

None of the presented fatigue tests was carried out to the final failure of the specimen, because after a significant stress drop of more than 75%, the specimens still did not fail. Instead, all specimens were subsequently loaded to fracture at room temperature, after the test. This needs to be considered for the analyses of the fracture surfaces.

In Figure 15 micrographs of the fracture surfaces for both the LCF- and RF-specimens, tested at different total strain ranges at 620 °C, are shown. The surfaces were oxidized and, therefore, appear dark grey and dull. On all fracture surfaces, fracture paths can be seen. The surfaces of the tests conducted at the smaller total strain range (±0.2%) still exhibit a residual fracture surface. For the LCF-specimen, the residual fracture surface lies in the upper half and covers about one-third of the entire fracture surface. For the RF-specimen, the residual fracture surface of about 12% was represented by the bright and sharply demarcated area, left from the center of the fracture surface. This is a clear indication that, for the RF-test, cracks were initiated all around the specimen surface. In contrast to the LCF-specimen, several points of crack initiation could be identified. For the LCF-tests at ±0.2%, the fracture paths led back to one point, at the surface at the 6 o'clock-position, with one exception at the 11 o'clock position, where a small residual area with a semicircular shape existed. At higher total strain range (±0.4%), the crack paths originated all around the specimen surfaces, for both tests. Especially in case of the RF-tests, the cracks seemed to start from different planes, at all strain levels. This led to rugged fracture surfaces, whereas, for the LCF-tests, the fracture surfaces were comparably smooth, at all strain levels.

The fractured surfaces were also analyzed by SEM. In Figure 16, detailed views of the crack growth zone of some specimens are shown. They exhibit the typical failure characteristics of a fatigue fracture, including fatigue striations and sometimes secondary cracking, as well as steps. Since no straight edges of the crystallites were visible, the crack propagation was assumed to be transgranular. Figure 16a,b,d shows the central part of the fractured surface. In general, in all specimens, the fracture surface edges were more severely coated with oxide layers than the center, which meant that the respective cracks were opened long before the final failure took place. An example is given in Figure 16c, in which an image taken at the edge of a fractured surface is shown. Here, the surface appeared to be very rough and coated. Furthermore, the visible steps in Figure 16c indicates that multiple cracks of the LCF-test started on different planes. On the surfaces of the samples with high total strain ranges (Figure 16b,d), periodic striations were observed. With the hold time, the distance between these striations seems to increase, as was systematically observed in different locations, all over the fracture surface. In the specimens from tests at a small total strain range, the striations could not be clearly identified, since, for example, the fracture surface in Figure 16a appeared to be more inhomogeneous and was not as smooth as the surfaces of tests with a higher total strain range. There were more steps visible indicating that the crack diverted during the propagation. All typical failure characteristics appeared to be diminished and finer.

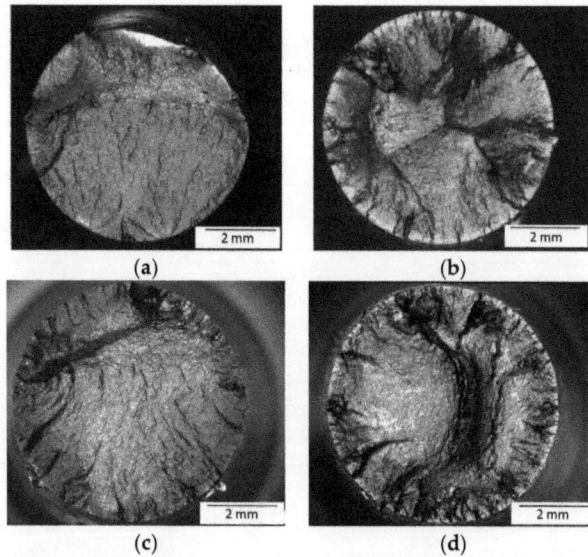

Figure 15. Optical micrographs of the fractured surfaces of the different LCF- and RF-tests, at 620 °C (a) LCF ±0.2%, (b) RF ±0.2%, (c) LCF ±0.4%, and (d) RF ±0,4%.

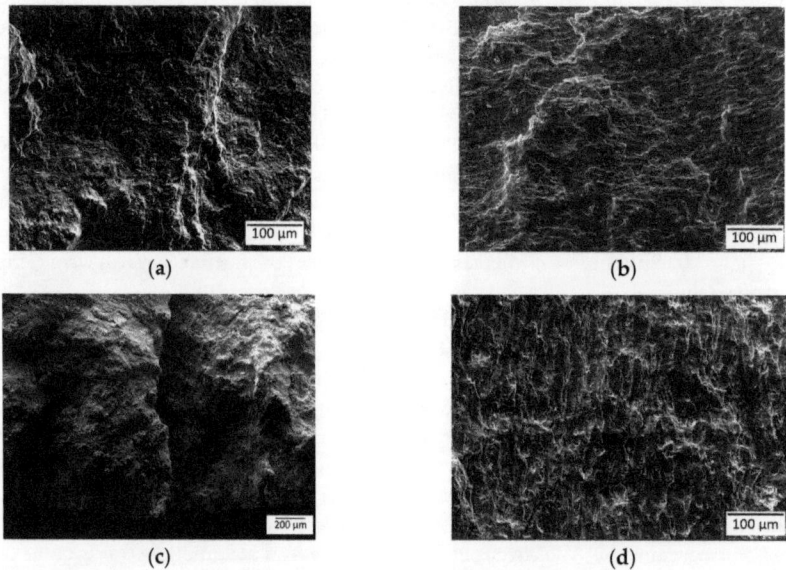

Figure 16. SEM micrographs of the fracture surfaces for different total strain ranges at 620 °C (a) LCF ±0.23%, (b,c) LCF ±0.4%, (d) RF ±0.4%; (a,b,d) were taken at the central part of the fracture surface, (c) was taken at the edge of the fracture surface.

In Figure 17, longitudinal cross-sections of the LCF- and RF-specimens tested at different strain ranges at 620 °C, are plotted. The micrographs show a representative section of the gauge length of the failed specimens; the sketch in the middle of Figure 17 indicates in which area of the specimen the micrographs were taken. The images show the polished specimens surrounded by mounting resin (in black). Based on these images, it could be concluded that higher the total strain range, higher was the number of cracks that developed. Additionally, as usual in fatigue testing, it could be concluded that the cracks initiated at the surface of the specimens. Moreover, in Figure 17c, the remaining pieces of the spot-welded thermocouple could be seen at the specimen surface, near the fracture edge. There were small cracks that initiated from the thermocouple, but obviously they did not accelerate or initiate the failure of the specimen. Comparing the LCF- and RF-specimens (see Figure 17b,d), it was obvious that in the RF-specimen, longer cracks had developed, compared to the LCF-specimen.

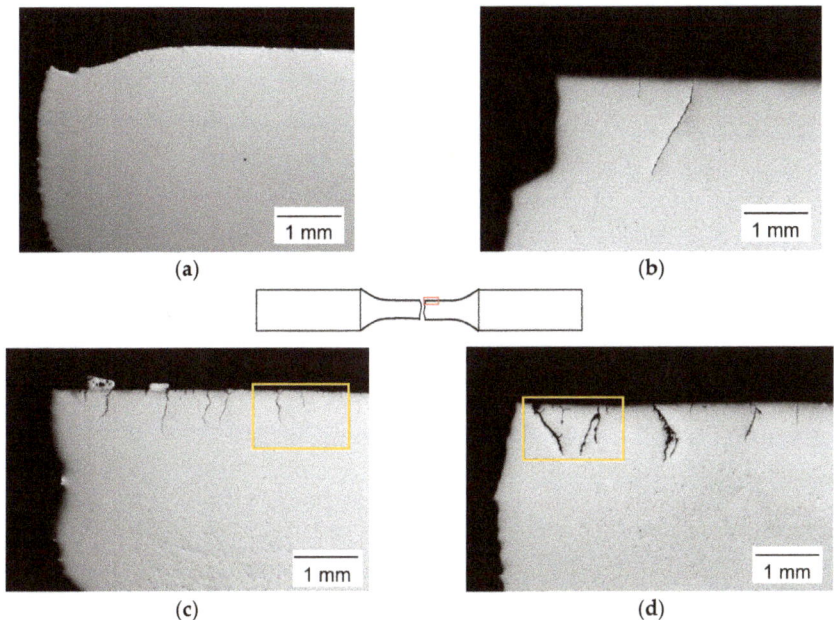

Figure 17. Optical micrograph of the longitudinal sections for the different total strain ranges at 620 °C (**a**) LCF ±0.2%, (**b**) LCF ±0.5%, (**c**) RF ±0.2%, and (**d**) RF ±0.4%. Fracture surfaces are left. Yellow rectangles indicate the regions which are shown in a higher magnification in Figure 18a,b.

More detailed micrographs of these sections are shown in the upper part of Figure 18. The micrographs show that there were short cracks (up to 250 μm), as well as the long cracks (up to 700 μm). In the detailed view, it is possible to see that the cracks of the LCF-test are filled with oxides. Additionally, the oxide layer at the outer specimen surface partially spalled off. The possibility that this happened during the sample preparation could not be excluded, but the SEM micrographs below the longitudinal sections in Figure 18c,d, which show the shell surface of the specimens, indicate that the oxide layer was already missing, immediately after the LCF-test and probably spalled off during the fatigue test. Moreover, both micrographs indicate that the oxides breached out of the cracks and the surface, especially for the RF-test. The cracks of the LCF-specimen were narrow and straight with slight branching. In the RF-specimen, in addition to some short and straight cracks, very broad long and partially fork shaped cracks were present. From the two big cracks at the RF micrographs, secondary cracks had formed. These side cracks probably formed due to a strong oxidation inside the cracks, during the tensile hold time. During compressive loading, these oxides were pressed together. It seems possible that the side cracks have developed during this step. Generally, the broader cracks of the RF specimens reflected the more pronounced oxide formation in these tests of longer duration. However, the images in Figure 18, especially the area which is marked by an arrow, show that it is impossible to say if the smaller cracks were fatigue cracks which directly formed at the specimen's surface or cracks that had formed out of an oxidation pit.

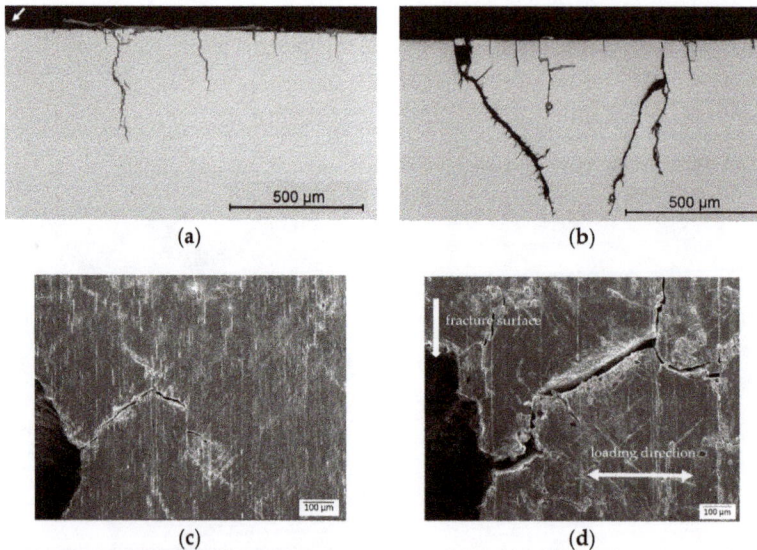

Figure 18. Optical micrographs of the longitudinal sections showing the regions marked by yellow rectangles in Figure 17c,d in a higher magnification. (a) LCF ±0.5% and (b) RF ±0.4%. SEM micrographs of the shell surface of the LCF and RF specimens, at the edge of the main crack, (c) LCF ±0.5% and (d) RF ±0.4%.

4. Discussion

4.1. Softening Behavior

In Figure 19, the softening ratio values (cf. Equation (8)) for all conducted LCF- and RF-tests at 500 °C and 620 °C have been plotted as a function of the applied strain amplitude. In the case of the LCF-tests (filled symbols), the softening ratio clearly increased with increasing strain amplitude, but with varying intensity. While a strong dependence was obvious at the lower applied strains, only a little further variation was obtained at the total strain amplitudes of 0.4% or higher, after reaching a softening ratio of about 20% (at 500 °C) and 30% (at 620 °C), respectively. As indicated by the cyclic stress response curves in Figure 5c,d, the tests at this intermediate strain amplitude level were characterized by quite high initial stress levels. In fact, any further increase in strain amplitude led to little or no change of the observed initial maximum stress, suggesting a direct relation between this initial value and the subsequent stress evolution. This phenomenon of nearly similar maximum stresses at different (but rather high) strain amplitudes could be rationalized on the basis of the hysteresis loops plotted in Figure 7. They show that higher amplitudes forced the specimens into plastic deformation, and that this plastic flow occurred at only moderately growing stresses.

When adding a hold time (RF-tests), the softening was more pronounced, as was reflected in the softening ratio-values which were up to 20 percentage points higher than that in the continuous (LCF) tests. The RF results in Figure 19 (open symbols) suggest that, within the range of strain amplitudes employed here, the softening ratio was nearly independent of the total strain. Especially in case of the tests at 620 °C, no clear dependence could be inferred from the data, anymore. However, it should not be overlooked that the phenomenon of initial hardening (as outlined in Section 3.3.2) contributed to the softening ratio values of these tests, since the overall maximum stress was taken for the calculation of the ratio in all cases (Equation (8)). The magnitude and varying impact of the hardening effect at different strain amplitudes is best demonstrated in Figure 11. At $N = 1$ the maximum stress was very different for the RF-tests performed at the same temperature. At 620 °C, for example, it was 258 MPa for a total strain range of ±0.2% and 345 MPa for ±0.4%. Due to the initial hardening during the first cycle of the test performed at 620 °C and a total strain range of ±0.2%, the maximum stress level of the test was raised to 295 MPa and approached the stress level of the other RF-tests performed at higher total strain ranges. From there on, a nearly similar softening trend was apparent for all tests. In other studies on ferritic–martensitic steels, e.g., [15,23], the softening ratio was often calculated on the basis of the maximum stress of the first cycle, neglecting the possibility that the global maximum stress was reached in a later cycle. If the maximum stress of the first cycle is taken for calculation, the softening ratio, in the present work, would result in values of 29.9% (for the RF-test at 620 °C and ±0.2%) and 31.1% (at 620 °C and ±0.23%). Returning to Figure 19, these values would represent a softening to strain amplitude dependence that would be more similar to the curves of the LCF-tests. The real material behavior under a combined static and cyclic loading, which is obviously a result of a competing hardening phenomenon, Figure 11, and pronounced softening, Figure 13a, would however, not have been recognized.

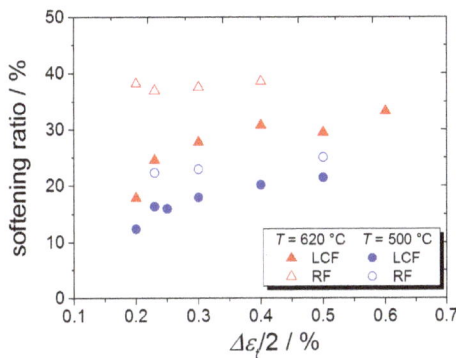

Figure 19. Softening ratio as a function of strain amplitude of all LCF- and RF-tests (3 min hold time) at 500 °C and 620 °C.

This leads to the question of why the initial hardening took place at smaller strain amplitudes. The hardening already occurred within the first cycle—after the first (tensile) hold time, a higher stress value in the compression was reached, Figure 13b. A hardening also occurs during the first stage of a creep test on ferritic–martensitic steels [30]. In these tests, the specimen was subjected to a constant stress and through solid solution hardening (W, Mo) and precipitation hardening ($M_{23}C_6$, MX, Laves-phase) the microstructure was stabilized [31,32]. Moreover, during the primary creep, an increase of the dislocation density due to deformation takes place, which results in a hardening. Similar mechanisms might also cause the short initial hardening in the RF-tests. The hold time resembles a creep deformation (elastic strains are substituted by creep strains, see discussion below) and, therefore, similar microstructural changes could take place. After the first tensile hold time, the strain at $\sigma = 0$ was 0.1% for the RF-test with ±0.2%, at 620 °C. To compensate for this extension, the material underwent a stronger deformation when switching into compression. Consequently, a higher stress value was reached. In the RF-test with ±0.4%, the specimen also exhibited a remaining deformation (0.28%) when unloading to $\sigma = 0$. In addition to that, at higher strain amplitudes (>0.3%) hardly no dependence of the maximum stress on the applied strain existed anymore (see Figures 12b and 13b). Therefore, in Figure 13b, despite the following strong deformation during the compression, plastic deformation occurred only at moderate growing stress. Consequently, the LCF- and RF-tests with ±0.4% reached the same minimum stress level, in contrast to the LCF- and RF-tests with ±0.2%.

During the hold time of an RF-test, a stress relaxation occurs and elastic strains are substituted by creep strains. With the development of creep strains during the hold time, the inelastic strain, being the sum of plastic strain and creep strain (Equation (6)), also increases. This increase in inelastic strain has been illustrated in Figure 20, where the inelastic strain amplitude for the first cycle (solid lines) and the cycle at 50% of lifetime (dashed lines) was plotted as a function of the total strain amplitude for all the LCF- and RF-tests, at 620 °C. Both in the LCF- (full symbols) and the RF-tests (open symbols) the inelastic strain amplitude increased between the first cycle and half-lifetime. The reason for this was the pronounced softening that occurred in both test types. When assuming constant elastic material properties in a first-order approximation, the continuously decreasing stresses resulted in similarly lower elastic strain contributions while the total strain limit was kept constant. The material, therefore, underwent more and more plastic deformation during a cyclic test. This also meant that the softening ratio and the increasing inelastic strain were closely linked, which could be shown by further evaluating the data in Figure 20. For the LCF-tests, the increase of inelastic strain amplitude between the first cycle and the cycle at 50% lifetime was constant, with an absolute value of 0.07%, for a strain amplitude of 0.3% and higher, whereas, at the lower strain amplitudes of 0.23% and 0.2%, it was considerably smaller. The same applied for the values of the softening ratio (Figure 19) as described in the previous paragraphs. For the RF-test, the increase of inelastic strain amplitude during the first half of lifetime was smaller than that of the LCF-test, although the inelastic strain level was generally higher. The increase was identical (0.03%-points) for all RF-tests, at different strain amplitudes, which was again in line with the constant softening ratios given in Figure 19. No larger increase was obtained, since the initial inelastic strain amplitude for the first cycle of the RF-tests already included the increase of inelastic strain, which was created during the hold time. Between $N = 1$ and $N = 50\%$ N_f, the composition of the inelastic strain changed for the RF-test. The creep strain which was accumulated during each hold period decreased (due to the generally lower stress level), while the plastic strain increased due to the softening process, as outlined before. Therefore, the inelastic strain did not increase as strongly as could be expected, due to the high softening ratio values of almost 40% (Figure 19). In other words, the increase of plastic strain was compensated by the decrease of the creep strain.

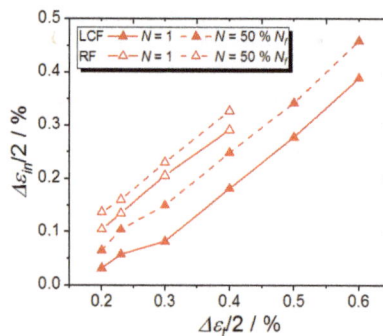

Figure 20. Inelastic strain amplitude at $N = 1$ and at $N = 50\%$ N_f of the LCF- and RF-tests at 620 °C.

The results regarding the lifetime, softening ratio, and maximum stresses of the LCF-tests were in a good agreement with the results obtained by Zhang et al. [23] and Wang et al. [15]. It should be mentioned that the tests results of the LCF-tests at 300 °C, in this work, were similar to those obtained at room temperature by Zhang et al., which showed in the values of the softening ratio and the constants of the Basquin Manson-Coffin relationship. Literature findings on the behavior of P92 under combined static and cyclic loading, and especially regarding the softening characteristics under these conditions, are rather scarce. Moreover, they appear to be partly in disagreement with the current findings. In fact, the cyclic softening results recently published by Gopinath et al. [19] explicitly suggest similar softening behavior during the LCF- and RF-tests at 600 °C, while the present work showed that clear differences in the softening behavior could be identified, e.g., in Figure 13a. However, the results of Gopinath et al. were obtained at only one rather high strain amplitude level (±0.6%). The results obtained at higher strain amplitudes, in this study, had also shown no dramatic influence of hold times on lifetime. Only when also considering small strain amplitudes, the influence on softening behavior can be studied in detail. Therefore, our results clearly demonstrate how important it is to consider small strains, and also because they are more realistic during the operation of a real power plant.

The cyclic softening of the ferritic–martensitic steels is attributed to the changes in the microstructure. After the initial heat treatment, a high dislocation density exists in the initial material condition. The rapid softening during the first loading cycles is, therefore, often attributed to a decrease of free dislocation density [6,17,33]. Especially at high temperatures, a microstructural recovery due to thermally activated deformation mechanisms, such as dislocation climb, cross-slip, and thermal recovery, may be expected [23]. The recovery results in the formation of a sub-grain structure, where the dislocations get arranged in cell walls [34]. Consequently, the dislocation density inside the grains declines. Moreover, upon mechanical cycling, the low-angle boundaries might disappear and, therefore, the initial lath structure coarsens [35]. However, this trend is partly counterbalanced by opposite effects—subgrains are typically not free of dislocations since they may get pinned by small and homogeneously distributed precipitates inside the subgrains [19,23]. Shankar et al. [33] reported that the application of a hold time promotes the formation of subgrains. Moreover, hold time tests also promote coarsening and morphological changes of carbides. This subgrain and carbide modification underlines that a unique and fast microstructure evolution might be obtained in the case of RF-testing, thereby, justifying the present finding of different softening behaviors in the LCF and RF experiments.

The microstructural changes which possibly represent the origin of the softening phenomenon might also be characterized by means of (micro) hardness measurements. The hardness was measured at the longitudinal sections of the failed specimens. In Figure 21 the course of the hardness from the fracture surface of the fatigue specimen to the end of its gauge length has been plotted for the different LCF- and RF-tests. For the first 4–5 mm distance, the hardness values were constantly low, except for the LCF-specimen tested at ±0.2%. It is very likely that the largest softening had taken place in this region. Then, with an increasing distance from the fracture surface, the hardness increased, eventually reaching a common value which corresponded to the original hardness in all specimens, at a position close to the end of the gauge length, right before the onset of the transition zone to the specimen head. The general shape of the hardness profiles was probably related to the corresponding temperature profiles in the specimens. The temperature was controlled at the center of the specimen and up to ±5 mm distance from the middle. After 5 mm, i.e., outside the controlled region, the temperature continuously decreased, due to the heat flow into the applied water-cooled grips. With decreasing temperature, the softening also decreased, as might be assumed from the data in Figure 19. The hardness level of the RF-tests was the lowest within the first 5 mm distance, which was consistent with the more pronounced softening in the tests with hold time. Additionally, for both RF-tests with higher and lower strain amplitudes, not only the hardness levels but also the softening ratio (Figure 19) were similar. This indicated that in all RF-specimens of the present study the microstructure might have evolved towards a similar condition. Moreover, the lower hardness of the RF- vs. LCF-specimens indicated that a coarser microstructure with less subgrains might have been present in the RF-specimen, since the subgrain walls, which consisted of arranged dislocations, were one of the hardest regions of the microstructure and would reduce the plastic flow during the hardness testing. For the LCF-tests, the hardness values seemed to depend on the applied strain amplitudes, with a trend towards a softer material condition, at a higher applied strain. For the test with a total strain range of ± 0.2%, the hardness first decreased and then increased after passing the 4 mm position. To ensure validity of this particular result, the hardness profile measurements were repeated both at a different place on the same surface, as well as after re-polishing the cross-section's surface. The same trend of hardness was obtained in all measurements, indicating that the microstructure softening might be rather inhomogeneous at the smaller strain amplitude. The need for a more detailed microstructure characterization to fully rationalize this finding was acknowledged.

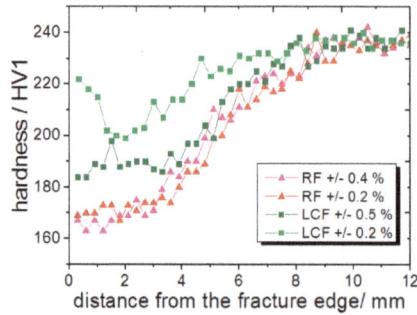

Figure 21. Hardness profiles measured on longitudinal gauge sections of specimens from different LCF- and RF-tests.

4.2. Lifetime Behavior

In Figure 22, the lifetimes of all conducted tests were plotted as a function of the applied strain amplitude. Like generally known, lifetime decreased with increasing strain range and temperature. The addition of a hold time might reduce the lifetime drastically. The reduction of lifetime, firstly, depends on the temperature. According to the rule of Tammann, a transition from time-independent to time-dependent processes takes place at a temperature of $T > 0.4 \cdot T_m$ (T_m = melting temperature in K) [36]. At 300 °C, no lifetime reduction due to holds could be identified, based on the limited available data. Additionally, nearly no stress relaxation occurred during the hold time, Figure 12b, which meant that nearly no additional inelastic strain, in the form of creep strain, developed. In contrast, at 500 °C and 620 °C, a stress relaxation associated with the development of creep strains was detected. Interestingly, there was only a slight lifetime reduction due to hold periods for tests at higher total strains. This observation suggests that the development of creep strains did not cause a reduction in lifetime under all loading conditions. An extension of the hold time from 3 minutes to 10 minutes only led to a small but further reduction in lifetime.

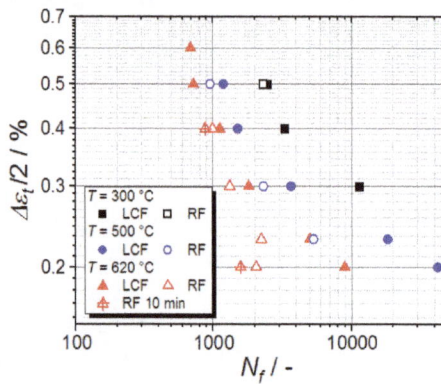

Figure 22. Lifetimes of all LCF- and RF-tests vs. the applied strain amplitude as a function of temperature.

The stronger reduction of lifetime at small strain amplitudes was possibly related to the more pronounced softening, as described in the previous chapter. At small strain amplitudes (620 °C), the softening ratio increased the most, compared to the LCF-tests (cf. Figure 19) and, similarly, the greatest influence on lifetime was obtained at ±0.2%, in Figure 19. The following considerations might explain the high hold-time-related reduction of lifetime at low strain amplitudes. In Figure 23. the composition of strain amplitude has been illustrated for the first cycle and the stabilized cycle (which is at half-lifetime), for the LCF- and RF-tests, at a small and a large strain amplitude. As a consequence of the hold time, all RF-tests exhibited a higher inelastic strain fraction than the corresponding LCF-test. Without a hold time at a total strain range of $\pm 0.2\%$, there was a fraction of 16% of inelastic strain at $N = 1$, whereas, the fraction was 52% in the RF-test with hold time. This means that due to the hold time, the inelastic strain was 3.25 times higher in the RF-test than in the LCF-test. For the total strain range of $\pm 0.4\%$, the inelastic strain fraction for the LCF-test was 45%, in the first cycle, significantly greater than that for $\pm 0.2\%$. The hold time resulted in a 1.6 times higher inelastic strain for the RF-test in the first cycle—which was a much lower variation—than that for $\pm 0.2\%$. This more prominent increase of the fraction of inelastic strain at small strain amplitudes could be a reason for the strong lifetime reduction due to hold time, at low strain amplitudes. Since inelastic strain is directly related to microstructural recovery (as discussed above), hold times at small amplitudes result in a more pronounced/accelerated microstructure modification. The respective values obtained at 50% N_f suggest that the higher level of inelastic strain in the RF-tests was indeed maintained throughout the tests, until fracture. The share of the inelastic strain grew with cycle number in the four experiments displayed in Figure 23. Nevertheless, the inelastic strains of the LCF-tests at 50% N_f did not even reach the level of the first RF cycle, despite their much higher number of accomplished reversals at half-lifetime.

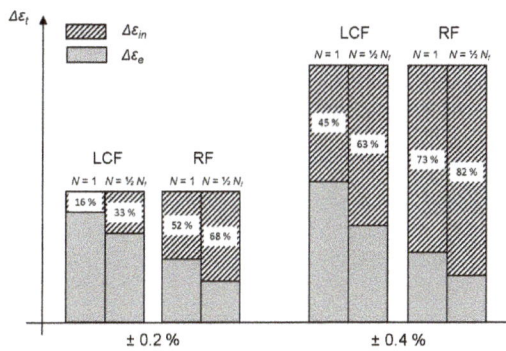

Figure 23. Calculated fractions of the composition of strain amplitude for LCF- and RF-tests at 620 °C.

The fracture surfaces of the specimen did not show any evidence for a creep damage, such as intergranular crack growth or the formation of pores. Additionally, no mix of transgranular (fatigue damage) and intergranular (creep damage) crack growth was found. A possible explanation is that the strain rate of the conducted tests was too high. According to Skelton et al. [37], only at small enough strain rates, creep damage might become predominant. However, a significantly higher number of cracks was found in the RF-specimens which were exposed to static loading phases. With more cracks, it might be expected that the RF-specimens would already fail earlier at a higher stress level than the LCF-specimens. This did not agree with the observation that in RF-tests, the transition from gradual lowering of the maximum stress to a sudden decrease at the end of the cyclic softening curves (Figure 13) occurred at a lower stress level than that in the LCF-tests. Therefore, it is tempting to assume that the more pronounced softening is a direct mechanical consequence of more cracks that developed during the RF-tests which reduce the load-carrying cross section of the specimen.

On the other hand, there are some facts that seem to disagree with this interpretation, First, the micrographs of Figure 17 indicate that more cracks developed during the RF-tests at higher strain amplitudes, but since the softening ratio of all RF-tests at 620 °C (Figure 19) was the same, the number of cracks should be similar. Second, the hardness profiles in Figure 21 suggest a lower hardness for the RF-specimen. This lower hardness could not result from cracking, since an unaffected central cross-section (not the outer shell) was investigated. Third, if more cracks would reduce the load-carrying cross-section of the specimen, an asymmetry in tension and compression (when the cracks were closed) would be the result, which was not the case. Therefore, the lower hardness at the RF-tests was apparently related to microstructure modification and not to the formation of multiple cracks.

In conclusion, the more pronounced softening under the RF conditions can mainly be attributed to changes in the microstructure. The dislocations are rearranged in a lower energy configuration, such as cells and subgrains [38]. It is important to consider that the test duration of the RF-tests was longer, although lifetime in terms of number of cycles was less than that in the LCF-tests, Table 3. During these far longer tests (e.g., T = 620 °C, ±0.2%: LCF = 19.8 h, RF = 209.6 h), the enhanced (time-dependent) dislocation rearrangement result in a more pronounced softening. Considerable mechanical stress is, however, a second prerequisite to allow for this microstructure evolution in the ferritic–martensitic steels, since it acts as the primary driving force. The finding that 10 min hold times did not considerably change the material reaction, compared to the tests with only 3 min hold times (Figure 14), indicates that the remaining stresses after 3 min were already insufficient to stimulate further short-term microstructure evolutions even though they still exceed the 150 MPa level, Figure 14b.

5. Conclusions

A series of strain-controlled LCF tests in the temperature range of 300 °C to 620 °C and total strain ranges ±0.2%–±0.6% were carried out on the ferritic–martensitic grade P92 steel to assess its fatigue behavior. In a second step, hold times under strain control were included into the test procedures to simulate superimposed static and cyclic loading. The following conclusions can be drawn:

1. P92 exhibits pronounced cyclic softening during fatigue loading. Its amount is increasing with increasing temperature and total strain. The softening depends on the maximum stress. It reaches a saturation at higher total strain amplitudes (>0.3%).
2. Hold times enhance cyclic softening. At small total strain amplitudes (<0.3%), tests with a hold time show an initial hardening. This results in the same pronounced softening as in RF-tests with high total strains (>0.3%). A new definition for the softening ratio was therefore proposed in which the initial maximum stress σ_0 is replaced by the overall maximum stress σ_{max}.
3. The influence of hold times on lifetime depends on temperature and strain amplitude. At 300 °C, no significant influence on lifetime is observed. At 500 °C and 620 °C, the influence on lifetime depends on the strain amplitude. At smaller and technically most relevant total strains of 0.2%, the lifetime is strongly reduced.
4. Hold times modify the damage mechanisms of ferritic-martensitic steels; a clear trend to multiple cracking and oxidation-related crack propagation is obtained. However, the number of cracks could not be related to the softening and lifetime behavior. Instead, these characteristic features depend on fatigue-related modifications of the ferritic-martensitic microstructure.

To extend the investigations towards even more realistic service conditions for flexible power plant operation including cyclic temperature variation, the thermomechanical fatigue (TMF) behavior of P92 will be studied in future work.

Author Contributions: Conceptualization, M.J., J.O., and B.F.; Funding acquisition, J.O.; Investigation, M.J.; Writing—original draft, M.J. and J.O.; Writing—review and editing, B.F., and B.S.

Funding: This research was funded by the German Federal Ministry of Education and Research, Grant Number 03SF0474.

Acknowledgments: The authors wish to acknowledge the support of M. Gerloff (metallographic preparation and microscopy of longitudinal sample sections), M. Buchheim (SEM images), C. Krimmling (hardness measurements), and M. Finn (Young's modulus measurements). The authors would like to thank P. Uhlemann, O. Kahlcke, and W. Wedell for their experimental support. We wish to thank B. Rehmer, D. Bettge, and G. Nolze for fruitful discussions. We are grateful to Vallourec Deutschland GmbH and Lausitz Energie Bergbau AG (LEAG) for providing the testing material and technical information, respectively.

Conflicts of Interest: The authors declare no conflict of interest.

References

1. Farragher, T.P.; Scully, S.; O'Dowd, N.P.; Leen, S.B. Development of life assessment procedures for power plant headers operated under flexible loading scenarios. *Int. J. Fatigue* **2013**, *49*, 50–61. [CrossRef]
2. Masuyama, F. History of Power Plants and Progress in Heat Resistant Steels. *ISIJ Int.* **2001**, *41*, 612–625. [CrossRef]
3. Viswanathan, R.; Bakker, W. Materials for ULTRASUPERCRITICAL coal power plants-boiler materials: Part 1. *J. Mater. Eng. Perform.* **2001**, *10*, 81–95. [CrossRef]
4. Hald, J. Metallurgy and creep properties of new 9-12%Cr steels. *Steel Res.* **1996**, *76*, 369–374. [CrossRef]
5. Kim, S.; Weertman, J.R. Investigation of Microstructural Changes in a Ferritic Steel Caused by High Temperature Fatigue. *Metall. Mater. Trans. A* **1988**, *19A*, 999–1007. [CrossRef]
6. Pineau, A.; Antolovich, S.D. High temperature fatigue: behaviour of three typical classes of structural materials. *Mater. High Temp.* **2015**, *32*, 298–317. [CrossRef]
7. Saad, A.A.; Sun, W.; Hyde, T.H.; Tanner, D.W.J. Cyclic softening behaviour of a P91 steel under low cycle fatigue at high temperature. *Procedia Eng.* **2011**, *10*, 1103–1108. [CrossRef]
8. Mishnev, R.; Dudova, N.; Kaibyshev, R. Low cycle fatigue behavior of a 10Cr-2W-Mo-3Co-NbV steel. *Int. J. Fatigue* **2016**, *83*, 344–355. [CrossRef]
9. Masuyama, F.; Komai, N. Evaluation of Long-Term Creep Rupture Strenght of Tungsten-Strengthened Advanced 9–12%Cr Steels. *Key Eng. Mater.* **1999**, *171–174*, 179–188. [CrossRef]
10. Earthman, J.C.; Eggeler, G.; Ilschner, B. Deformation and Damage Processes in a 12-Percent-Cr-Mo-V Steel under High-Temperature Low-Cycle Fatigue Conditions in Air and Vacuum. *Mater. Sci. Eng. A* **1989**, *110*, 103–114. [CrossRef]
11. Fournier, B.; Dalle, F.; Sauzay, M.; Longour, J.; Salvi, M.; Caes, C.; Tournie, I.; Giroux, P.F.; Kim, S.H. Comparison of various 9-12%Cr steels under fatigue and creep-fatigue loadings at high temperature. *Mater. Sci. Eng. A* **2011**, *528*, 6934–6945. [CrossRef]
12. Abe, F. Creep Behavior, Deformation Mechanisms and Creep Life of Mod.9Cr-1Mo Steel. *Metall. Mater. Trans. A* **2015**, *46A*, 16.
13. National Institute for Material Science. MatNavi—NIMS Materials Database. Available online: http://mits.nims.go.jp/index_en.html. (accessed on 15 July 2018).
14. Penalba, F.; Gomez-Mitxelena, X.; Jimenez, J.A. Effect of Temperature on Mechanical Properties of 9% Cr Ferritic Steel. *ISIJ Int.* **2016**, *56*, 1662–1667. [CrossRef]
15. Wang, X.; Jiang, Y.; Gong, J. Characterization of Low Cycle Fatigue of Ferritic–martensitic P92 Steel: Effect of Temperature. *Steel Res. Int.* **2015**, *87*, 761–771. [CrossRef]
16. Zhang, Z.; Hu, Z.; Fan, L. Low Cycle Fatigue Behavior and Cyclic Softening of P92 Ferritic–martensitic Steel. *J. Iron Steel Res. Int.* **2015**, *22*, 534–542. [CrossRef]
17. Giroux, P.F.; Dalle, F.; Sauzay, M.; Caes, C.; Fournier, B.; Morgeneyer, T.; Gourgues-Lorenzon, A.F. Influence of strain rate on P92 microstructural stability during fatigue tests at high temperature. *Procedia Eng.* **2010**, *2*, 2141–2150. [CrossRef]
18. Wang, X.W.; Gong, J.M.; Zhao, Y.P.; Wang, Y.F.; Yu, M.H. Characterization of Low Cycle Fatigue Performance of New Ferritic P92 Steel at High Temperature: Effect of Strain Amplitude. *Steel Res. Int.* **2015**, *86*, 1046–1055. [CrossRef]
19. Gopinath, K.; Gupta, R.K.; Sahu, J.K.; Ray, P.K.; Ghosh, R.N. Designing P92 grade martensitic steel header pipes against creep-fatigue interaction loading condition: Damage micromechanisms. *Mater. Des.* **2015**, *86*, 411–420. [CrossRef]
20. Zhang, W.; Wang, X.; Jiang, Y. Thermal-mechanical fatigue behaviour of P92 T-piece and Y-piece pipe. *Mater. High Temp.* **2017**, *33*, 609–616. [CrossRef]
21. Scholz, A.; Berger, C. Deformation and life assessment of high temperature materials under creep fatigue loading. *Materialwiss. Werkstofftech.* **2005**, *36*, 722–730. [CrossRef]
22. DIN. *Nahtlose Stahlrohre für Druckbeanspruchungen—Technische Lieferbedingungen—Teil 2: Rohre aus Unlegierten und Legierten Stählen mit Festgelegten Eigenschaften bei Erhöhten Temperaturen*; Deutsche Fassung EN-10216-2:2013; Beuth Verlag GmbH: Berlin, Germany, 2013.
23. Zhang, Z.; Hu, Z.; Schmauder, S. Low-Cycle Fatigue Properties of P92 Ferritic–martensitic Steel at Elevated Temperature. *J. Mater. Eng. Perform.* **2016**, *25*, 1650–1662. [CrossRef]

24. Fournier, B.; Sauzay, M.; Pineau, A. Micromechanical model of the high temperature cyclic behavior of 9–12%Cr martensitic steels. *Int. J. Plast.* **2011**, *27*, 1803–1816. [CrossRef]
25. Klueh, R.L. Elevated temperature ferritic and martensitic steels and their application to future nuclear reactors. *Int. Mater. Rev.* **2005**, *50*, 287–310. [CrossRef]
26. Shibli, A.; Gostling, J.; Starr, F. *Damage to Power Plants Due to Cycling*; Product ID: 1001507; EPRI: Palo Alto, CA, USA, 2001.
27. *ISO 12106: Metallic Materials—Fatigue Testing—Axial-Strain-Controlled Method*; International Organization for Standardization: Geneva, Switzerland, 2017.
28. *ASTM E 1875: Standard Test Method for Dynamic Young's Modulus, Shear Modulus, and Poisson's Ratio by Sonic Resonance*; ASTM International: West Conshohocken, PA, USA, 2013.
29. Suresh, S. *Fatigue of Materials*; Press Syndicate of the University of Cambridge: Cambridge, UK, 1998.
30. Sklenicka, V.; Kucharova, K.; Svoboda, M.; Kloc, L.; Bursik, J.; Kroupa, A. Long-term creep behavior of 9–12%Cr power plant steels. *Mater. Charact.* **2003**, *51*, 35–48. [CrossRef]
31. Prat, O.; Garcia, J.; Rojas, D.; Carrasco, C.; Inden, G. Investigations on the growth kinetics of Laves phase precipitates in 12% Cr creep-resistant steels: Experimental and DICTRA calculations. *Acta Mater.* **2010**, *58*, 6142–6153. [CrossRef]
32. Rojas, D.; Garcia, J.; Prat, O. Design and Characterization of Microstructure evolution during creep of 12%Cr heat resistant Steels. *Mater. Sci. Eng, A* **2010**, *527*, 3864–3876. [CrossRef]
33. Shankar, V.; Bauer, V.; Sandhya, R.; Mathew, M.D.; Christ, H.J. Low cycle fatigue and thermo-mechanical fatigue behavior of modified 9Cr-1Mo ferritic steel at elevated temperatures. *J. Nucl. Mater.* **2012**, *420*, 23–30. [CrossRef]
34. Nagesha, A.; Kannan, R.; Sastry, G.V.S.; Sandhya, R.; Singh, V.; Rao, K.B.S.; Mathew, M.D. Isothermal and thermomechanical fatigue studies on a modified 9Cr-1Mo ferritic–martensitic steel. *Mater. Sci. Eng. A* **2012**, *554*, 95–104. [CrossRef]
35. Sauzay, M.; Brillet, H.; Monnet, I.; Mottot, M.; Barcelo, F.; Fournier, B.; Pineau, A. Cyclically induced softening due to low-angle boundary annihilation in a martensitic steel. *Mater. Sci. Eng. A* **2005**, *400*, 241–244. [CrossRef]
36. Meetham, G.W.; Van de Voorde, M.H. *Materials for High Temperature Engineering Applications*; Springer: Berlin, Germany, 2000.
37. Skelton, R.P.; Gandy, D. Creep-fatigue damage accumulation and interaction diagram based on metallographic interpretation of mechanisms. *Mater. High Temp.* **2008**, *25*, 27–54. [CrossRef]
38. Shankar, V.; Valsan, M.; Rao, K.B.S.; Kannan, R.; Mannan, S.L.; Pathak, S.D. Low cycle fatigue behavior and microstructural evolution of modified 9Cr-1Mo ferritic steel. *Mater. Sci. Eng. A* **2006**, *437*, 413–422. [CrossRef]

metals

MDPI

Article

The Effect of Normalizing Temperature on the Short-Term Creep Rupture of the Simulated HAZ in Gr.91 Steel Welds

Hao-Wei Wu [1], Tai-Jung Wu [1,2], Ren-Kae Shiue [3] and Leu-Wen Tsay [1,*]

[1]   Institute of Materials Engineering, National Taiwan Ocean University, Keelung 20224, Taiwan;
      howard22322@gmail.com (H.-W.W.); gordon@iner.gov.tw (T.-J.W.)
[2]   Division of Nuclear Fuels and Materials, Institute of Nuclear Energy Research, Taoyuan 32546, Taiwan
[3]   Department of Materials Science and Engineering, National Taiwan University, Taipei 10617, Taiwan;
      rkshiue@ntu.edu.tw
*   Correspondence: b0186@mail.ntou.edu.tw; Tel.: +886-2-24622192 (ext. 6405)

Received: 14 November 2018; Accepted: 14 December 2018; Published: 16 December 2018

Abstract: As-received Gr.91 steel tube was normalized at either 940 or 1060 °C for 1 h, followed by Ar-assisted cooling to room temperature, then tempered at 760 °C for 2 h. Those samples were designated as 940NT or 1060NT samples. An infrared heating system was used to simulate HAZ microstructures in the weld, which included over-tempering (OT) and partial transformation (PT) zones. The results of short-term creep tests showed that normalizing at higher temperature improved the creep resistance of the Gr.91 steel. By contrast, welding thermal cycles would shorten the creep life of the Gr.91 steel. Among the tested samples in each group, the PT samples had the shortest life to rupture, especially the 940NT-PT sample. The microstructures of the PT samples comprised of fine lath martensite and ferrite subgrains with carbides decorating the grain and subgrain boundaries. Excessive dislocation recovery, rapid coalescence of refined martensite laths, and growth of ferrite subgrains were responsible for the poorer creep resistance of the PT samples relative to those of the other samples.

Keywords: Gr.91; normalizing; simulate HAZ

1. Introduction

The increase in steam temperature and pressure can increase the thermal efficiency of fossil fuel steam power plants, which enforce the development of creep-resistant alloys. Gr.91 steel, developed by Oak Ridge National Laboratory, is designed to be applied in fast breeder reactors. The combination of good creep resistance [1,2] and high temperature oxidation resistance has made Gr.91 steel a dominant material for supercritical and ultra-supercritical fossil power plants [3,4]. $M_{23}C_6$, MC, and MCN precipitates in a tempered martensite matrix can effectively strengthen Gr.91 steel [5]. The increase in strength and creep rupture life of P91 steel with increasing the normalizing temperature can be attributed to fine precipitates and low inter-particle spacing, which hinder dislocation motion [6]. Increasing the normalization temperature reduces the Charpy impact toughness of P92 steel, which is related to the increase in grain size [7].

Continuous phase transformation can occur within the heat-affected zone (HAZ) of a steel weld. The intercritical heat-affected zone (ICHAZ) or partial transformation zone (PTZ) [8] experiences a peak temperature between A_{C1} and A_{C3}, which leads to partial transformation or incomplete austenite formation therein. Heating around the A_{C1} of Gr.91 steel may result in no austenite formation. Short-time over-tempering causes the recovery of dislocations, break-down of the lath structure, polygonization, and carbide spheroidizing as well as coarsening [9]. Premature failure of 9–12 Cr creep-strength-enhanced ferritic steel welds after long-term service at elevated temperature is known as Type IV cracking [10–14]. The fine-grained HAZ (FGHAZ) has the lowest creep strength among the different regions of a P91 steel weld [15]. In creep tests of a simulated HAZ, the FGHAZ heated to near the A_{C3} temperature of Gr.91 steel shows the lowest creep rupture strength [11]. The creep crack growth life of the FGHAZ of a Gr.91 weld at 600 °C is about 45% of that the base metal [16]. The low creep growth resistance of FGHAZ in a Gr.91 steel weld at 625 °C is due to the presence of fine grains and suppression of the lath structure [17]. The rapid growth of undissolved $M_{23}C_6$ carbides in a simulated FGHAZ during PWHT at 760 °C for 4 h accounts for the shorter creep life of Gr.91 steel with a higher preweld tempering temperature [18]. It is also reported that the ICHAZ (or PTZ) is the main causes of Type IV cracking of Gr.91 steel welds [14]. Preferential deformation of Cr-depleted grains accelerates grain boundary sliding and the nucleation of creep cavities, which result in Type IV cracking [19].

As-received T91 steel tubes were normalized (N) at either 940 or 1060 °C for 1h, followed by tempering (T) at 760 °C for 2 h. The aim of this study was to investigate the effects of various prior normalizing treatments on simulated HAZ microstructures of T91 steel tube. Gr.91 steel pipes are designed to be used mainly in the boiler at the operation temperature around 600 °C for long term service in ultra-supercritical fossil power plants. However, the uneven temperature distribution in the boiler or concentrated burner fire on the pipes may cause local temperature surge much higher than designed operation temperature, thus obviously deteriorating service life of the boiler. In this work, the short-term creep life of the simulated samples was determined under constant stress at elevated temperature and compared with the original substrates. Detailed microstructures of the samples were inspected by transmission electron microscope (TEM, JEOL, Tokyo, Japan). Electron backscatter diffraction (EBSD, Oxford Instruments, Abingdon, UK) was applied to reveal the grain orientation, high-angle, and subgrain boundaries of the specimens.

2. Materials and Experimental Procedures

The chemical composition (wt.%) of the T91 steel tube used in this study was 0.09 C, 0.36 Mn, 0.40 Si, 0.013 P, 0.003 S, 8.75 Cr, 0.90 Mo, 0.13 Ni, 0.20 V, 0.07 Nb, 0.040 N, and a balance of Fe. The as-received steel tube (T91) was subjected to normalizing treatment at 940 or 1060 °C for 1 h in high vacuum, followed by Ar-assisted cooling to room temperature. The normalized (N) specimens were then tempered (T) at 760 °C for 2 h. The normalized and tempered substrates were named as 940NT and 1060NT, accordingly. The transformation temperatures of the Gr. 91 steel were determined by a dilatometer at heating and cooling rates of 15 °C/sec. An infrared heating system with rapid heating and controlled cooling was used to simulate the HAZ microstructures of the weld. The HAZ microstructures were simulated by heating the NT substrates to 860 or 900 °C for 1 min, which was either slightly below the A_{C1} temperature (over-tempering, OT) or a little below the A_{C3} temperature (partial transformation, PT). After infrared heating, the simulated specimens were tempered at 760 °C for 2 h. To emphasize thermal history of the sample, the 940NT-PT sample, representing the 940NT substrate, was further heated to 900 °C for 1 min by infrared heating, followed by tempering at 760 °C for 2 h. The 1060NT substrate heated at 860 °C for 1 min by infrared, and then further tempered at 760 °C for 2 h, was designated as 1060NT-OT sample.

Figure 1 shows the tested samples sectioned from the tempered Gr. 91 steel tubes. The cuboidal samples (Figure 1a) with the dimensions of about 10 mm L × 10 mm W × 3 mm T were used for microhardness measurements, microstructural examinations, and structural identifications. A Vickers microhardness tester was applied with a load of 300 g and duration time of 15 sec, and 8 measurements were obtained from cuboidal specimens under different treated conditions. The microstructures of various cuboidal specimens were inspected by optical microscope (OM, Olympus, Tokyo, Japan) and scanning electron microscope (SEM, JEOL, Tokyo, Japan). The detailed microstructures of the specimens sliced from the cuboidal specimens in different states were inspected by transmission electron microscope (TEM, JEOL, Tokyo, Japan). Thin foils were prepared by twin-jet polishing in an electrolyte consisting of 75% ethanol, 20% $C_3H_5(OH)_3$, and 5% $HClO_4$ acid at −20 °C.

Moreover, the cuboidal specimens of various microstructures were also examined with an SEM equipped with an electron backscatter diffraction (EBSD) to reveal the grain size, high-angle and subgrain boundaries of the specimens. To realize the influence of prior thermal history on the short-term creep rupture of various specimens, dog-bone samples, as shown in Figure 1b, were loaded with dead weight under different stresses and temperatures [20]. The changes in microstructures in the fracture zone of crept samples were analyzed by using EBSD, as shown in Figure 1c. In case of simulated samples, the cuboidal and dog-bone samples wire-cut from the NT tubes were subjected to infrared treatment, then tempered at 760 °C for 2 h.

(a)   (b)   (c)

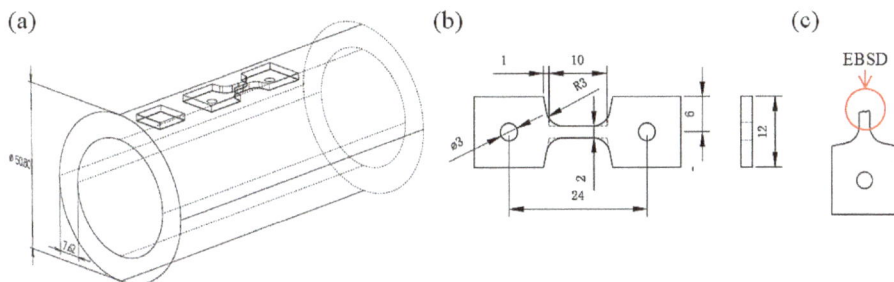

Figure 1. Schematic diagrams showing (**a**) tested samples cut from the Gr.91 steel tube, (**b**) the specimen dimensions for creep test, (**c**) the fractured zone cut from crept sample for EBSD analysis.

3. Results

The A_{C1}, A_{C3}, M_s, and M_f temperatures determined by a dilatometer in this work at heating and cooling rates of 15 °C/sec were 870, 915, 425, and 258 °C, respectively. The A_{C1} and A_{C3} temperatures of T91 and T92 steels will moderately increase with increasing the heating rates [9,20]. The in situ martensitic transformation of 201 SS under straining can be determined by synchrotron X-ray [21]. It is reported [22] that synchrotron X-ray diffraction and laser dilatometry are a powerful tool to measure the transformation temperatures of a martensitic stainless steel. The in situ tracking of the reversion kinetics even enables the construction of a reverse Transformation-Temperature-Time diagram [22]. After normalizing and tempering treatments, the micro-hardnesses of the 1060NT and 940NT samples were HV 223 ± 4 and HV 226 ± 3, respectively. The hardness of the 940NT sample was slightly higher than that of the 1060NT sample, as shown by multiple measurements. Short-time over-tempering reduced sample hardness, regardless of the prior normalizing condition. The micro-hardnesses of the 940NT-OT and 1060NT-OT samples were HV 205 ± 3 and HV 200 ± 4, respectively. Regarding the samples heated to below the A_{C3} temperature, the hardness of the 1060NT-PT sample (HV 211± 3) was slightly higher than that of the 940NT-PT sample (HV 205 ± 3). As a whole, welding thermal cycles obviously reduced the hardness of all simulated samples.

Figure 2 presents SEM micrographs showing the typical microstructures of various samples after tempering at 760 °C for 2 h. As shown in Figure 2a,b, the normalized and tempered microstructure of Gr.91 steel showed lath martensite packets and austenite boundaries decorated by $M_{23}C_6$ carbides. Stable fine MX carbides or carbonitrides were dispersed in the martensite matrix, which resisted coarsening at elevated temperature [23]. The prior austenite grain sizes (PAGSs) of the samples were determined by means of the line-intercept method and further confirmed by EBSD crystallographic analysis, as described in the following section. The PAGSs of the 940NT and 1060NT samples were about 7.4 and 17.0 μm, respectively. It has been reported that increasing the normalizing temperature from 1050 to 1150 °C also causes a decrease in $M_{23}C_6$ carbide size, which is associated with an increase in tensile strength but a reduction in ductility of P91 steel [6].

For the over-tempered specimens, it seemed that a decrease in carbide density and an increase in carbide size were observed (Figure 2c,d). The lath morphologies of the OT samples were also not as prominent as those of the NT ones. Moreover, the distribution of all precipitates was not uniform in the OT samples. A few areas in the OT samples lacked precipitates, which were more obvious in the 940NT-OT sample. As shown in Figure 2e,f, the grain structure of the PT samples was even more refined than those of the NT and OT samples. The lath morphology was obscure in two PT samples, especially in the 940NT-PT one. In previous work [8,9], the ICHAZ in the as-welded condition comprises of fine ferrite subgrains and fresh fine martensite, indicating a refined structure. Lath martensite in the PT samples was too fine to be revealed in the SEM observation. It was deduced that the coalescence of carbide-free lath martensite in the NT sample during the heating cycle contributed to the formation of ferrite subgrains in the PT specimen.

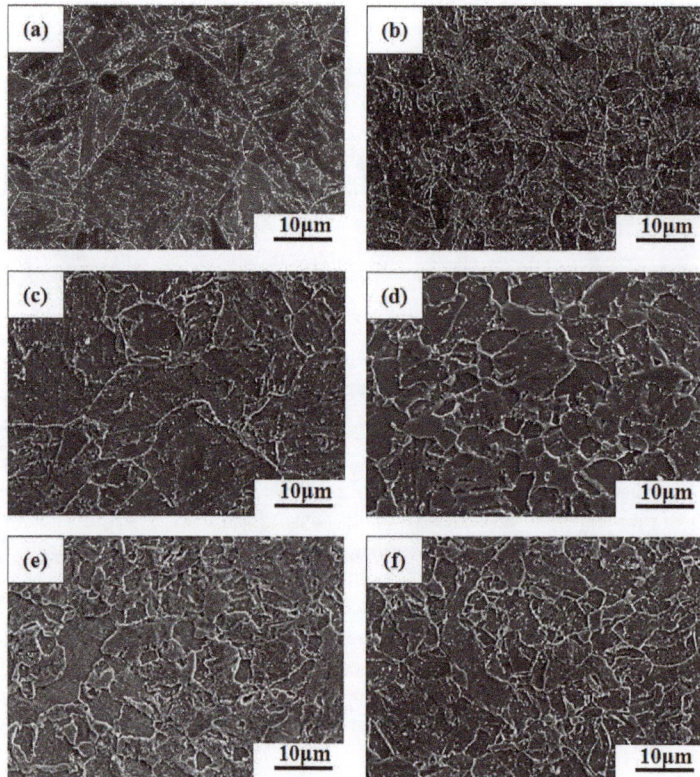

Figure 2. Scanning electron microscope (SEM) micrographs of the (**a**) 1060NT, (**b**) 940NT, (**c**) 1060NT-OT, (**d**) 940NT-OT, (**e**) 1060NT-PT, and (**f**) 940NT-PT samples.

SEM micrographs and the inverse pole figures (IPFs) showing the individual grain orientation in different colors of 1060NT and 940NT are displayed in Figure 3. The great changes in color represent great differences in grain orientations. Within one grain, martensite packets oriented in nearly the same direction are of the same color, whereas different color zones are related to the lath martensite in different orientations. As shown in Figure 3a,b, the PAGSs and martensite packets of the 1060NT sample were obviously larger and coarser than those of the 940NT sample (Figure 3c,d). The coarser martensite packets in the 1060NT sample also made the lath structure much easier to be distinguished in this sample, in comparison with those in the 940-NT sample. In addition, the PAGSs of the 1060NT and 940NT samples were measured by EBSD maps to be about 7 and 17 µm, accordingly.

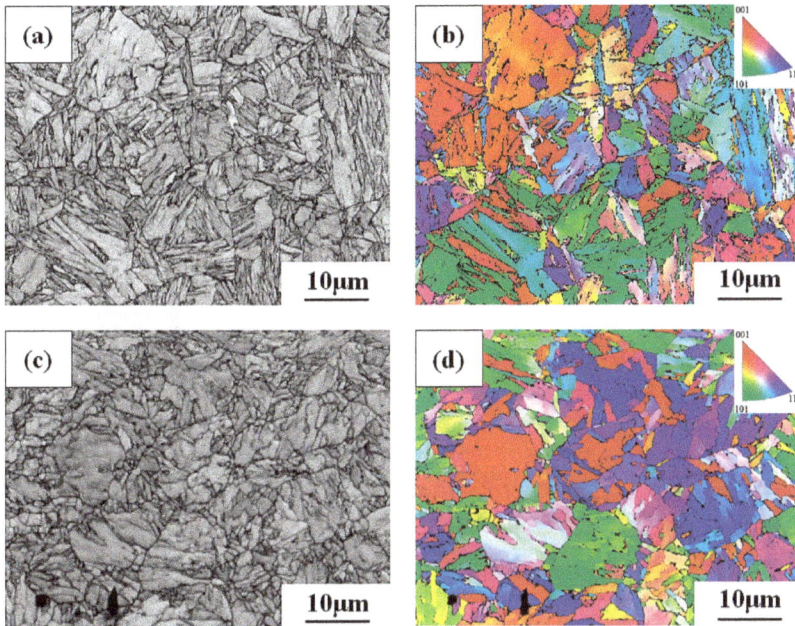

Figure 3. (a) SEM micrographs and (b) inverse pole figure (IPF) map of the 1060NT sample; (c) SEM micrograph and (d) IPF map of the 940NT sample.

Figure 4a,b present the SEM micrograph and IPF map of the 1060NT-OT sample. The lath morphology was still distinguishable in the 1060NT-OT sample (Figure 4a). The PAGSs and martensite packet sizes of the 1060NT-OT sample were similar to those of the 1060NT sample. As indicated by the arrow in Figure 4a, a few fine grains were found in the 1060NT-OT sample. The IPF map (Figure 4b) showed that those fine grains indicated by the arrow had nearly the same orientation. It was expected that the short time of over-heating would cause disruption of the lath structure and enhance the polygonization process locally. An SEM micrograph and an IPF map of 1060NT-PT sample are shown in Figure 4c,d. The SEM micrograph (Figure 4c) revealed numerous fresh nucleated fine grains in the 1060NT-PT sample. The lath structure could hardly be observed in the 1060NT-PT sample, which was partially related to the extremely fine martensite [9]. The IPF map (Figure 4d) displayed some local areas, indicated by the arrow, which were oriented in almost the same direction but divided by sub-boundaries. Based on Figure 4d, the calculated PAGS of the 1060NT-PT sample was approximately 5.3 μm. The 940-OT sample also exhibited similar PAGS and microstructure of 940NT sample. Grain refinement was also observed in the 940NT-PT sample as illustrated in Figure 2f.

Figure 5 shows TEM micrographs of Gr.91 steel in the normalized and normalized-tempered conditions. In the normalized condition, lath martensite with a high dislocation density was observed (Figure 5a,b). Complete dissolution of Cr-rich carbides during normalization caused the prior austenite grain boundaries (PAGBs) of the two samples to be barely decorated by precipitates (Figure 5a,b). Normalizing at 940 °C assisted the formation of fine lath martensite therein (Figure 5b). After tempering, the detailed microstructures of the 1060NT and 940NT samples were similar (Figure 5c,d), consisting of lath martensite packets along with prior austenite grain boundaries decorated by precipitates. During tempering, the dislocation density was decreased and carbide precipitation was enhanced. The predominant precipitates formed in the NT specimens were $M_{23}C_6$ carbides, as confirmed by the diffraction pattern. A few fine MX precipitates, which could be carbides or carbonitrides (NbC, VC, and (NbV)(CN)), were dispersed in the martensite matrix [5]. Non-uniform carbide distributions were observed in the tempered martensite; some lath packages were free of precipitate along the lath boundaries. It is reported that increasing the normalizing temperature from 1050 to 1150 °C reduces the size of $M_{23}C_6$ carbides in P91 steel [6]. In this work, the carbide sizes in the two samples did not show significant differences between them. However, the lath width and packet size of the martensite in the 1060NT sample were larger than those in the 940NT sample.

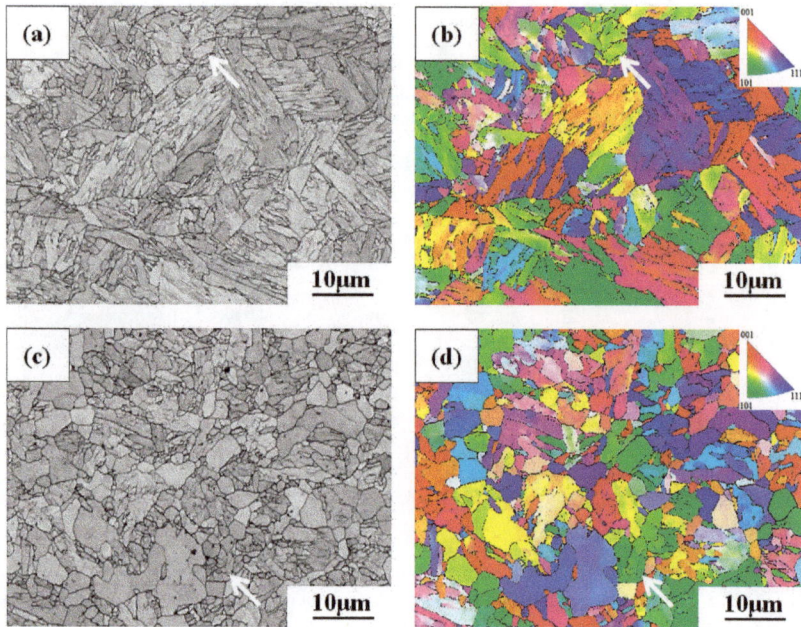

Figure 4. (**a**) SEM micrographs and (**b**) IPF map of the 1060NT-OT sample; (**c**) SEM micrographs and (**d**) IPF map of the 1060NT-PT sample.

TEM micrographs of the simulated microstructures after tempering are shown in Figure 6. Short-time over-tempering of tempered Gr.91 steel is expected to annihilate excess dislocations and cause carbide spheroidizing and coarsening [9]. The degraded lath structure and developed subgrains with low dislocation density in the over-tempered Gr.91 steel were responsible for the decrease in hardness, as compared with the NT substrates. It was seen that the parallel lath structure in the 1060NT sample was prone to be replaced by numerous cell structures or subgrains in the 1060NT-OT sample (Figure 6a). Similar morphologies were observed in the 940NT-OT sample (Figure 6b). It was observed that the carbide sizes seemed to be coarser in the 940NT-OT sample relative to those in the 1060NT-OT sample. The coexistence of fresh refined lath martensite with a high dislocation density and carbide-free ferrite subgrains caused incomplete hardening of the as-simulated PT or ICHAZ specimens [9]. In the as-simulated condition, most $M_{23}C_6$ carbides heated to slightly below the A_{C3} temperature dissolved into the matrix, and only few residual carbides remained [9]. After tempering, carbides precipitated mainly along lath boundaries and in ferrite subgrains. As shown in Figure 6c, tempered fine lath martensite was decorated by precipitates along the PAGBs and lath boundaries of the PT specimen. The fine laths in the PT sample were also comprised of many subgrains. A few coarse carbides in the PT specimen could be related to the growth of residual carbides during tempering of the Gr.91 steel. In addition, the aligned carbides in the 1060NT-PT specimens also showed the fact of migrated sub-boundaries after tempering.

Metals 2018, 8, 1072

Figure 5. Transmission electron microscope (TEM) micrographs of the (**a**) 1060N, (**b**) 940N, (**c**) 1060NT, and (**d**) 940NT samples.

Figure 7 presents grain boundary crystallographic maps showing the detailed austenite grain boundaries and sub-boundaries, identified by measuring the misorientation between the adjacent grains. Subgrain boundaries include the lath and blocks/packets boundaries within the matrix. Prior austenite grain boundaries (PAGBs) are identified as high-angle grain boundaries with a misorientation above 15 degrees. Subgrain boundaries are indicated by red and green lines, whereas high-angle grain boundaries are indicated by blue lines. The grain boundary characteristics of the OT and PT samples normalized at different temperatures were evaluated. The high-angle grain boundaries delineated PAGB profiles of the 1060NT-OT sample (Figure 7a), which had the same grain boundary characteristics as the 1060NT sample. Subgrain boundaries (red and green) indicated the over-tempered martensite substructures. For the 1060NT-PT sample (Figure 7b), the grains profiled by high-angle grain boundaries were finer and more irregular than those of the 1060NT-OT sample (Figure 7a). Moreover, some fine grains (white zones in the figure), which were associated with the lack of subgrain boundaries inside, were deduced to be newly formed ferrite subgrains. It was seen that some long-island-like white zones were divided by subgrain boundaries in green lines in the 1060NT-PT sample, implying that those white zones could have been formed by excessive annihilation of dislocations and coalescence of martensite laths during infrared heating and tempering. Those white zones were expected to be the weak points in the samples.

157

Figure 6. TEM micrographs of the (**a**) 1060NT-OT, (**b**) 940NT-OT and (**c**) 1060NT-PT samples.

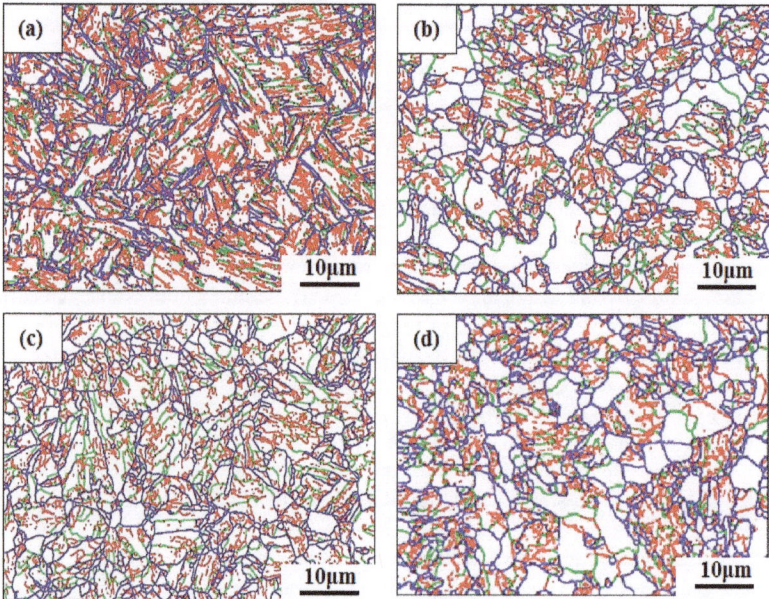

Figure 7. Grain boundary maps of the (**a**) 1060NT-OT, (**b**) 1060NT-PT, (**c**) 940NT-OT and (**d**) 940NT-PT samples. ━:1°~5°, ━:5°~15°, ━:15°~62.5°.

Figure 7c,d display the grain boundary maps of the 940NT-OT and 940NT-PT samples. As compared with the counterpart samples normalized at 1060 °C, the 940NT-OT sample consisted of much finer austenite grain sizes. Similar to the 1060NT-PT sample, the 940NT-PT sample comprised of a fine-grained structure and a certain amount of white zones (Figure 7d). Those white zones obviously comprised of fewer subgrain boundaries. It was deduced that the fast coalescence of fine laths in the 940NT sample during infrared heating enhanced the formation of ferrite subgrains. The presence of fine ferrite subgrains, which were weak relative to the tempered lath martensites, was obviously harmful to the creep resistance of the PT samples. In addition, these fine laths in the 940NT sample were prone to combine and grow into ferrite subgrains during subsequent infrared heating and tempering.

Figure 8 shows the results of short-term creep tests for the specimens loaded under condition of 630 °C /120 MPa or 660 °C /80 MPa. Stress-rupture tests were terminated if the tested specimens did not rupture after 1000 h duration. The specimen elongation was also measured as an index of creep strength of the specimen at elevated temperature. The results indicated that the creep resistance of the samples normalized at 1060 °C was higher than the counterpart samples normalized at 940 °C, regardless of testing condition. Under the 630 °C /120 MPa condition, the stress-rupture life of the 1060NT sample (804 h) was much longer than that of the 940NT sample (244 h), as shown in Figure 8a. It was noticed that the stress-rupture lifetimes of the thermally treated samples were shorter than those of the NT samples in each group. The rupture life of the 1060NT-OT sample (256 h) was longer than that of the 940NT-OT sample (192 h). Among the tested samples in each group, the PT samples had the lowest resistance to creep rupture, especially the 940NT-PT sample. The 940NT-PT sample ruptured in a very short time (28 h) after applying stress. The high elongation of the PT samples before rupture indicated their inherent low strength. Figure 8b shows the creep samples tested under the condition of 660 °C/80 MPa. The PT samples also had a shorter time-to-rupture than did the NT and OT samples in each group. The OT samples were more likely to deform and rupture than were the NT samples. It was obvious that normalizing at higher temperature was helpful to improve the creep resistance of Gr.91 steel. By contrast, welding thermal cycles shortened the rupture time of the Gr.91 steel. Inevitably, the PTHAZ or ICHAZ in a Gr.91 steel weld was the zone most sensitive to creep rupture among the distinct zones in the weld.

The SEM fractographs of all the samples after short-term creep tests showed ductile dimple fracture. The IPF maps and grain boundary maps next to the fractured zones in the ruptured samples are displayed in Figure 9. All the samples exhibited a fine-grained structure and texture under the 630 °C/120 MPa loading condition (Figure 9). The fracture life of OT samples was longer than that of the PT samples in each group, as displayed in Figure 8. During straining samples at elevated temperature, dynamic recrystallization resulted in the formation of refined grains (Figure 9a,b). Extensive nucleation of new grains occurred in the 1060NT-OT samples (Figure 9a), revealing that the original coarse grains were replaced by fine grains around the fractured zone in the 1060NT-OT sample. The corresponding grain boundary maps of the tested samples are shown in Figure 9c,d. It was observed that the low-angle grain boundaries (lath and packet boundaries) and PAGBs were less dense in the 940NT-PT sample (Figure 9d) than those in the 1060NT-OT one (Figure 9c). The time-to-fracture of the 940NT-PT sample was quite short (28 h). It was deduced that rapid combination and coalescence of refined substructures during straining at elevated temperature could explain the lower stress-rupture life of the PT samples at elevated temperature.

Figure 10 presents the IPF maps and grain boundary maps around the fractured zone, showing the distribution of grain size and grain boundary characteristics of the samples tested in the 660 °C/80 MPa condition. The PT samples had shorter fracture lifetimes than the OT ones (Figure 8), particularly the 940NT-PT sample (61h). As displayed in Figure 10a, grain growth after dynamic recrystallization occurred in the 940NT-PT sample. It was noticed that grain growth after recrystallization became much faster in the samples strained in the 660 °C/80 MPa condition. Similar results were reported that both sub grains and particle sizes increase during short term creep of Gr.91 steel [24]. The 1060NT-PT sample exhibited behaviors of variation in microstructures similar to those of the 940NT-PT one.

By contrast, predominantly dynamic recrystallization still dominated the microstructural changes of the OT samples strained in the 660 °C/80 MPa condition (Figure 10b). Moreover, the grain boundary map of the 940NT-PT sample (Figure 10c) revealed a lack of substructures within the grain interior, which meant the great loss of strengthening substructures under the testing condition. By contrast, the 940NT-OT sample consisted of fine grains and a certain number of sub-boundaries (Figure 10d), as compared with the 940NT-PT sample (Figure 10c). It was also seen that the white grained zones were coarser relative to the neighboring grains, possibly due to the removal of the sub-structure during grain growth. Softening during creep could result from many causes, e.g., the growth of subgrains and the recovery of dislocations, coarsening of fine $M_{23}C_6$ carbides to hinder the growth of subgrains and boundary migrations. It was deduced that the fast coalescence of ferrite subgrains during straining at 660 °C assisted the formation of the coarse-grained structure of the PT samples. By contrast, the dynamic recrystallization retained the fine grains close to the fractured zone of the 940NT-OT and 1060NT-OT samples.

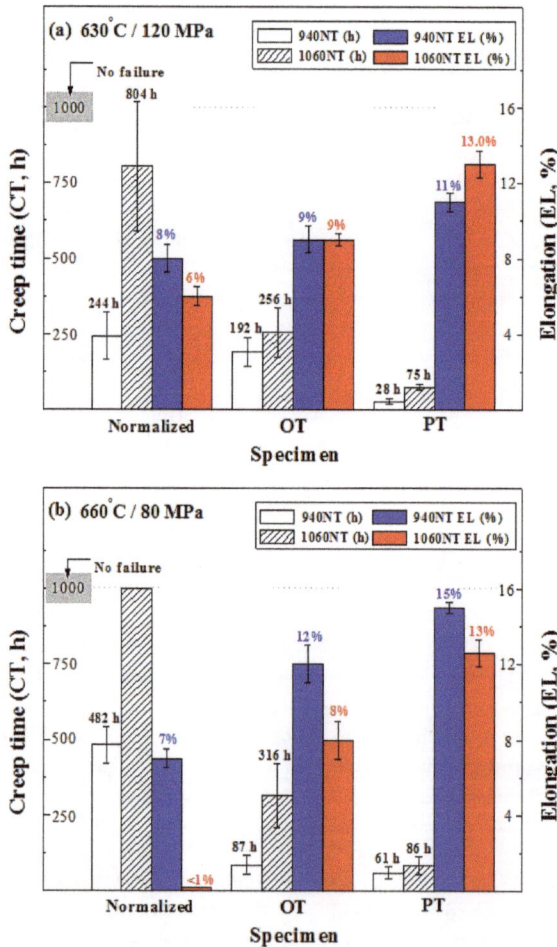

Figure 8. The short-term creep tests of various samples at (**a**) 630 °C/120 MPa, (**b**) 660 °C/80 MPa.

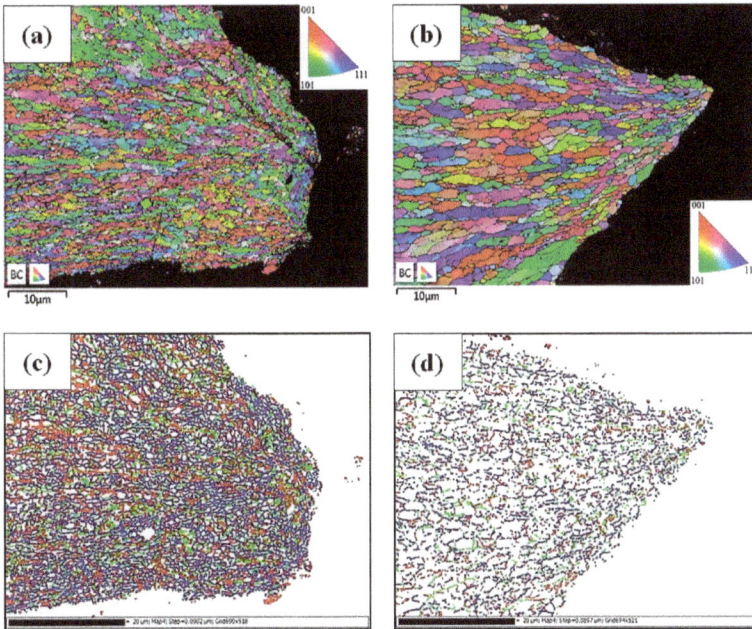

Figure 9. IPF maps of the (**a**) 1060NT-OT, (**b**) 940NT-PT samples; grain boundary maps of the (**c**) 1060NT-OT, (**d**) 940NT-PT samples strained at 630 °C/120 MPa. ▬:1°~5°, ▬:5°~15°, ▬:15°~62.5°.

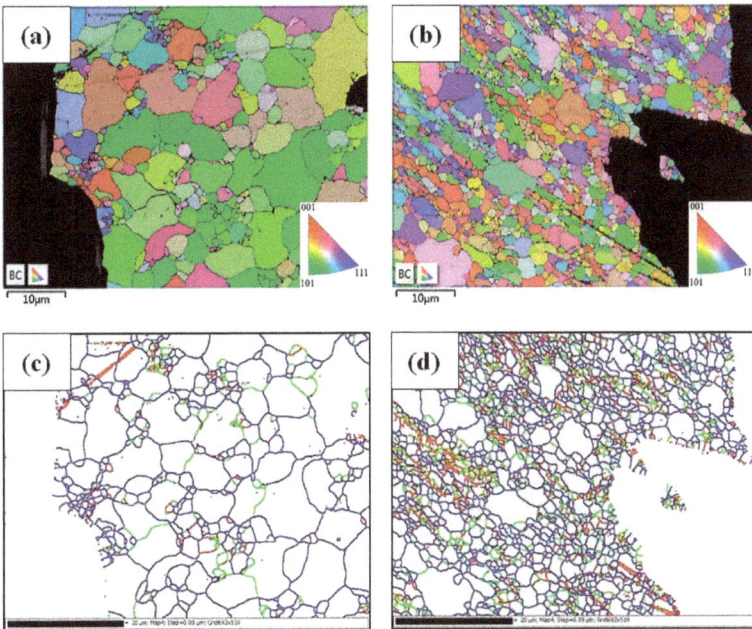

Figure 10. IPF crystallographic maps of the (**a**) 940NT-PT, (**b**) 940NT-OT samples; grain boundary crystallographic maps of the (**c**) 940NT-PT, (**d**) 940NT-OT samples strained at 660 °C /80 MPa. ▬:1°~5°, ▬:5°~15°, ▬:15°~62.5°.

4. Discussion

Increasing the normalizing temperature from 940 to 1060 °C increased the martensite lath and packet sizes, as well as the PAGSs. The measured PAGSs of the 940NT and 1060NT samples were about 7.4 and 17.0 μm, respectively. The short-term over-tempering of the NT substrates had less effect on changing the PAGS but induced microstructural evolution. The TEM micrographs showed that the degraded lath structures and developed subgrains with low dislocation densities in the OT samples were associated with a decrease in hardness, as compared with the NT substrates. The tempered intercritical (IC) or partial transformation (PT) HAZ, which had been heated to below the A_{C3} temperature, exhibited refined grain sizes of less than 6.0 μm, regardless of the prior normalizing temperatures. The microstructures of PT samples comprised mainly of ferrite subgrains and refined tempered laths with carbides decorating the boundaries. It was expected that in the NT substrates, the original martensite packets with fewer carbides would assist the coalescence of fine laths into ferrite subgrains during infrared-heating and tempering. Similar results have indicated that preferential austenitization initiates from the prior austenite grain boundaries in the PTHAZ of Gr.91 steel welds during the heating cycle of welding [14]. The formation of subgrains in a lath with low Cr concentration in the tempered lath microstructure have a lower potential to transform to austenite during welding, thus, leading to the formation of over-tempered martensite [14]. On cooling, fine grains with large numbers of Cr-carbides are more likely to transform into fine laths or tiny packet-sized martensite with un-dissolved carbides [14]. Therefore, the mixed microstructures in the PT samples exhibited a fine-grained structure, as displayed in Figures 4 and 7.

It is reported a finer PAGS, as in the fine-grained HAZ, leads to a higher rate of recovery of dislocations, higher coarsening rate of the subgrain structure and intergranular Cr-rich carbides, and consequently, to lower creep resistance [2]. The applied strain will enhance the growth of precipitates during creep of T91 steel [25]. In addition, preferential deformation of Cr-depleted grains accelerates grain boundary sliding and nucleation of creep cavities, which result in Type IV cracking [14,19]. In this work, the PT sample had the lowest rupture life among the tested samples in each group, regardless of the test condition. The rapid coalescence and growth of ferrite subgrains in the PT samples during straining at elevated temperature accounted for the deteriorated short-term creep resistance.

5. Conclusions

An infrared heating system was used to simulate the HAZ microstructures in Gr.91 steel welds, which included over-tempering (OT) and partial transformation (PT) zones. The short-term creep life of the simulated samples was determined under constant load at elevated temperature, and compared with the original substrates.

(1) Increasing the normalizing temperature from 940 to 1060 °C improved the creep resistance of Gr.91 steel substrate, whereas the time-to-rupture of infrared-heated samples were shorter than those of the NT samples in each group. Among the tested samples in each group, the PT samples had the lowest resistance to creep rupture, especially the 940NT-PT sample.

(2) Martensite packet sizes and PAGSs of the 1060NT sample were larger than those of the 940NT sample. The degraded lath structure and developed subgrains with low dislocation density in the over-tempered Gr.91 steel were responsible for a decrease in hardness, as compared with that of the NT substrates. Overall, the PAGSs and martensite packet sizes of the OT sample were similar to those of the NT substrate in each group; by contrast, refined grains were found in the PT samples. The microstructures of the PT samples comprised of fine lath martensite, ferrite subgrains, and carbides decorated the grain and sub-grain boundaries.

(3) Excessive dislocation recovery, rapid coalescence of refined martensite laths and growth of ferrite subgrains were responsible for the deteriorated creep resistance of the PT samples as compared with other samples.

Author Contributions: L.-W.T. designed and planned the experiment. R.-K.S. assisted EBSD and microstructural characteristics evaluation. H.-W.W. and T.-J.W. carried out the experimental tests. All authors were involved in completing the manuscript.

Funding: This research was funded by the Ministry of Science and Technology, ROC, grant number MOST 106-2221-E-019 -060 -MY3, and the APC was funded by National Taiwan Ocean University.

Conflicts of Interest: The authors declare no conflict of interest.

References

1. Fujio, A.B. Creep behavior, deformation mechanisms, and creep life of modified 9Cr-1Mo steel. *Metall. Mater. Trans. A* **2015**, *46*, 5610–5625.
2. El-Azim, M.E.; Ibrahim, O.H.; El-Desoky, O.E. Long term creep behaviour of welded joints of P91 steel at 650 °C. *Mater. Sci. Eng. A* **2013**, *560*, 678–684. [CrossRef]
3. Viswanathan, R.; Bakker, W. Materials for ultrasupercritical coal power plants- boiler materials: Part 1. *J. Mater. Eng. Perform.* **2001**, *10*, 81–95. [CrossRef]
4. Viswanathan, R.; Bakker, W. Materials for ultrasupercritical coal power plants- turbine materials: Part II. *J. Mater. Eng. Perform.* **2001**, *10*, 96–101. [CrossRef]
5. Shrestha, T.; Alsagabi, S.F.; Charit, I.; Potirniche, G.P.; Glazoff, M.V. Effect of heat treatment on microstructure and hardness of Grade 91 steel. *Metals* **2015**, *5*, 131–149. [CrossRef]
6. Das, C.R.; Albert, S.K.; Swaminathan, J.; Raju, S.; Bhaduri, A.K.; Murty, B.S. Transition of crack from type IV to Type II resulting from improved utilization of boron in the modified 9Cr-1Mo steel weldment. *Metall. Mater. Trans. A* **2012**, *43*, 3724–3741. [CrossRef]
7. Saini, N.; Pandey, C.; Mahapatra, M.M. Effect of normalizing temperature on fracture characteristic of tensile and impact tested creep strength-enhanced ferritic P92 steel. *J. Mater. Eng. Perform.* **2017**, *26*, 5414–5424. [CrossRef]
8. Xu, X.; West, G.D.; Siefert, J.A.; Parker, J.D.; Thomson, R.C. Microstructural characterization of the heat-affected zones in Grade 92 steel welds: double-pass and multipass welds. *Metall. Mater. Trans. A* **2018**, *49*, 1211–1230. [CrossRef]
9. Hsiao, T.H.; Chen, T.C.; Jeng, S.L.; Chung, T.J.; Tsay, L.W. Effects of simulated microstructure on the creep rupture of the modified 9Cr-1Mo steel. *J. Mater. Eng. Perform.* **2016**, *25*, 4317–4325. [CrossRef]
10. Abson, D.J.; Rothwell, J.S. Review of type IV cracking of weldments in 9–12%Cr creep strength enhanced ferritic steels. *Int. Mater. Rev.* **2013**, *58*, 437–473. [CrossRef]
11. Hongo, H.; Tabuchi, M.; Watanabe, T. Type IV creep damage behavior in Gr. 91 steel welded joints. *Metall. Mater. Trans. A* **2012**, *43*, 1163–1173. [CrossRef]
12. Divya, M.; Das, C.R.; Albert, S.K.; Goyal, S.; Ganesh, P.; Kaul, R.; Bhaduri, A.K. Influence of welding process on Type IV cracking behavior of P91 steel. *Mater. Sci. Eng. A* **2014**, *613*, 148–158. [CrossRef]
13. Laha, K.; Chandravathi, K.S.; Parameswaran, P.; Rao, K.B.S. Type IV cracking susceptibility in weld joints of different grades of Cr-Mo ferritic steel. *Metall. Mater. Trans. A* **2009**, *40*, 386–397. [CrossRef]
14. Wang, Y.; Kannan, R.; Li, L. Correlation between intercritical heat-affected zone and type IV creep damage zone in Grade 91 steel. *Metall. Mater. Trans. A* **2018**, *49*, 1264–1275. [CrossRef]
15. Spigarelli, S.; Quadrini, E. Analysis of the creep behaviour of modified P91 (9Cr–1Mo–NbV)welds. *Mater. Des.* **2002**, *23*, 547–552. [CrossRef]
16. Yatomi, M.; Yoshida, K.; Kimura, T. Difference of creep crack growth behaviour for base, heat-affected zone and welds of modified 9Cr–1Mo steel. *Mater. High Temp.* **2011**, *28*, 109–113. [CrossRef]
17. Kumar, Y.; Venugopal, S.; Sasikala, G.; Albert, S.K.; Bhaduri, A.K. Study of creep crack growth in a modified 9Cr–1Mo steel weld metal and heat affected zone. *Mater. Sci. Eng. A* **2016**, *655*, 300–309. [CrossRef]
18. Yu, X.; Babu, S.S.; Terasaki, H.; Komizo, Y.; Yamamoto, Y.; Santella, M.L. Correlation of precipitate stability to increased creep resistance of Cr–Mo steel welds. *Acta Mater.* **2013**, *61*, 2194–2206. [CrossRef]
19. Wang, Y.; Kannan, R.; Li, L. Identification and characterization of intercritical heat-affected zone in as-welded Grade 91 weldment. *Metall. Mater. Trans. A* **2016**, *47*, 5680–5684. [CrossRef]
20. Peng, Y.Q.; Chen, T.C.; Chung, T.J.; Jeng, S.L.; Huang, R.T.; Tsay, L.W. Creep rupture of the simulated HAZ of T92 steel compared to that of a T91 steel. *Materials* **2017**, *10*, 139. [CrossRef]

21. Gauss, C.; Souza Filho, I.R.; Sandim, M.J.R.; Suzuki, P.A.; Ramirez, A.J.; Sandim, H.R.Z. In situ synchrotron X-ray evaluation of strain-induced martensite in AISI 201 austenitic stainless steel during tensile testing. *Mater. Sci. Eng. A* **2016**, *651*, 507–516. [CrossRef]
22. Escobar, J.D.; Faria, G.A.; Wu, L.; Oliveira, J.P.; Mei, P.R.; Ramirez, A.J. Austenite reversion kinetics and stability during tempering of a Ti-stabilized supermartensitic stainless steel: Correlative in situ synchrotron x-ray diffraction and dilatometry. *Acta Mater.* **2017**, *138*, 92–99. [CrossRef]
23. Pandey, C.; Giri, A.; Mahapatra, M.M. Effect of normalizing temperature on microstructural stability and mechanical properties of creep strength enhanced ferritic P91 steel. *Mater. Sci. Eng. A* **2016**, *657*, 173–184. [CrossRef]
24. Cerri, E.; Evangelista, E.; Spigarelli, S.; Bianchi, P. Evolution of microstructure in a modified 9Cr–1Mo steel during short term creep. *Mater. Sci. Eng. A* **1998**, *245*, 285–292. [CrossRef]
25. Nakajima, T.; Spigarelli, S.; Evangelista, E.; Endo, T. Strain Enhanced Growth of Precipitates during Creep of T91. *Mater. Trans.* **2003**, *44*, 1802–1808. [CrossRef]

![metals logo] *metals*

MDPI

Article

Creep Behaviour and Microstructural Characterization of VAT 36 and VAT 32 Superalloys

Vagner João Gobbi [1], Silvio José Gobbi [1], Danieli Aparecida Pereira Reis [2],
Jorge Luiz de Almeida Ferreira [1], José Alexander Araújo [1] and Cosme Roberto Moreira da Silva [1,*]

[1]   Faculty of Technology, University of Brasilia—UnB, Brasilia 70910-900, Brazil;
       vagnergobbi@yahoo.com.br (V.J.G.); silviogobbi2@gmail.com (S.J.G.); jorge@unb.br (J.L.d.A.F.);
       jaaunb@gmail.com (J.A.A.)
[2]   Science and Technology Department, Universidade Federal de São Paulo—UNIFESP,
       São José dos Campos 12231-280, Brazil; danielireis@gmail.com
*   Correspondence: cosmeroberto@gmail.com; Tel.: +55-61-31071144

Received: 2 October 2018; Accepted: 23 October 2018; Published: 27 October 2018

Abstract: Superalloys are used primarily for the aerospace, automotive, and petrochemical industries. These applications require materials with high creep resistance. In this work, evaluation of creep resistance and microstructural characterization were carried out at two new nickel intermediate content alloys for application in aerospace industry and in high performance valves for automotive applications (alloys VAT 32 and VAT 36). The alloys are based on a high nickel chromium austenitic matrix with dispersion of intermetallic $L1_2$ and phases containing different (Nb,Ti)C carbides. Creep tests were performed at constant load, in the temperature range of 675–750 °C and stress range of 500–600 MPa. Microstructural characterization and failure analysis of fractured surfaces of crept samples were carried out with optical and scanning electron microscopy with EDS. Phases were identified by Rietveld refinement. The results showed that the superalloy VAT 32 has higher creep resistance than the VAT 36. The superior creep resistance of the alloy VAT 32 is related to its higher fraction of carbides (Nb,Ti)C and intermetallic $L1_2$ provided by the amount of carbon, titanium, and niobium in its chemical composition and subsequent heat treatment. During creep deformation these precipitates produce anchoring effect of grain boundaries, hindering relative slide between grains and therefore inhibiting crack formation. These volume defects act also as obstacles to dislocation slip and climb, decreasing the creep rate. Failure analysis of surface fractures of crept samples showed intergranular failure mechanism at crack origin for both alloys VAT 36 and VAT 32. Intergranular fracture involves nucleation, growth, and subsequent binding of voids. The final fractured portion showed transgranular ductile failure, with dimples of different shapes, generated by the formation and coalescence of microcavities with dissimilar shape and sizes. The occurrence of a given creep mechanism depends on the test conditions. At creep tests of VAT 32 and VAT 36, for lower stresses and higher temperature, possible dislocation climb over carbides and precipitates would prevail. For higher stresses and intermediate temperatures shear mechanisms involving stacking faults presumably occur over a wide range of experimental conditions.

Keywords: creep; superalloy VAT 36; superalloy VAT 32; high temperature

1. Introduction

Creep is the slow and continuous deformation of a solid with time, under the influence of mechanical stresses [1–4]. Superalloys play a role in the development of jet engine technology [5–8]. Over the past 20 years, the thrust of jet engines has increased by more than 60% whereas the fuel consumption has fallen by 15–20%, and these improvements are, in part, the result of improvements in the high-temperature properties of superalloys [9].

A variety of high-performance materials is used in modern jet engines. Aluminum and carbon-fibers composites are used in the coolest sections of engines (operating at temperatures under 150 °C), such as the fan and inlet casing, to reduce weight. Titanium ($\alpha + \beta$ and β) alloys are used in engine components with operating temperatures around 550 °C, which includes parts in the fan and compressor sections. Superalloys are used for components that operate above 550 °C, such as the blades, discs, vanes, and other parts found in the combustion chamber and other high-temperature engine sections [10].

Materials used in the hottest engine components, such as high-pressure turbine blades and discs, must have high strength, fatigue life, fracture toughness, creep resistance, hot-corrosion resistance, and low thermal expansion properties. Nickel-based superalloys are the appropriate material for these engine components bearing in mind their capability to operate at high temperatures for long periods of time [10].

Nickel-based superalloys used in jet engines have high concentration of alloying elements (up to about 50% by weight) to provide strength, creep resistance, fatigue endurance, and corrosion resistance at high temperature. The types and concentration of alloying elements determines whether the superalloy is a solid solution-hardened or precipitation-hardened material. Precipitation-hardened superalloys are used in the hottest engine components, with their high-temperature strength and creep resistance improved by presence of $L1_2$ and other precipitates that have high thermal stability [10,11].

Iron-nickel superalloys are used in jet engines for their high-temperature properties and low thermal expansion. These superalloys, which contain 15–60% iron and 25–45% nickel, are used in blades, discs, and engine casings that require low thermal expansion properties. Other alloying elements such as Nb, Mo, W, Ti, Al, and Cr are also added deliberately to the formation of new phases with ability to support exposure to oxidizing environments at high temperatures for reasonable periods of time [11,12].

However, there is a growing interest in new superalloys with good creep resistance and low cost. For this purpose two new superalloys have been developed by Villares Metals, the VAT 36 and VAT 32, both with intermediate nickel content but different amounts of carbon, iron, titanium, and niobium. The VAT 32 and VAT 36 poly-grained alloys can replace superalloys in turbines but its main application is for high temperature automotive valves. The VATs alloys intend to replace UNS N07751 and UNS N07080 (Nimonic 80A). These alloys are commonly applied in high temperature components such as those for automotive valves production for high performance internal combustion engines, considering its high hot strength. However, the high cost is still a problem due to the high nickel contents. The new developed alloys (VAT 32 and VAT 36) present economic advantages caused by its lower nickel content and are based on a high nickel-chromium austenitic matrix with dispersion of $L1_2$ phases and different carbides. In this work, these alloys were subjected to creep tests at constant load, at a temperature range of 675–750 °C and a stress range of 500–600 MPa. The creep tests were conducted in accordance with ASTM E139. The study is complemented by microstructural characterization with optical microscopy (OM), Scanning Electron Microscopy (SEM), Energy Dispersive Spectroscopy (EDS), and Rietveld refinement. The VAT 32 alloy showed higher creep resistance and these results were correlated to formation of higher amount of carbides and intermetallic phases caused by higher mass percentage of carbon and different amounts of alloying elements.

2. Materials and Methods

Cylindrical bars of superalloys VAT 36 and VAT 32 were provided by Villares Metals Company (Sumaré, Brazil). Specimens for creep tests were prepared with shapes and dimensions depicted in Figure 1.

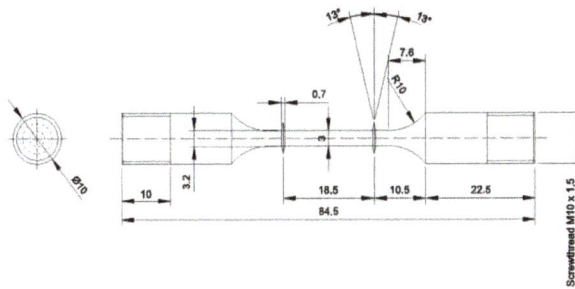

Figure 1. Dimensions of the specimen for creep tests in mm.

The specimens were subjected to heat treatment comprising two steps: (a) solution, held at 1050 °C during 30 min with a heating rate of 20 °C/min, followed by air cooling; (b) aging, carried out at 750 °C during 4 h, with a heating rate of 10 °C/min followed by air cooling. The creep tests were accomplished according to ASTM E-139-11 [13]. Electrical systems and controllers (Denison Mayes Group, Leeds, UK) were adapted in the furnace (Denison Mayes Group, Leeds, UK). A linear variable differential displacement transducer (LVDT, Denison Mayes Group, Leeds, UK) was used to obtain elongation measurements, and a Cromel-Alumel thermocouple was used for temperature control as detailed in [14]. The creep tests were performed at temperatures of 675, 700 and 750 °C and in constant load mode at stresses of 500, 550, and 600 MPa. A diffractometer Philips model PW 3710 (Philips, Best, The Netherlands) was used for Rietveld refinement, operated with Cuk α radiation (λ = 1.54184 Å). The specimens were scanned at 2θ in a range of 10° ≤ 2θ ≤ 125° and step increment of 0.02°.

The samples preparation for analysis via Optical Microscopy (Zeiss, Jena, Germany) and Scanning Electron Microscopy (SEM) (Jeol, Tokyo, Japan) followed the standard grinding procedure using silicon carbide sandpapers. Polishing was accomplished with diamond paste Codemaq with 0.3 nm. Etching was carried out with the kalling's reagent, with the following composition: 33 mL HCl, 33 mL ethanol, 33 mL water, and 1.5 g of $CuCl_2$. The scanning electron microscope was used for microstructural and energy dispersive spectroscopy (EDS) analyses. The same microscope was operated aiming to evaluate fracture surfaces of crept samples and to identify the dominant fracture mechanism in each case.

3. Results and Discussion

3.1. Chemical Composition

The chemical compositions of VAT 36 and VAT 32 alloys are shown in Table 1.

Table 1. Hemical compositions: VAT 36 and VAT 32. (Mass.%).

Alloys/Elem.	Ni	Fe	Cr	C	Ti	Nb	Al
VAT 36	35.8	40.21	18.6	0.05	1.14	2.0	1.90
VAT 32	32.0	44.14	15.5	0.26	2.0	3.90	1.90

The VAT 32 alloy has about five times more carbon and approximately two times more titanium and niobium, when compared to VAT 36. Titanium and niobium preferably combine with carbon to form carbides. The excess of titanium and niobium unreacted with carbon combines with nickel giving rise to intermetallic phases $L1_2$ and increases the creep resistance.

3.2. Microstcuctural Evaluation after HT

Images obtained with optical and scanning electron microscopy showed both alloys (VAT 32 and VAT 36) with heterogeneous microstructures, with twinned areas, and grains of average size of 71.83 μm to 32 VAT and 70.75 μm to 36 VAT.

3.2.1. Microstructural Evaluation of Superalloy VAT 36

The microstructures for VAT 36 alloy are depicted in Figures 2 and 3.

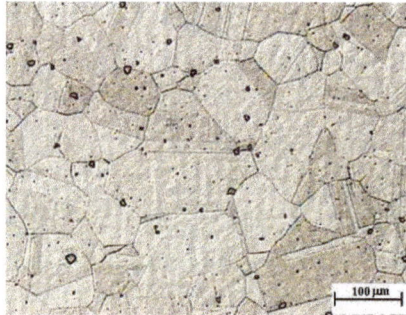

Figure 2. Optical micrograph of heat treated superalloy VAT 36. Magnification: 100×. Heterogeneous grain sizes and small amount of precipitates. Small twinned areas are observed.

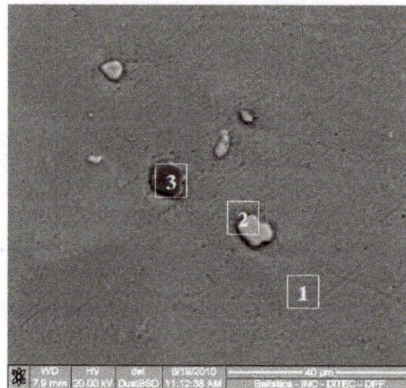

Figure 3. Scanning Electron Microscopy (SEM) micrograph of heat treated superalloy VAT 36 in backscattered electron image. Magnification: 3000×. Matrix (1) and two different precipitates (2 and 3).

Matrix chemical composition (1) was already presented in Table 1 for this alloy. Precipitates highlighted (2 and 3) were evaluated via EDS with semi quantitative analysis depicted in Table 2.

Table 2. Semi quantitative analysis of precipitates 2 and 3 of VAT 36, showed in Figure 3.

ZAF—Method Standardless Quantitative Analysis		
Element	Mass. % (Precipitate 2)	Mass. % (Precipitate 3)
C K	8.66	5.67
Fe K	2.28	4.65
Ti K	21.72	44.41
Nb K	67.24	18.69
Cr K	0.04	23.22
Ni K	0.06	3.36
TOTAL	100	100

From the values observed in Table 2 we can infer a possible formation of (Nb,Ti)C carbides in precipitate 2 and (Nb,Ti,Cr) carbides in precipitate 3.

3.2.2. Microstructural Evaluation of Superalloy VAT 32

Figures 4 and 5 depict microstructures for VAT 32 alloy. Precipitates were observed distributed throughout the microstructures, with higher amount of these precipitates for VAT 32 alloy (Figure 4) when compared with VAT 36 (Figure 2).

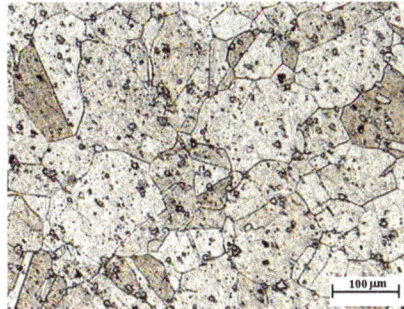

Figure 4. Optical micrograph of heat treated superalloy VAT 32. Magnification: 100×. Higher amount of precipitates is observed, in comparison with VAT 36 microstructure. Negligible twinned areas are observed.

Figure 5. SEM micrograph of heat treated superalloy VAT 32 in backscattered electron mode in which matrix (1) and precipitates (2 and 3) are highlighted. Magnification: 3000×.

Matrix chemical composition of VAT 32 (1) is available in Table 1 for this alloy. Precipitates highlighted (2 and 3) were evaluated via EDS with compositions depicted in Table 3 as follows.

Table 3. Semi quantitative analysis of precipitates 2 and 3 observed in VAT 32, highlighted in Figure 5.

ZAF—Method Standardless Quantitative Analysis		
Element	Mass % (Precipitate 2)	Mass % (Precipitate 3)
C K	5.23	2.23
Fe K	2.64	1.78
Ti K	23.51	31.30
Nb K	65.34	62.39
Cr K	1.32	1.34
Ni K	1.96	0.96
Total	100	100

Both precipitates are quite similar for this alloy, possibly (Nb,Ti)C carbides.

The chemical analysis of both alloys VAT 36 and VAT 32 (Table 1) showed high mass percentage of iron, nickel, and chromium, as expected for Fe-Ni-Cr based superalloys. Other elements such as carbon, aluminum, titanium, and niobium are also present, however at lower concentrations. These elements form carbides and intermetallic compounds and, as a result, cause the alloy precipitation hardening as described in [15]. Evaluation of EDS semi quantitative analysis, obtained from precipitates observed in Figures 3 and 5 showed considerably higher concentration of niobium and titanium at precipitates for VAT 32 alloy, indicative of (Nb,Ti)C carbides formation and possibly intermetallic $L1_2$. For VAT 36, EDS analysis showed possibly formation of (Nb,Ti,Cr) carbides at precipitate 3. The weight percentage of precipitates is about 1.2% to VAT 36 and 5.5% to the VAT 32, as calculated in GSAS software in Rietveld refinement. The largest amount of precipitates for the alloy VAT 32 is due to higher concentration of carbon, titanium, and niobium in its chemical composition.

3.3. Rietveld Refinement

The peaks of the intermetallic $L1_2$ phases [Ni_3Al, $Ni_3(Al_{0.5}Ti_{0.5})$] are superimposed with the highest peak of γ austenitic matrix. A refinement of the crystal structure was accomplished by the Rietveld method (Figures 6 and 7), due to the difficulty of distinguishing the phase γ and other precipitates with similar lattice parameters as described in [16,17]. Table 4 contains the lattice parameters and calculated weight percentages of the phases for the alloys VAT 36 and VAT 32.

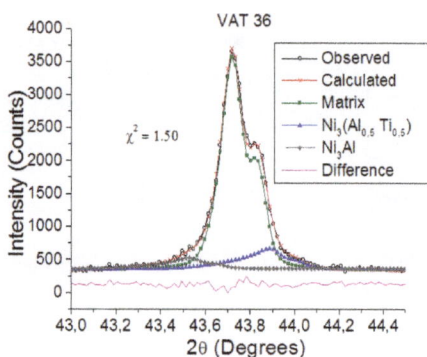

Figure 6. Rietveld refinement carried out at 2θ = 44° for the alloy VAT 36.

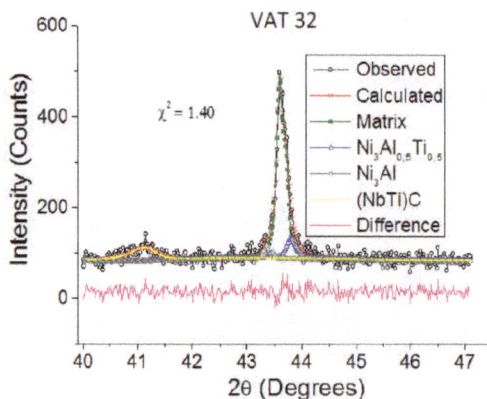

Figure 7. Rietveld refinement carried out in 40° ≤ 2θ ≤ 47° (VAT 32).

Table 4. Lattice parameters and percentages of the phases for the alloys VAT 36 and VAT 32.

VAT 36		
Phases	**wt.% Calculated**	**Lattice Parameters (Å)**
Matrix (γ)	92	$a = b = c = 3.581$
L1$_2$ Phase	5	$a = b = c = 3.596$
L1$_2$ Phase	3	$a = b = c = 3.568$
VAT 32		
Phases	**wt.% Calculated**	**Lattice Parameters (Å)**
Matrix (γ)	64	$a = b = c = 3.590$
L1$_2$ Phase	9	$a = b = c = 3.611$
L1$_2$ Phase	5	$a = b = c = 3.581$
(NbTi)C	22	$a = b = c = 4.384$

3.4. Creep Tests

Figures 8–10 show the creep curves corresponding to deformation ε as a function of the time t for VAT 36 and VAT 32 alloys in creep tests carried out for a temperature range of 675–750 °C and stress range of 500–600 MPa.

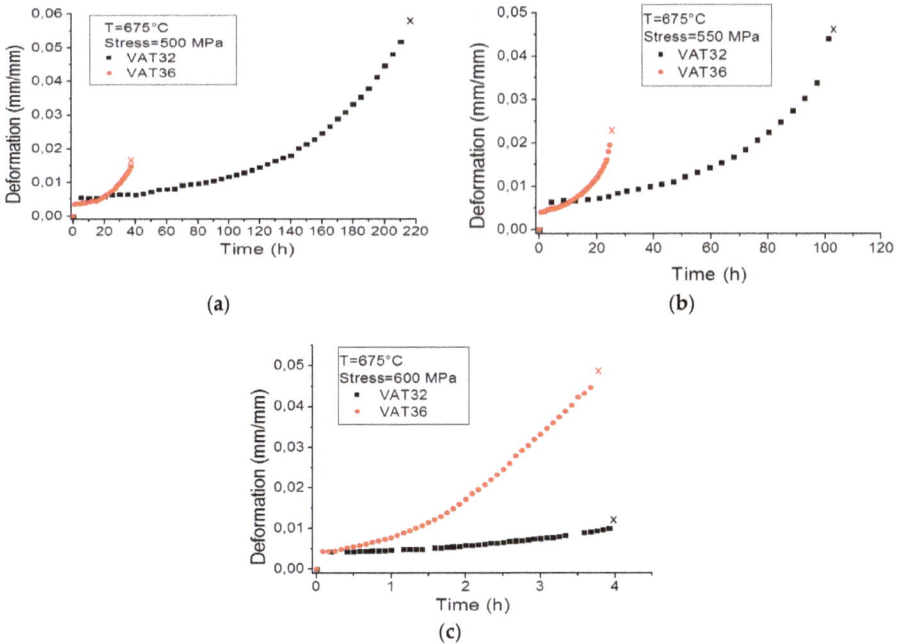

(a)

(b)

(c)

Figure 8. Creep curves of the alloy VAT 36 and VAT 32 at 675 °C at the stress of (**a**) 500. (**b**) 550 and (**c**) 600 MPa.

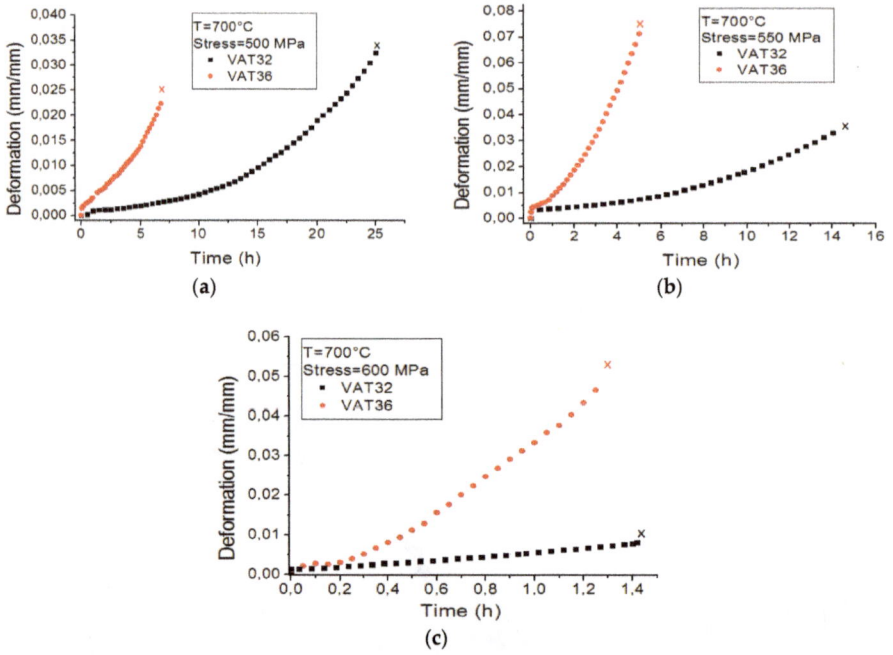

Figure 9. Creep curves of the alloys VAT 36 and VAT 32 at 700 °C at the stress of: (**a**) 500. (**b**) 550 and (**c**) 600 MPa.

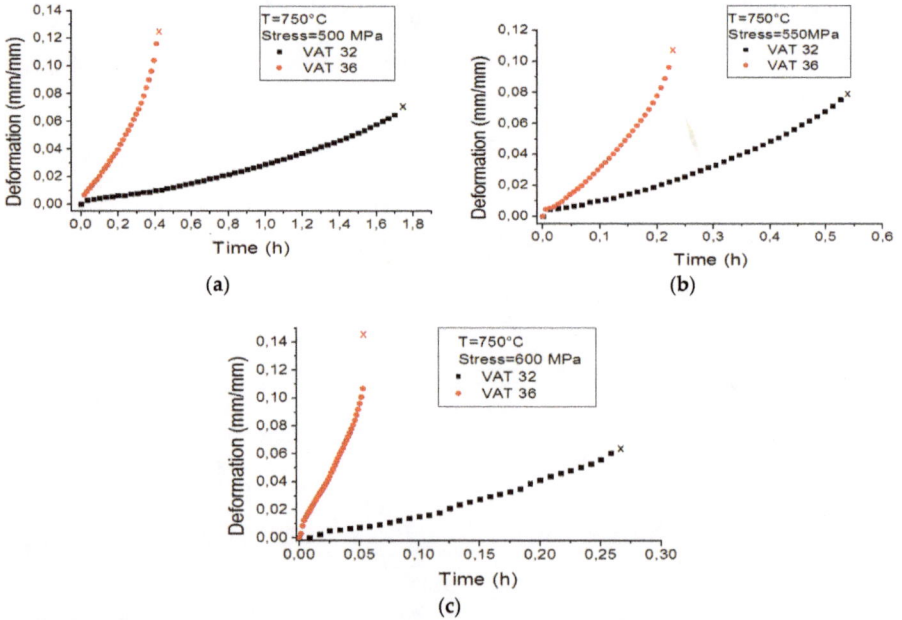

Figure 10. Creep curves of the alloys VAT 36 and VAT 32 the 750 °C at the stress of: (**a**) 500. (**b**) 550 and (**c**) 600 MPa.

Table 5 shows the creep data of all the creep curves.

Table 5. Creep tests data of the alloys VAT 36 and VAT 32 at temperatures (T) of 675, 700, and 750 °C and stresses (σ) of 500, 550, and 600 MPa, showing creep rate ($\dot{\varepsilon}_s$), time of failure (t_f) and reduction of area (RA).

T (°C)	Material	σ (MPa)	$\dot{\varepsilon}_s$ (1/h)	t_f (h)	RA (%)
		500	3.25×10^{-4}	37.25	3.31
	VAT 36	550	9.08×10^{-4}	25.27	3.30
675		600	4.01×10^{-3}	3.76	5.91
		500	5.08×10^{-5}	214.58	3.30
	VAT 32	550	1.75×10^{-4}	102.33	2.00
		600	7.56×10^{-4}	3.96	1.32
		500	2.36×10^{-3}	6.95	14.12
	VAT 36	550	7.89×10^{-3}	5.12	11.64
700		600	2.58×10^{-2}	1.30	12.26
		500	4.4×10^{-4}	25.16	4.61
	VAT 32	550	1.55×10^{-3}	14.57	3.96
		600	5.93×10^{-3}	1.43	3.26
		500	1.71×10^{-1}	0.41	19.29
	VAT 36	550	3.87×10^{-1}	0.23	16.45
750		600	14.65×10^{-1}	0.054	17.40
		500	2.15×10^{-2}	1.73	5.91
	VAT 32	550	8.43×10^{-2}	0.53	5.26
		600	2.38×10^{-1}	0.26	4.61

Table 5 shows smaller reduction of area after creep tests for VAT 32 alloy. Therefore this alloy has lower ductility than the VAT 36. The creep rate is also lower for VAT 32 for all temperatures and stresses by, at least, one order of magnitude. This higher creep resistance for this alloy (VAT 32) is attributed to a larger amount of MC carbides and intermetallic compound L1$_2$. The VAT 32 showed also higher times to failure at lower stresses.

At 675 °C, when the stress is increased from 500 to 600 MPa, the time to rupture decreases 54.2 times for the VAT 32 and to 9.9 times for the VAT 36. At 700 °C, this reduction is of 17.6 times for the VAT 32 and 5.3 times for the VAT 36. At a temperature of 750 °C, t_f decreases 6.6 times for the VAT 32 and 7.6 times for the VAT 36. Increase of creep tests temperatures results in higher ductility and the material becomes less sensitive to stress, causing lower differences between the times of rupture for higher stresses for both compositions. In general, both alloys showed typical behavior in creep tests, with decrease in the creep rate and increase of the rupture time with reduction of the stress or temperature. This indicates that the alloys exhibits creep rate sensitivity to both temperature and applied stress. For metals and alloys, the greater the temperature and higher strain rate, the earlier the stage of cavities and cracks formation will be [18].

The higher creep resistance of the alloy VAT 32 in comparison to VAT 36 is caused by the largest fraction of carbides (Nb,Ti)C and intermetallic L1$_2$ in VAT 32 alloys. These precipitates originated from the higher amount of carbon, titanium, and niobium in the VAT 32 chemical composition. These precipitates are stable and have a low rate of coalescence. As a result dislocations sliding are hindered by these obstacles during creep deformation [19]. Carbides also produce an anchoring effect at grain boundaries, hence delaying relative sliding between grains, affecting the ease with which vacancies can be generated in these regions [20,21]. Dependence of the steady state creep rate with stress and temperature could be expressed as an Arrhenius type relationship, and represented by the Equation (1) [21].

$$\dot{\varepsilon}_s = B_0\, \sigma^n \exp\left(\frac{-Q_C}{RT}\right) \tag{1}$$

where B_0 is a factor dependent on the stress and the structure of the material, σ is the stress applied, R is the gas constant, and T is the absolute temperature. The concept of a mechanism responsible for creep phenomenon can be characterized by different values of stress exponent n combined with the activation energy Q_c. Therefore, values of n and Q_c were determined in this work, based on Equation (1) and the data in Table 3, as can be seen in Figures 11 and 12.

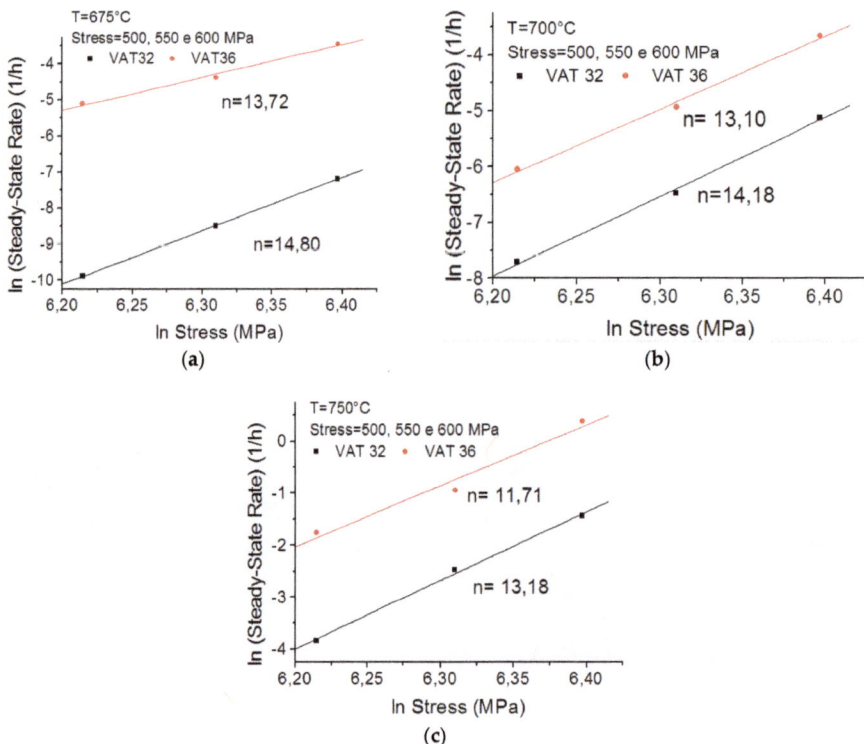

Figure 11. Dependence of the steady-state creep rate with the applied stress to the alloys VAT 32 and VAT 36 at: (a) 675 °C. (b) 700 °C and (c) 750 °C.

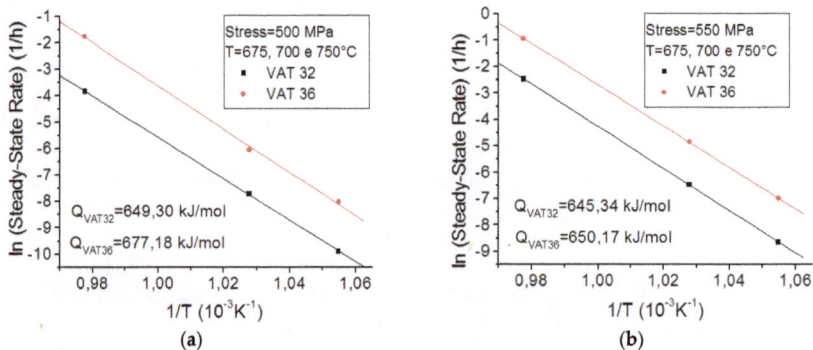

Figure 12. *Cont.*

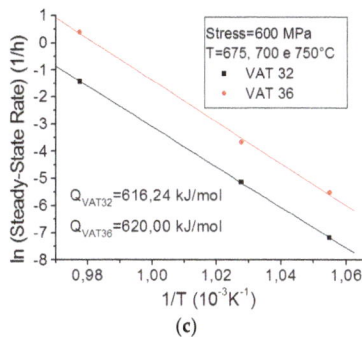

(c)

Figure 12. Steady-state creep rate dependence with the temperature for the alloys VAT 32 and VAT 36 at: (**a**) 500 MPa. (**b**) 550 MPa and (**c**) 600 MPa.

At 675 °C a stress exponent of 14.80 was obtained for the VAT 32 and 13.72 for the VAT 36. At 700 °C, *n* values reached 14.18 for the VAT 32 and 13.10 for the VAT 36. At 750 °C the stress exponents were 13.18 for the VAT 32 and 11.71 for the VAT 36.

Activation energies values at 600 MPa were near 616 kJ/mol for the VAT 32 and 620 kJ/mol for the VAT 36. At 550 MPa these values reached 645 kJ/mol for the VAT 32 and 650 kJ/mol for the VAT 36. At 500 MPa activation energies were 649.30 kJ/mol for the VAT 32 to 677 kJ/mol for the VAT 36.

The occurrence of a given creep mechanism depends on the test conditions. For lower stresses and high temperatures, possibly dislocation climb over carbides and precipitates would prevail. For higher stresses and lower temperatures shear mechanisms involving stacking fault occurs over a wide range of experimental conditions [22,23]. Previous research showed the dependence of velocity of dislocation movement with carbides volumetric fraction [24,25].

Dislocations movement is hindered by carbide particles. Reduced creep rate of VAT 32 is strongly affected by higher amounts of carbides (Nb,Ti)C and intermetallic $L1_2$, when compared to VAT 36. Those particles behave as obstacles to dislocations movement, with the creep rate dependent on their size and distribution.

3.5. Failure Analysis

Fracture surfaces of crept samples are depicted in fractographs showed in Figures 13–15. Figure 13 shows the fractured surface of sample VAT 36 after creep test. The crack starts with intergranular mode (region A). After reaching critical size, crack propagation occurs at transgranular ductile mode (region B), with dimples.

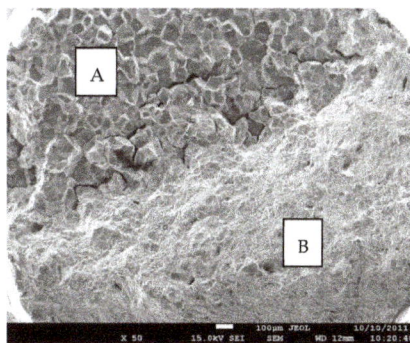

Figure 13. Fracture surface of crept VAT 36 sample, in creep test at 700 °C and stress of 600 MPa, with intergranular (A) and transgranular ductile (B) regions.

Figure 14. Surface of fracture initiation sites of VAT 36 (**a**) and VAT 32 (**b**), crept at 700 °C and 500 MPa, with an intergranular mode for both cases. Considerable amount of precipitated carbides can be seen at the VAT 32 surface.

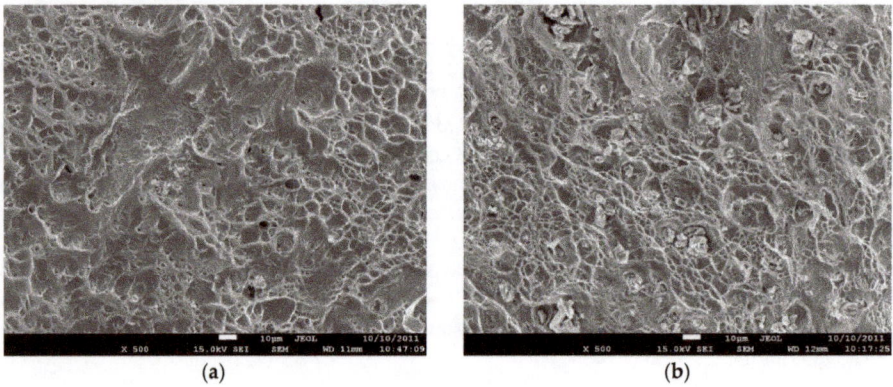

Figure 15. Final portions of fractured surfaces of VAT 36 (**a**) and VAT 32 (**b**) alloys, subjected to creep tests at 750 °C showing dimples of ductile fracture.

Detailed examination of the crack initiation of both compositions showed intergranular mode at these regions, as depicted in Figure 14. Higher amount of precipitated carbides can be observed for VAT 32 composition.

At the final region of fractured surfaces, ductile fractures prevailed, with a large amount of dimples. Particulate carbides remained inside dimples for VAT 32 alloy, as depicted in Figure 15b.

The fractographs showed intergranular fracture at creep initial stages for both alloys. Steps of creep intergranular fracture are nucleation, growth, and subsequent binding of voids at triple points. The formation of cavities is controlled by diffusion processes. The stacking of dislocations in regions near the grain boundaries during creep can increase the stress concentration favoring the initiation and propagation of intergranular microcracks [22].

At the final surface cracks, ductile fracture prevails (transgranular ductile fracture) with formation and coalescence of microcavities of variable shapes and sizes. The size and shape of the dimples are governed by the number and distribution of nucleated microcavities and by the level of internal stresses present in the material [26].

The alloy VAT 36 showed more areas of transgranular ductile fracture, considering its higher ductility caused by lower weight percentage of carbides and intermetallics.

4. Conclusions

1. Rietveld refinement allowed qualitative and quantitative evaluation of carbides and intermetallic precipitates in both alloys. The VAT 36 has approximately 8 wt.% of intermetallic $L1_2$ and VAT 32, 14 wt.%. The precipitates hindered dislocation slip and grain boundaries slip, mainly in VAT 32 alloy, increasing its creep resistance.

2. The VAT 32 alloy showed smaller reduction in area after creep tests in relation to VAT 36. This result is caused by lower ductility of VAT 32 originated from higher mass percentage of carbon in its composition, favoring formation of phases based on carbides.

3. The higher creep resistance of the alloy VAT 32 is related to its substantial fraction of carbides $(Nb,Ti)C$ and intermetallic $L1_2$, provided by its larger carbon content. The excess of titanium and niobium unreacted with carbon combines with nickel giving rise to intermetallic phases $L1_2$. These precipitates are stable and have low rate of coalescence. As a result, during creep deformation these precipitates produce anchoring effect of grain boundaries hindering relative slide between grains and therefore causes crack formation delay. These volume defects act also as obstacles to dislocation slip and climb, decreasing the creep rate.

4. Failure analysis of surface fractures of crept samples showed intergranular failure mechanism at crack origin for both alloys VAT 36 and VAT 32. Intergranular fracture involves nucleation, growth, and subsequent binding of voids. The final fractured portion showed transgranular ductile failure, with dimples of different shapes and sizes, typical of ductile failure. Transgranular ductile fracture involves the formation and coalescence of microcavities with dissimilar shape and sizes. The VAT 32 showed smaller areas of intragranular failure mechanism (dimples). This behavior is caused by its lower ductility.

5. Stress exponents obtained is this work were in a range of 14.80–11.71 and activations energies between 677–616 kJ/mol. The occurrence of a given creep mechanism depends on the test conditions. At creep tests of VAT 32 and VAT 36, for lower stresses and high temperatures, dislocation climb over carbides and precipitates possibly prevail. For higher stresses and intermediate temperatures shear mechanisms involving stacking fault can occur over a wide range of experimental conditions

Author Contributions: Conceptualization: C.R.M.d.S., V.J.G., S.J.G., D.A.P.R., J.L.d.A.F., J.A.A.; Validation: C.R.M.S., V.J.G., S.J.G., D.A.P.R., J.L.d.A.F., J.A.A.; Formal Analysis: C.R.M.d.S., V.J.G; Investigation: C.R.M.d.S., D.A.P.R., V.J.G; Writing-Review and Editing: C.R.M.d.S., V.J.G.; Supervision: C.R.M.d.S.

Funding: This research received no external funding.

Acknowledgments: We would like to show our gratitude to Villares Metals who provided the samples for creep tests.

Conflicts of Interest: The authors declare no conflict of interest.

References

1. Sawada, K.; Kimura, K.; Abe, F. Mechanical response of 9% Cr heat-resistant martensitic steels to abrupt stress loading at high temperature. *Adv. Mater. Sci. Eng. A* **2003**, *358*, 52–58. [CrossRef]
2. Furtado, H.C.; de Almeida, L.H.; Le May, I. Precipitation in 9Cr–1Mo steel after creep deformation. *Mater. Charact.* **2007**, *58*, 72–77. [CrossRef]
3. Ennis, P.J.; Quadakkers, W.J. *9–12% Chromium Steels: Application Limit and Potential for Further Development in Parsons. Advanced Materials for 21st Century Turbines and Power Plants*; The Institute of Materials: London, UK, 2000; pp. 265–275.
4. Berneti, J.; Brada, B.; Kosec, G.; Bricelj, E.; Kosec, B.; Vodopivec, F.; Kosec, L. Centreline Formation of Nb(C, N) eutectic in structural steel. *Metall* **2010**, *49*, 29–32.
5. Schafrik, R.E.; Ward, D.D.; Groh, J.R. Application of alloy 718 in GE aircraft engines: Past, present and next five years. *Superalloys* **2001**, *718*, 1–11.

6. Schafrik, R.; Christodoulou, L.; Williams, J.C. Collaboration isan essential part of materials development. *JOM* **2005**, *57*, 14–16. [CrossRef]
7. Hohmann, M.; Brooks, G.; Spiegelhaue, C. production methods and applications for high-quality metal powders and sprayformed products. *Acta Metall. Sin. (Eng. Lett.)* **2005**, *18*, 15–23.
8. Walston, S.; Cetel, A.; MacKay, R.; O'Hara, K.; Duhl, D.; Dreshfield, R. Joint development of a fourth generation single crystal superalloy. *Superalloys* **2004**, 15–24.
9. Clarke, D.; Bold, S. Materials Developments in aeroengine gas turbines. In *Aerospace Materials*; Institute of Physics Publishing: Bristol, UK, 2001; pp. 71–80.
10. Smith, G.D.; Patel, S.J. The role of niobium in wrought precipitation-hardened nickel-base alloys. *Superalloys* **2005**, *718*, 625–706.
11. Tresa, M.P.; Sammy, T. Nickel-based superalloys for advanced turbine engines: Chemistry, microstructure and properties. *J. Propul. Power.* **2006**, *22*, 361–374.
12. Edmonds, D.V.; Cochrane, R.C. The effect of alloying on the resistance of carbon steel for oilfield applications to CO_2 corrosion. *Mat. Res.* **2005**, *8*, 377–385. [CrossRef]
13. American Society for Testing and Materials (ASTM). *ASTM E139-11: Standard Test Methods for Conducting Creep, Creep Rupture, and Stress Rupture Tests of Metallic Materials*; ASTM International: West Conshohocken, PA, USA, 2011.
14. Almeida, G.F.C.; Couto, A.A; Reis, D.A.P.; Massi, M.; Sobrinho, A.S.S.; Lima, N.B. Effect of plasma nitriding on the creep and tensile properties of the Ti-6Al-4V alloy. *Metals* **2018**, *8*, 618. [CrossRef]
15. Donachie, M.J; Donachie, S.J. *Superalloys: A Technical Guide*, 2nd ed.; ASM International: Materials Park, OH, USA, 2002.
16. Xu, Y.; Jin, O.; Xiao, X.; Cao, X.; Jia, G.; Zhu, Y.; Yin, H. Strengthening mechanisms of carbon in modified nickel-based superalloy Nimonic 80A. *Mater. Sci. Eng. A* **2011**, *528*, 4600–4607. [CrossRef]
17. Tian, S.; Wang, M.; Yu, H.; Yu, X.; Li, T.; Qian, B. Influence of element Re on lattice misfits and stress rupture properties of single crystal nickel-based superalloys. *Mater. Sci. Eng. A* **2010**, *527*, 4458–4465.
18. Kassner, M.E. *Fundamentals of Creep in Metals and Alloys*, 3rd ed.; Elsevier Ltd.: Amsterdam, The Netherlands, 2015.
19. He, L.Z.; Zheng, Q.; Sun, X.F.; Guan, H.R.; Hu, Z.Q.; Tieu, A.K.; Lu, C.; Zhu, H.T. Effect of carbides on the creep properties of a Ni-base superalloy M963. *Mater. Sci. Eng. A* **2005**, *397*, 297–304. [CrossRef]
20. Ha, V.T.; Jung, W.S. Effects of heat treatment processes on microstructure and creep properties of a high nitrogen 15Cr–15Ni austenitic heat resistant stainless steel. *Mater. Sci. Eng. A* **2011**, *528*, 7115–7123. [CrossRef]
21. Evans, R.W.; Wilshire, B. *Introduction to Creep*; The Institute of Materials: London, UK, 1993; p. 115.
22. Tian, S.; Xie, J.; Zhou, X.; Qian, B.; Lun, J.; Yu, L.; Wang, W. Microstructure and creep behavior of FGH95 nickel-base superalloy. *Mater. Sci. Eng. A* **2011**, *528*, 2076–2084.
23. Sajjadi, S.A.; Nategh, S.A. High temperature deformation mechanism map for the high performance Ni-base superalloy GTD-111. *Mater. Sci. Eng. A* **2001**, *307*, 158–164. [CrossRef]
24. Society of Automotive Engineers. *SAE J775: Engine Poppet Valve Information Report*; SAE International: Warrendale, PA, USA, 2004.
25. American Society for Testing and Materials Standards. ASTM B637: Standard Specification for Precipitation-Hardening Nickel Alloy Bars, Forgings and Forging Stock for High Temperature Service. ASTM: West Conshohocken, PA, USA, 2011.
26. Brooks, C.R.; Choudhury, A. Fracture mechanisms and microfractographic features. In *Metallurgical Failure Analysis*; McGraw-Hill: New York, NY, USA, 1993; pp. 119–211.

metals **MDPI**

Article

Influence of Excess Volumes Induced by Re and W on Dislocation Motion and Creep in Ni-Base Single Crystal Superalloys: A 3D Discrete Dislocation Dynamics Study

Siwen Gao *, Zerong Yang, Maximilian Grabowski, Jutta Rogal, Ralf Drautz and Alexander Hartmaier

Interdisciplinary Centre for Advanced Materials Simulation, Ruhr-Universität Bochum, 44801 Bochum, Germany; zerong.yang@rub.de (Z.Y.); maximilian.grabowski@rub.de (M.G.); jutta.rogal@rub.de (J.R.); ralf.drautz@rub.de (R.D.); alexander.hartmaier@rub.de (A.H.)
* Correspondence: siwen.gao@ruhr-uni-bochum.de; Tel.: +49-234-32-22612; Fax: +49-234-32-14984

Received: 20 March 2019; Accepted: 25 May 2019; Published: 1 June 2019

Abstract: A comprehensive 3D discrete dislocation dynamics model for Ni-base single crystal superalloys was used to investigate the influence of excess volumes induced by solute atoms Re and W on dislocation motion and creep under different tensile loads at 850 °C. The solute atoms were distributed homogeneously only in γ matrix channels. Their excess volumes due to the size difference from the host Ni were calculated by density functional theory. The excess volume affected dislocation glide more strongly than dislocation climb. The relative positions of dislocations and solute atoms determined the magnitude of back stresses on the dislocation motion. Without diffusion of solute atoms, it was found that W with a larger excess volume had a stronger strengthening effect than Re. With increasing concentration of solute atoms, the creep resistance increased. However, a low external stress reduced the influence of different excess volumes and different concentrations on creep.

Keywords: superalloy; excess volume; solute atom; dislocation dynamics; creep; DFT

1. Introduction

The new generation Ni-base single crystal superalloys that are used to make turbine blades in aero engines show a significant improvement of the mechanical properties at high temperatures when adding specific solid solution elements, such as Re and W [1–3]. The cuboidal γ' precipitates with a high volume fraction separated by the narrow γ matrix channels form the typical microstructure of superalloy single crystals [4]. Both the γ/γ' microstructure and dislocation motion inside are influenced by these elements. An increasing content of Re and W increases the volume fraction of cuboidal γ' phase, and Re decreases the γ' particle size, while W has no effect on the size of γ' [5,6]. During creep, Re decelerates the coarsening and rafting of γ' precipitates [7,8], as well as Re and W retard the dislocation motion through solid solution strengthening mechanisms [9,10]. Due to the atomic size difference and shear modulus mismatch between solute and solvent, the solute atoms can impede dislocation motion by parelastic and dielastic interactions, respectively [11–13]. Moreover, the stacking fault energy can be reduced by increasing solute concentration, such that more solute atoms segregate to the core of the dissociated dislocation and therefore drag dislocation glide [14]. The mobility of vacancies is hardly influenced by the solute atoms within the dilute limit [15], but the low stacking fault energy induced by the solute atoms can raise the migration energy of vacancies to hinder dislocation climb [9]. Although discussions about the effect of solute atoms in superalloys, especially Re, have been proposed in a view of experiments [2,8,16–19], the debate still exists. Furthermore, in complex superalloys, many possible effects of solute atoms can occur simultaneously, thus it is difficult to clarify how each individual mechanism relates to the creep properties. A 3D discrete dislocation dynamics (DDD) model is used in the present work to exclusively elucidate the effect of excess volume due to the size mismatch between solute atoms (Re and W) and the Ni matrix on the dislocation motion and creep.

Extensive experimental studies have shown that the creep mechanisms in single crystal superalloy involve the dislocation motion. The creep deformation generally starts with the dislocations filling γ matrix channels, and continues with the dislocation climb along γ/γ' interfaces and the shearing of γ' precipitates [4,20,21]. For different temperatures and different loading conditions, the microstructural and dislocation evolution during creep is diverse. At high temperatures above 900 °C and low stresses below 400 MPa, the rafting of γ' precipitates happens due to the γ/γ' lattice mismatch and the heterogeneity of γ matrix plasticity [22,23]. The shearing of γ' precipitates by pairs of superdislocations combined with anti-phase boundary (APB) is usually observed at the temperature higher than 850 °C [24–27]. When the temperature is lower than 850 °C and the stress is higher than 500 MPa, the shearing of γ' precipitates by superlattice stacking faults is found, and the microtwins appear for some non-$\langle100\rangle$-oriented superalloys [28–32]. The DDD model, as an outstanding tool to simulate the dislocation motion, has been used to investigate the deformation behavior in superalloys widely. The dislocation glide in γ channels, the shearing of γ' precipitates, the effect of dislocation climb, and the interaction of interfacial dislocations are well discussed by using different DDD methods [33–39]. However, the DDD study on the influence of substitutional atoms in superalloys is rare thus far. In addition, some authors incorporate the elastic stress induced by the concentration of interstitial atoms in the DDD simulation to quantify the effect of hydrogen on dislocations [40,41]. For the large substitutional atoms which form neither the cluster nor the cloud, we introduced the individual solute atoms in specific positions of our DDD simulation cell. In this work, the DDD model captures all the important deformation mechanisms of superalloys, including the dislocation glide and dislocation climb associated with the vacancy diffusion in the γ/γ' microstructure [39]. We performed simulations at a temperature of 850 °C, at which the morphology change of γ' precipitates is absent. According to a previous study [15], the diffusion coefficients of Re and W at 850 °C are approximately 1.0×10^{-18} m^2/s and 1.0×10^{-17} m^2/s, respectively. In the DDD time scale, such slow diffusion can be neglected. Moreover, it is stated that the mobility of vacancies is only minimally influenced by the solute atoms in Ref. [15]. Therefore, the diffusion of solute atoms and their interaction with vacancies were not taken into account. Furthermore, the segregation of Re around the dislocation line in γ' precipitates by the pipe diffusion at the later creep stage observed at 750 °C and 800 MPa could not be simulated by our current model [19]. For the shear modulus, we utilized a general value of Ni-base superalloys [42]. The excess volumes of Re and W solute atoms in pure Ni were obtained by density functional theory (DFT) calculations, which were applied in the DDD simulation to determine the back stresses on dislocations. Some experimental evidence shows that Re and W mainly partition to the γ matrix and no Re cluster is found in the undeformed microstructure [43–47]. Therefore, in this work, Re and W were only introduced into the γ channels to qualitatively study the influence of their respective excess volumes on the dislocation motion, and the corresponding dislocation motion controlled creep behavior was compared for different concentrations of Re and W.

2. Simulation Method

2.1. Implementation of Excess Volume Effect in Dislocation Dynamics

In our previously developed 3D discrete dislocation dynamics model [39], the shear stress of dislocation glide and the mechanical force of dislocation climb are generally determined from the main stresses by the Peach–Köhler equation in the glide and climb directions, respectively. The main stresses include the external stress, the internal stress from other dislocations, and the misfit stress due to the γ/γ' lattice mismatch [48]. The total effective stress on dislocation glide is also influenced by the line tension and the Peierls stress. In addition, the local chemical potential of vacancies gives rise to an osmotic force, which varies with the temperature and the local concentration of vacancies, affecting the dislocation climb. The vacancy diffusion is solved in a "meso" scale by a combination of finite difference method and fast Fourier transformation method. The velocity of dislocation motion is calculated by a linear relation from the effective stress with a mobility. The mobility of dislocation glide is determined by the drag coefficient, while the mobility of dislocation climb is based on the bulk diffusion such as other studies, in which the pipe diffusion along the dislocation line is not considered [49–52]. Because the mobility of dislocation climb is much smaller than the dislocation glide, we used a larger time step for dislocation climb. Initially, only dislocation glide with a small time step starts. When the plastic strain produced by dislocation glide is smaller than a threshold value, one step of dislocation climb with a large time step is activated, and afterwards dislocation climb step changes back to the glide step. The criterion of dislocation glide step changing to dislocation climb step is defined as $|1 - \epsilon_{eq}^t / \epsilon_{eq}^{t-\Delta t}| < 10^{-4}$, where ϵ_{eq}^t and $\epsilon_{eq}^{t-\Delta t}$ are the equivalent plastic strains produced by dislocation glide in the present and previous time step, respectively. The dislocation glide dominates the primary creep stage. The alternating dislocation glide and climb governs the secondary creep stage. For the shearing of γ' precipitates, we implemented the mechanism of the creation of APB by a single cutting dislocation and the destruction of APB by a second dislocation on the same slip plane behind the first dislocation, which is explained in our previous paper [48]. A single dislocation has to overcome a high APB stress to move in the γ' precipitate further. Without another dislocation coming from the same slip plane to delimit the APB, the cutting event rarely occurs. If we set the periodical boundary condition as an exact cubic periodicity for the dislocation motion in the simulation, the dislocation going out of the simulation box would return back on the exact same slip plane. This would increase the probability of two dislocations forming the pair of superdislocations artificially. To avoid the spurious self-cutting, an orthorhombic periodicity was used.

The dislocation line is discretized into an alternating sequence of pure straight edge and screw segments in our DDD code [53–55]. Each edge or screw dislocation segment was characterized by its line direction and glide direction precisely. The stress field of dislocation segment is calculated by a sum of stress fields of two semi-infinite dislocations [55,56]. The effective stress was applied on the middle point of each dislocation segment. On the basis of this edge-screw structure, we introduced additional back stresses due to the parelastic interaction between solute atoms and dislocations. If a solute atom with a different size from the matrix atoms were incorporated into the crystal, the generated excess volume would result in an interaction energy [13] of

$$\Delta E = -p\Delta V \left(3\frac{1-\nu}{1+\nu} \right) \quad , \tag{1}$$

where p is the hydrostatic stress field of a dislocation, ΔV is the excess volume, and ν is the Poisson's ratio. From a cross-section of an edge dislocation segment at the middle point perpendicular to its line direction, we found the stress field of the dislocation segment has the same feature as that of an infinite long straight edge dislocation [57]. For the qualitative study, we used the hydrostatic stress from an infinite long straight edge dislocation to calculate the back stress approximately, which is given by

$$p = \frac{1}{3}(\sigma_{xx} + \sigma_{yy} + \sigma_{zz}) = -\frac{\mu b}{3\pi}\frac{1+\nu}{1-\nu}\frac{y}{x^2+y^2} \quad , \tag{2}$$

where σ_{xx}, σ_{yy}, and σ_{zz} are the normal stress components of the elastic stress field of an edge dislocation, μ is the shear modulus, and b is the magnitude of the Burgers vector. The dislocation line is along z-direction, while x and y represent the coordinates with respect to the glide and climb directions, respectively. By inserting Equation (2) into Equation (1), we obtain

$$\Delta E = \frac{\mu b \Delta V}{\pi}\frac{y}{x^2+y^2} \quad . \tag{3}$$

Thus, the parelastic interaction forces due to a solute atom can be determined by the derivatives of ΔE with respect to x in the dislocation glide direction

$$F_{gl}^{P} = -\frac{\partial \Delta E}{\partial x} = \frac{\mu b \Delta V}{\pi} \frac{2xy}{(x^2 + y^2)^2} \quad , \tag{4}$$

and y in the dislocation climb direction

$$F_{cl}^{P} = -\frac{\partial \Delta E}{\partial y} = \frac{\mu b \Delta V}{\pi} \frac{y^2 - x^2}{(x^2 + y^2)^2} \quad , \tag{5}$$

respectively. For a dislocation segment of length l, the back stress due to the parelastic interaction force is given by

$$\tau_{bk} = \frac{F^{P}}{bl} \quad . \tag{6}$$

The back stress has the opposite sign of the effective stress on the dislocation, which is only added on edge dislocation segments, because screw dislocations have no hydrostatic stress field. When edge segments in a dislocation line are hindered by the back stress, their connected screw segments will be also hindered due to the local line tension.

2.2. Calculation of Excess Volumes

To obtain accurate values for the excess volume of the solute atoms, we employed DFT calculations. We used the projector augmented wave (PAW) method [58,59] as implemented in the Vienna Ab initio Simulation Package (VASP 5.4) [59–62] with the gradient corrected PBE exchange-correlation functional [63]. All calculations were carried out spin-polarized in $(3 \times 3 \times 3)$ face-centered cubic (fcc) supercells with a $[5 \times 5 \times 5]$ Monkhorst–Pack k-point mesh [64]. The lattice constant before relaxation is around 3.518 Å, which is close to the equilibrium lattice constant of Ni. The convergence of the electronic self-consistency was set to 10^{-7} eV and ionic positions were relaxed until all forces were below 0.01 eV Å^{-1}.

The equilibrium volumes of the supercells are determined in two different ways: by direct relaxation of the ionic positions, the cell shape, and the cell volume; and by fitting energy-volume curves using the Birch–Murnaghan equation of state [65]. In the former approach, the change in volume during the calculation leads to Pulay stresses [66], which result in an underestimation of the equilibrium volume. Furthermore, this approach is more sensitive with respect to the energy cutoff for the plane waves. To check the convergence of the excess volume as a function of the energy cutoff, calculations were performed for cutoffs of 450–700 eV. To record the energy–volume curves in the latter approach, supercells with ± 2 % change in the lattice constant around the equilibrium value are setup. For each volume, ionic positions are fully relaxed. The excess volume of atom X is calculated as

$$\Delta V_{X} = V_{Ni_{107}X} - V_{Ni_{108}} \quad , \tag{7}$$

where V is the volume of the $(3 \times 3 \times 3)$ fcc supercell with one solute atom ($V_{Ni_{107}X}$) and in pure Ni ($V_{Ni_{108}}$).

2.3. Dislocation Dynamics Setup and Solute Atom Arrangement

The same γ/γ' microstructure of Ni-base single crystal superalloys as in our previous study [39] was used in this work. As shown in Figure 1, 48 Frank–Read sources with an average length of 154 nm were randomly distributed on 12 slip systems in the γ channels with a uniform width of approximately 80 nm. The volume fraction of the cuboidal γ' precipitates was approximately 66%. The initial dislocation density was 1.8×10^{12} m^{-2}. The whole simulation box had a size of $1890 \times 1890 \times 1890$ nm^3, but the periodicity for the dislocation motion was set as 1890 nm $\times$ 1910 nm $\times$ 1930 nm. The material parameters used in the current study were the same as in Ref. [39]. The time steps of dislocation glide, climb and vacancy diffusion were 0.1 ns, 0.3 s, and 0.3 s, respectively.

Slip system	1	2	3	4	5	6	7	8	9	10	11	12
Slip plane	(111)	$(\bar{1}1\bar{1})$	$(\bar{1}1\bar{1})$	$(\bar{1}1\bar{1})$	(111)	$(\bar{1}\bar{1}1)$	$(\bar{1}1\bar{1})$	$(1\bar{1}\bar{1})$	$(1\bar{1}\bar{1})$	(111)	$(\bar{1}\bar{1}1)$	$(1\bar{1}\bar{1})$
Slip direction	$[\bar{1}01]$	$[\bar{1}01]$	$[011]$	$[011]$	$[1\bar{1}0]$	$[1\bar{1}0]$	$[\bar{1}\bar{1}0]$	$[\bar{1}\bar{1}0]$	$[0\bar{1}1]$	$[0\bar{1}1]$	$[101]$	$[101]$

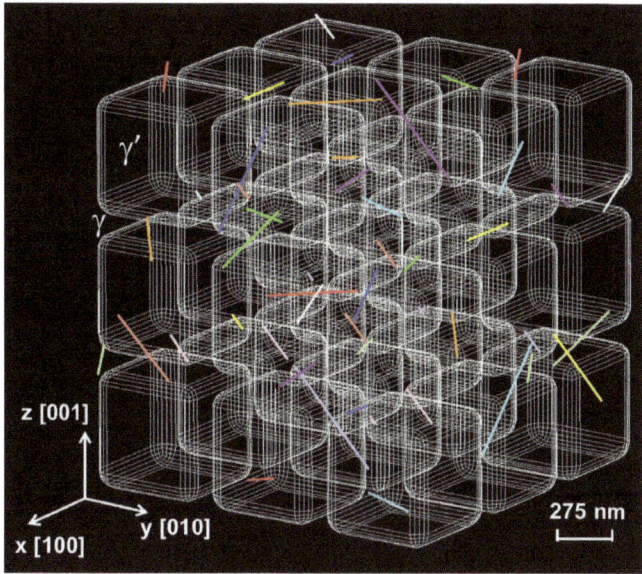

Figure 1. Representative γ/γ' microstructure with uniformly distributed cuboidal γ' precipitates surrounded by narrow γ matrix channels. Dislocation segments representing initial Frank–Read sources are distributed on 12 slip systems displayed by different colors as given in the color bar.

A strong partitioning of Re to the γ phase has been observed experimentally [43,44]. In addition, no Re clusters and no Re enrichment in the un-crept superalloys have been detected in 3D atom probe analysis and scanning transmission electron microscopy [45–47]. Although W is found in both the γ and γ' phase with similar concentrations [43,44], only W in the γ phase was considered in the present study to directly compare the effect of Re and W. Neglecting the influence of solute atoms in the γ' precipitates, we introduced homogeneously distributed solute atoms only in γ channels in our DDD simulations. The individual solute atoms were distributed on regular 3D grids with three different grid lengths of 5.4 nm, 10.8 nm, and 12.6 nm, which correspond to different solute concentrations of approximately 7×10^{-3} at.%, 9×10^{-4} at.%, and 6×10^{-4} at.%, respectively. These concentrations are much lower than the actual solute concentration in superalloys, but our simulations could provide qualitative trends with increasing solute content.

To determine the back stresses, the distances between the middle point of every edge dislocation segment and every solute atom in both glide and climb directions were calculated. The total back stress for one edge dislocation segment was the sum over the interaction with all solute atoms. We assumed that the minimum effective distance between the dislocation segment and the solute atom was 1.5 Å. The mechanism of the solute atom passing through the dislocation core was not taken into account, since there is no unequivocal theory to describe the interaction between a solute atom and a dislocation core thus far. Hence, when the distances in both glide and climb directions were smaller than 1.5 Å, the back stress was set to the value at the the minimum effective distance in the present simulations.

3. Results

3.1. Calculated Excess Volumes

The dependence of the excess volume of Re and W in Ni on the energy cutoff for the plane waves in the DFT calculations is shown in Figure 2. As discussed in Section 2.2, we employed two approaches to determine the equilibrium volumes, a complete relaxation of all degrees of freedom including the cell shape and volume (red lines in Figure 2) and a fit to the energy–volume curves using the Birch–Murnaghan equation of state (blue lines in Figure 2). Due to the Pulay stresses, the excess volume extracted from the complete relaxation converged only slowly with the plane wave cutoff, and even for a rather high energy cutoff of 700 eV the values were not fully converged. The values extracted from the Birch–Murnaghan fits converged much more quickly and were more robust. The excess volumes of both Re and W determined by the complete relaxation were underestimated, also due to the presence of Pulay stresses. For the DDD simulations, we correspondingly used the excess volumes calculated with our second approach yielding values of 4.34 Å^3 for Re and 7.21 Å^3 for W at an energy cutoff of 700 eV. Independent of the computational approach, the excess volume of Re was always smaller than the one of W.

The excess volume could also be estimated based on the atomic radii. The conventional covalent radii of Ni, Re, and W are $r_{Ni} = 1.24$ Å, $r_{Re} = 1.51$ Å, and $r_W = 1.62$ Å, respectively. The corresponding excess volumes are 6.43 Å^3 for Re and 9.82 Å^3 for W. Qualitatively, this simple estimate was in agreement with our DFT calculations, but quantitatively the excess volumes were significantly overestimated. The implications for the corresponding deformation resistance is discussed below.

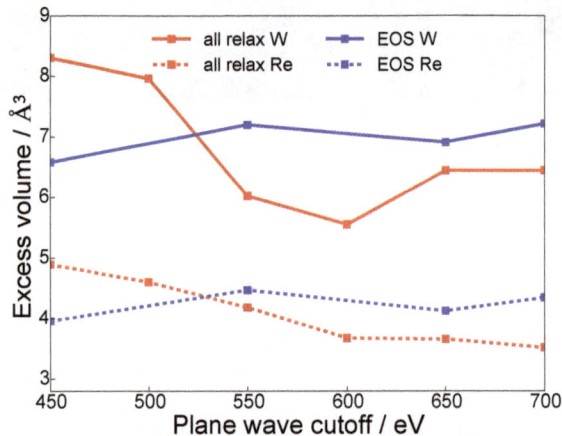

Figure 2. Calculated excess volumes for Re and W as a function of the plane wave cutoff for different methods determining the equilibrium volume.

3.2. Back Stress on Dislocations Due to Excess Volume

Assuming an edge dislocation segment with a length of 1 nm, the back stresses for its glide and climb due to the presence of a Re or W atom was calculated using the DFT excess volumes and Equations (4)–(6). Neglecting any influence along the dislocation line direction, the back stresses decreased with increasing distance between the atom and the dislocation core, as shown in Figure 3. When the distance was larger than 10 Å, the back stresses were almost zero. This demonstrated that only when the atom was very close to the dislocation core the effect of the back stresses became important. The back stresses for dislocation glide and climb showed a different dependence on the position of the atom in the plane. For instance, if a Re atom were located at $x = 2.0$ Å and $y = 2.0$ Å, the back stress for glide, τ_{bk}^{gl}, would be approximately 60 MPa (Figure 3a), whereas the back stress for climb, τ_{bk}^{cl}, would be lower than 4 MPa (Figure 3b). For some other configurations, τ_{bk}^{cl} could be larger than τ_{bk}^{gl}. However, the maximum value of the back stress for dislocation climb was much smaller than for dislocation glide. Moreover, because of the larger excess volume of W, the back stresses for W were higher than for Re.

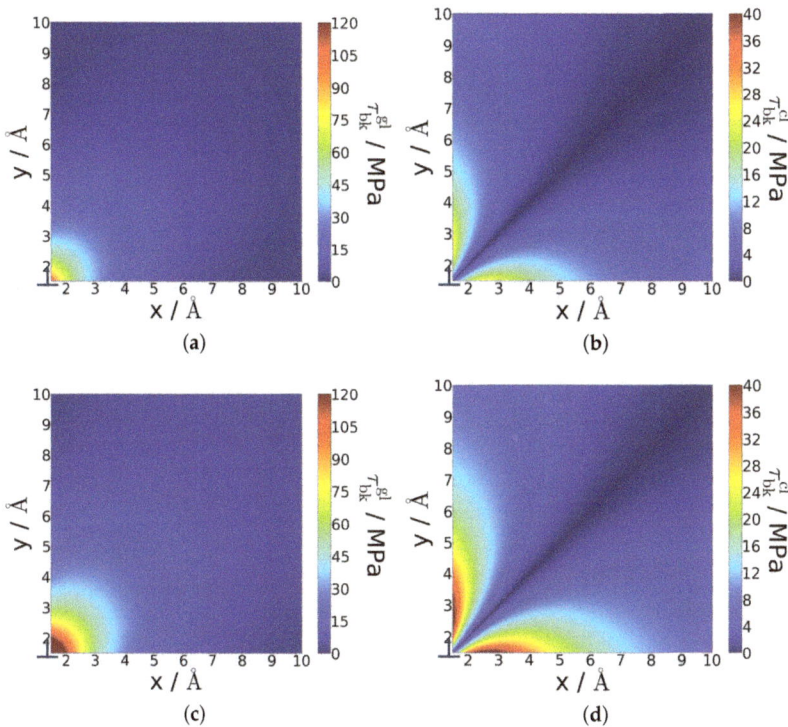

Figure 3. Back stresses for dislocation glide τ_{bk}^{gl} and dislocation climb τ_{bk}^{cl} with respect to the distance between one atom and an edge dislocation core. The x-axis and y-axis represent the glide and climb directions, respectively. The plotted range in x- and y-direction is from 1.5 Å to 10 Å: (**a,b**) for Re; and (**c,d**) for W. Note the different scales on the colorbars for glide and climb back stresses, respectively.

3.3. Influence of Excess Volume on Dislocation Motion and Creep

In the DDD simulation, two constant external tensile loads of 500 MPa and 200 MPa along the [001] direction were applied at 850 °C. At the high stress of 500 MPa, only dislocation glide in the microstructure was considered, while at the low stress of 200 MPa both dislocation glide and climb were included. The plastic strain discussed in this work was the equivalent plastic strain and the strain rate was its corresponding derivative. In the present study, it was found that only a few single dislocations slightly cut into the γ' precipitates and no shearing by superdislocations was observed, because we used a high APB energy of 250 mJ/m^2 [67] and there was no second dislocation coming on the exact same slip plane to destroy the APB. Hence, the cutting event is not the focus here.

In Figure 4, we show the evolution of the plastic strain, the strain rate, and the dislocation density in our DDD simulations at 500 MPa. The addition of both Re and W decreased the plastic strain and the strain rate. With increasing concentration of solute atoms, the deformation resistance was further improved. At the same concentration, W had a stronger retard effect than Re, which was due to its larger excess volume and correspondingly larger back stresses. However, an increase of concentration could compensate the insufficient resistance due to low excess volume, because the highest concentration of Re resulted in lower strain and strain rate than the medium concentration of W, as shown in Figure 4a,b. The analysis of the dislocation evolution in the simulation cell showed that the dislocation filling γ channels was decelerated by adding solute atoms. To avoid the visual confusion due to the full of dislocations in the microstructure, we take the dislocation pattern in one slip system as an example to show the influence of solute atoms on the dislocation evolution. It can be seen in Figure 5a that the dislocations in the [0$\bar{1}$1](111) slip system glided and multiplied easily in γ matrix channels without the solute atom, resulting in a high dislocation density. The existence of 7×10^{-3} at.% Re in γ matrix channels hindered the dislocation glide and multiplication evidently (Figure 5b). The 7×10^{-3} at.% W enhanced the retard effect even more, where the dislocations hardly propagated in γ matrix channels (Figure 5c). The solute atoms led to slow rises of dislocation densities, as shown in Figure 4c, corresponding to small plastic strains in Figure 4a.

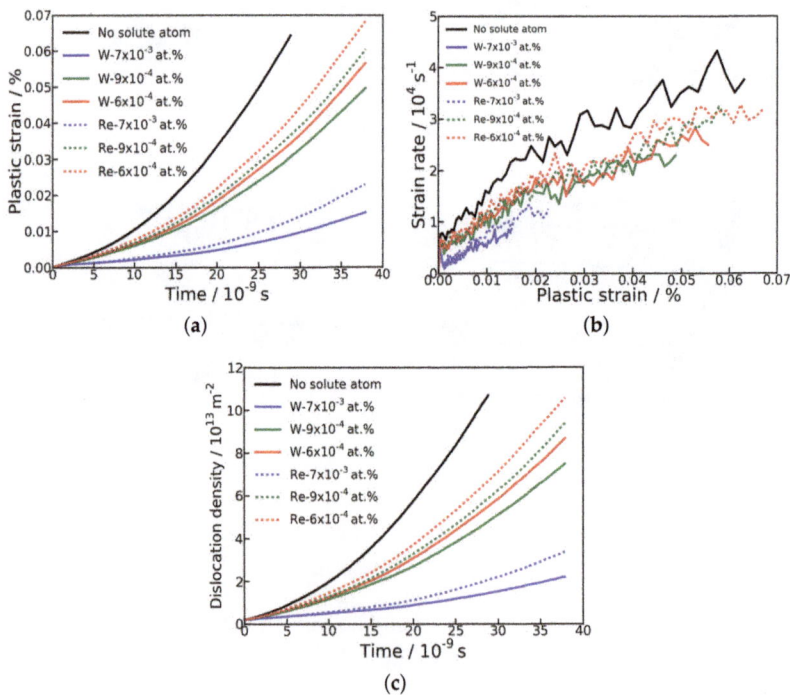

(a)

(b)

(c)

Figure 4. (a) Plastic strains as a function of time; (b) strain rates as a function of plastic strain; and (c) total dislocation densities as a function of time for the addition of Re and W with different concentrations under 500 MPa tensile load along [001] direction at 850 °C.

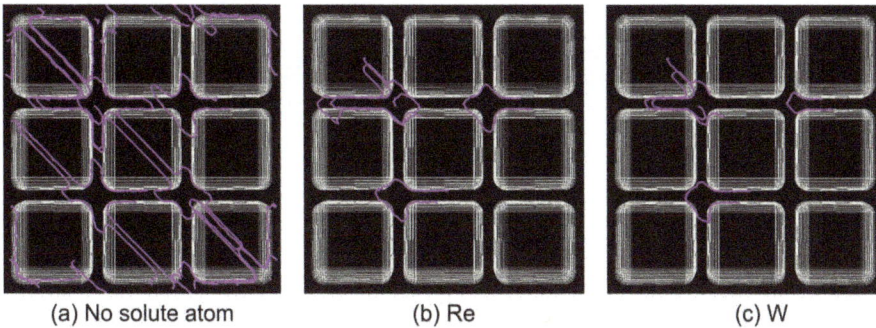

Figure 5. Dislocation patterns in the [0$\bar{1}$1](111) slip system in γ matrix channels after 30 ns under 500 MPa tensile load along [001] direction at 850 °C for different solute atoms. Dislocations are viewed in the projection on (100) crystallographic planes, corresponding to Figure 1: (**a**) no solute atom; and (**b**) 7×10^{-3} at.% Re; (**c**) 7×10^{-3} at.% W.

In the case of 200 MPa tensile load, the plastic strains and strain rates during creep for the addition of Re and W with different concentrations are compared in Figure 6. Because the small and large time steps were used for dislocation glide and climb, respectively, as explained in Ref. [39], the primary creep merely ascribed to the dislocation glide ended in a short time, which is hardly seen in Figure 6a but is partly displayed in Figure 6b. The high concentration of solute atoms restricted the dislocation glide and reduced the plastic strain at the primary creep stage, which was similar to the previous results at 500 MPa. According to the setting in our DDD simulation, when the dislocation glide was limited, the dislocation climb was activated, and the creep entered the secondary creep stage with alternating dislocation glide and climb. The more solute atoms triggered the dislocation climb early due to their obstruction on dislocation glide. However, the different excess volumes of Re and W did not lead to distinct discrepancies of plastic strains and strain rates at the primary stage, as can be seen in Figure 6b,c. At the secondary creep stage, the plastic strains in Figure 6a did not show a completely monotonous decreasing behavior with increasing concentration. The plastic strain for the highest concentration of Re (blue dashed line in Figure 6a) became higher than the lower concentration (green dashed line in Figure 6a) during the later deformation. The back stress on dislocation climb was essentially small compared to dislocation glide, and only atoms in particular positions near the dislocation core led to higher back stresses. If dislocation glide were retarded by a solute atom due to the local high back stress, but the back stress on dislocation climb were still small, the dislocation could climb to a new position where the glide resistance would be relatively low. Subsequently, the dislocation could glide again. In this case, dislocation climb assisted the dislocation to overcome the barriers from the solute atoms. Although a high concentration of solute atoms was substantial in achieving high deformation resistance, the dislocation climb could rearrange the dislocation configuration, which also played an important role on creep. Once the dislocation pattern changed, the local stress environment altered, which influenced the further deformation. Therefore, the highest concentration of solute atoms in our simulations may not have led to the best creep resistance. In addition, the influence of different excess volumes on deformation became obvious in a longer time. This indicated that the lower back stress on dislocation climb for smaller excess volume supplied more opportunities to let the dislocation move far away from the atom, so as to avoid the barrier. In general, the low external load reduced the difference of strains for different atoms and different concentrations. Furthermore, if the concentration of Re were 3 wt.%, as is the real case in CMSX-4 [68], the distance between Re atoms would be approximately 10 Å. Such a fine distribution of solute atoms would make the dislocations nearly have no chance to avoid a high resistance. Figure 7 displays a comparison of dislocation patterns in [101]($\bar{1}\bar{1}$1) and [101](1$\bar{1}\bar{1}$) slip systems in γ matrix channels for different solute atoms at 200 MPa. From this, we could find a similar phenomenon as shown in Figure 5, where the solute atoms hindered the dislocation motion, but the dislocation climb complicated the dislocation evolution.

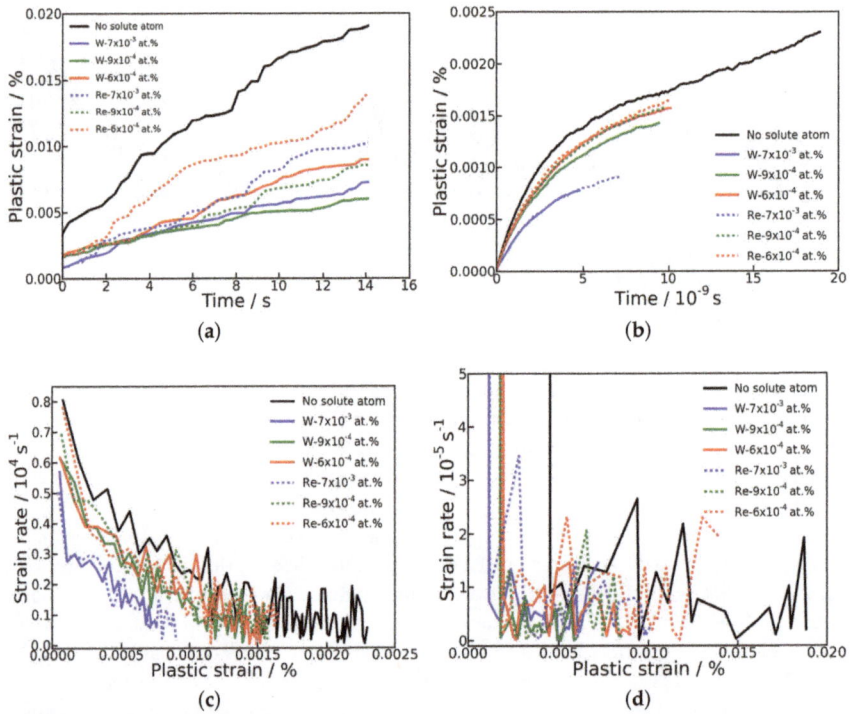

Figure 6. (**a**) Plastic strains as a function of time; (**b**) plastic strains in the primary stage of creep as a function of time; (**c**) strain rates in the primary stage of creep as a function of plastic strain; and (**d**) strain rates in the secondary stage of creep as a function of plastic strain for the addition of Re and W with different concentrations under 200 MPa tensile load along [001] direction at 850 °C.

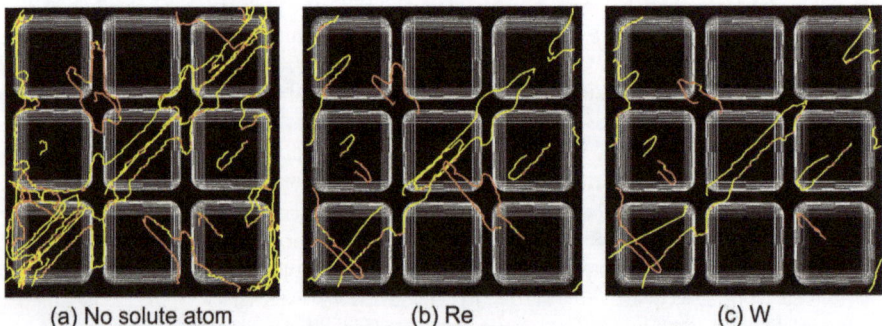

Figure 7. Dislocation patterns in [101]($\bar{1}\bar{1}$1) (yellow) and [101](1$\bar{1}\bar{1}$) (orange) slip systems in γ matrix channels after 14 s under 200 MPa tensile load along [001] direction at 850 °C for different solute atoms. Dislocations are viewed in the projection on (001) crystallographic planes, corresponding to Figure 1. One dislocation line containing some dislocation jogs with another color indicates the occurrence of dislocation climb: (**a**) no solute atom; (**b**) 7×10^{-3} at.% Re; and (**c**) 7×10^{-3} at.% W.

4. Discussion

In the present work, we focused on the effect of the excess volume of Re and W on creep deformation in Ni-base single crystal superalloys to elucidate its role in solid solution strengthening. Different from other works [11,12,40,41,69–71], in which only the value of concentration of atoms is concerned, we introduced the individual solute atoms in the DDD simulation to explicitly consider the interaction of atoms and dislocations. We only added a certain number of solute atoms in the γ channels, which corresponded to a much lower concentration compared to actual superalloys. If realistic concentrations of Re and W in Ni-base superalloys were considered, the number of atoms in the simulation cell would dramatically increase. Since the dislocation segments have to interact with all atoms in the microstructure, the DDD simulations would become computationally extravagant. Although the applied concentrations were low, the influence of different concentrations could be still revealed. The current setup represents a suitable compromise between meaningful results and affordable computational costs, but restricts our interpretation to qualitative trends.

This study was a qualitative analysis, in which the accurate quantitative comparison was not of importance, since the DDD simulation results vary with the initial dislocation sources. If the dislocation sources were assigned in different ways, the resulting specific strains and dislocation densities at the specific time step would be various. For instance, if we used different random dislocation sources with the same initial dislocation density in two simulations, we could obtain similar plastic strain curves as a function of time. However, the values of strain and dislocation density at each time step were not exactly same, but the difference was small for these two simulations. In addition, if we used different dislocation sources with different initial dislocation densities or the dislocation sources were not random, we would obtain different plastic strain curves as a function of time. Thus, at the same time step, the strains and dislocation densities would certainly be different.As long as we kept the same initial dislocation microstructure for different alloy elements and different alloy concentrations, we always found the same trend: the high excess volume and high concentration resulted in the small deformation, regardless of the accurate difference of values. Therefore, if we only discuss the influence of solute atoms on the deformation qualitatively, the choice of initial dislocation microstructure is not strict.

In the DDD simulations, we used the excess volumes for Re and W extracted from our DFT calculations. As discussed in Section 2.2, a simple estimate of the excess volume based on the covalent radii led to the same trend, but the absolute values were significantly larger. A comparison of the plastic strain for the different values of the excess volume with a solute concentration of 9×10^{-4} at.% under 500 MPa tensile load along the [001] direction at 850 °C is shown in Figure 8. Since the difference between excess volumes of Re and W in the simple estimate was nearly the same as that in our DFT calculations, the observed result in the reduction of plastic strains was similar. However, for a quantitative evaluation of the solid solution strengthening effect, an accurate determination of the excess volumes appears to be important.

Experimentally, it has been observed that solute atoms with a larger size mismatch to Ni exhibit a stronger strengthening effect at 800 °C, while Re with a smaller size has a more potent effect at temperatures higher than 1000 °C [10]. Without considering the diffusion of solute atoms in our present simulations at 850 °C, our results confirm the experimental observation at a lower temperature in Ref. [10] that W has a stronger hardening effect than Re, due to the larger excess volume. The slower diffusing Re may influence vacancy diffusion at non-dilute solute concentrations, thereby slowing down the dislocation motion at high temperatures. This will be investigated by DFT calculations first [72], and the findings will be applied in DDD simulation to clarify the diffusion effect of solute atoms on creep in the future work.

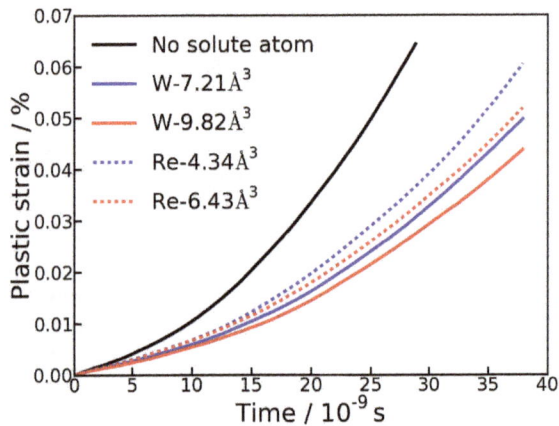

Figure 8. Plastic strains as a function of time for different excess volumes of Re and W for a concentration of 9×10^{-4} at.% under 500 MPa tensile load along the [001] direction at 850 °C.

5. Summary and Conclusions

The parelastic interaction between solute atoms and dislocations due to the excess volume of large atoms in the host matrix was implemented in a 3D discrete dislocation dynamics model. By using this model, the influence of excess volumes of Re and W on dislocation motion and creep in Ni-base single crystal superalloys under tensile loads of 200 MPa and 500 MPa along the [001] direction at 850 °C was studied. The excess volumes obtained from DFT calculations were used to determine the back stresses on dislocation glide and climb. The relative positions of dislocations and solute atoms governed the magnitude of back stresses. The excess volume affected dislocation glide more strongly than dislocation climb. The DDD simulation results show that W with a larger excess volume had a stronger strengthening effect than Re, ignoring the diffusion of solute atoms. An increase in the concentration of solute atoms resulted in a decrease of plastic strain associated with dislocation glide at a high stress lever, while the different concentrations did not lead to a pronounced difference in creep at a low stress level where both dislocation glide and climb were considered.

Author Contributions: Conceptualization, S.G. and A.H.; methodology, S.G., Z.Y. and M.G.; formal analysis, S.G., Z.Y. and M.G.; investigation, S.G., Z.Y. and M.G.; visualization, S.G.; writing–original draft preparation, S.G.; writing–review and editing, J.R., R.D. and A.H.; supervision, R.D. and A.H.

Funding: This research was funded by the Deutsche Forschungsgemeinschaft (DFG) through project C2 and project C4 of the collaborative research center SFB/Transregio 103 superalloy single crystals under grant number INST 213/747-2.

Acknowledgments: The authors are grateful for the funding by the Deutsche Forschungsgemeinschaft (DFG) through project C2 and project C4 of the collaborative research center SFB/Transregio 103 superalloy single crystals.

Conflicts of Interest: The authors declare no conflict of interest.

References

1. Geddes, B.; Leon, H.; Huang, X. *Superalloys: Alloying and Performance*; ASM International: Novelty, OH, USA, 2010.
2. Huang, M.; Zhu, J. An overview of rhenium effect in single-crystal superaloys. *Rare Metals* **2016**, *35*, 127–139. [CrossRef]
3. Erickson, G. *The Development and Application of* CMSX-10; Superalloys, The Minerals, Metals and Materials Society: Warrendale, PA, USA, 1996; pp. 35–44.
4. Reed, R. *The Superalloys: Fundamentals and Applications*; Cambridge University Press: Cambridge, UK, 2008.

5.   Wang, B.; Zhang, J.; Huang, T.; Su, H.; Li, Z.; Liu, L.; Fu, H. Influence of W, Re, Cr, and Mo on the microstructural stability of the third generation Ni-based single-crystal superalloys. *J. Mater. Res.* **2016**, *31*, 3381–3389. [CrossRef]

6.   Ritter, N.; Sowa, R.; Schauer, J.; Gruber, D.; Göhler, T.; Rettig, R.; Povoden-Karadeniz, E.; Körner, C.; Singer, R. Effects of Solid Solution Strengthening Elements Mo, Re, Ru, and W on Transition Temperatures in Nickel-Based Superalloys with High γ′-Volume Fraction: Comparison of Experiment and CALPHAD Calculations. *Metall. Mater. Trans. A* **2018**, *49*, 3206–3216. [CrossRef]

7.   Yoon, K.; Nöbe, R.; Seidman, D. Effects of rhenium addition on the temporal evolution of the nanostructure and chemistry of a model Ni-Cr-Al superalloy. II: Analysis of the coarsening behavior. *Acta Mater.* **2007**, *55*, 1159–1169. [CrossRef]

8.   Mottura, A.; Reed, R. What is the role of rhenium in single crystal superalloys? In *Proceedings of the 2nd European Symposium on Superalloys and Their Applications*; MATEC Web of Conferences: New York, NY, USA, 2014; pp. 1–6.

9.   Yu, X.; Wang, C. The effect of alloying elements on the dislocation climbing velocity in Ni: A first-principles study. *Acta Mater.* **2009**, *57*, 5914–5920. [CrossRef]

10.   Rehman, H.; Durst, K.; Neumeier, S.; Sato, A.; Reed, R.; Göken, M. On the temperature dependent strengthening of nickel by transition metal solutes. *Acta Mater.* **2017**, *137*, 54–63. [CrossRef]

11.   Fleischer, R. Substitutional solution hardening. *Acta Mater.* **1963**, *11*, 203–209. [CrossRef]

12.   Labusch, R. A statistical theory of solid solution hardening. *Phys. Status Solidi (B)* **1970**, *41*, 659–669. [CrossRef]

13.   Gottstein, G. *Physical Foundations of Materials Science*; Springer: Berlin/Heidelberg, Germany, 2004.

14.   Suzuki, H. Segregation of Solute Atoms to Stacking Faults. *J. Phys. Soc. Jpn.* **1962**, *17*, 322–325. [CrossRef]

15.   Schuwalow, S.; Rogal, J.; Drautz, R. Vacancy mobility and interaction with transition metal solutes in Ni. *J. Phys. Condens. Matter* **2014**, *26*, 485014. [CrossRef]

16.   Giamei, A.; Anton, D. Rhenium additions to a Ni-base superalloy: Effects on microstructure. *Metall. Trans. A* **1985**, *16*, 1997–2005. [CrossRef]

17.   Wollgramm, P.; Buck, H.; Neuking, K.; Parsa, A.; Schuwalow, S.; Rogal, J.; Drautz, R.; Eggeler, G. On the role of Re in the stress and temperature dependence of creep of Ni-base single crystal superalloys. *Mater. Sci. Eng. A* **2015**, *628*, 382–395. [CrossRef]

18.   Ding, Q.; Li, S.; Chen, L.; Han, X.; Zhang, Z.; Yu, Q.; Li, J. Re segregation at interfacial dislocation network in a nickel-based superalloy. *Acta Mater.* **2018**, *154*, 137–146. [CrossRef]

19.   Wu, X.; Makineni, S.; Kontis, P.; Dehm, G.; Raabe, D.; Gault, B.; Eggeler, G. On the segregation of Re at dislocations in the γ′ phase of Ni-based single crystal superalloys. *Materialia* **2018**, *4*, 109–114. [CrossRef]

20.   Zhang, J.; Murakumo, T.; Harada, H.; Koizumi, Y.; Kobayashi, T. Creep deformation mechanisms in some modern single-crystal superalloys. *Superalloys* **2004**, 189–195.

21.   Rae, C.; Reed, R. Primary creep in single crystal superalloys: Origins, mechanisms and effects. *Acta Mater.* **2007**, *55*, 1067–1081. [CrossRef]

22.   Reed, R.; Matan, N.; Cox, D.; Rist, M.; Rae, C. Creep of CMSX-4 superalloy single crystals: effects of rafting at high temperature. *Acta Mater.* **1999**, *47*, 3367–3381. [CrossRef]

23.   Kamaraj, M.; Mayr, C.; Kolbe, M.; Eggeler, G. On the influence of stress state on rafting in the single crystal superalloy CMSX-6 under conditions of high temperature and low stress creep. *Scr. Mater.* **1998**, *38*, 589–594. [CrossRef]

24.   Eggeler, G.; Dlouhy, A. On the formation of ⟨010⟩-dislocations in the γ′-phase of superalloy single crystals during high temperature low stress creep. *Acta Mater.* **1997**, *45*, 4251–4262. [CrossRef]

25.   Srinivasan, R.; Eggeler, G.; Mills, M. γ′-cutting as rate-controlling recovery process during high-temperature and low-stress creep of superalloy single crystals. *Acta Mater.* **2000**, *48*, 4867–4878. [CrossRef]

26.   Kostka, A.; Mälzer, G.; Eggeler, G. High-temperature dislocation plasticity in the single-crystal superalloy LEK94. *J. Microsc.* **2006**, *223*, 295–297. [CrossRef]

27.   Jácome, L.A.; Nörtershäuser, P.; Somsen, C.; Dlouhy, A.; Eggeler, G. On the nature of γ′ phase cutting and its effect on high temperature and low stress creep anisotropy of Ni-base single crystal superalloys. *Acta Mater.* **2014**, *69*, 246–264. [CrossRef]

28.   Rae, C.; Matan, N.; Reed, R. The role of stacking fault shear in the primary creep of [001]-oriented single crystal superalloys at 750 °C and 750 MPa. *Mater. Sci. Eng. A* **2001**, *300*, 125–134. [CrossRef]

29. Knowles, D.; Chen, Q. Superlattice stacking fault formation and twinning during creep in γ/γ' single crystal superalloy CMSX-4. *Mater. Sci. Eng. A* **2003**, *340*, 88–102. [CrossRef]

30. Caron, P.; Khan, T.; Nakagawa, Y. Effect of orientation on the intermediate temperature creep behaviour of Ni-base single crystal superalloys. *Scr. Metall.* **1986**, *20*, 499–502. [CrossRef]

31. Wu, X.; Wollgramm, P.; Somsen, C.; Dlouhy, A.; Kostka, A.; Eggeler, G. Double minimum creep of single crystal Ni-base superalloys. *Acta Mater.* **2016**, *112*, 242–260. [CrossRef]

32. Sass, V.; Glatzel, U.; Feller-Kniepmeier, M. Anisotropic creep properties of the nickel-base superalloy CMSX-4. *Acta Mater.* **1996**, *44*, 1967–1977. [CrossRef]

33. Huang, M.; Zhao, L.; Tong, J. Discrete dislocation dynamics modelling of mechanical deformation of nickel-based single crystal superalloys. *Int. J. Plast.* **2012**, *28*, 141–158. [CrossRef]

34. Yang, H.; Li, Z.; Huang, M. Modelling dislocation cutting the precipitate in nickel-based single crystal superalloy via the discrete dislocation dynamics with SISF dissociation scheme. *Comput. Mater. Sci.* **2013**, *75*, 52–59. [CrossRef]

35. Yang, H.; Li, Z.; Huang, M. Modeling of abnormal mechanical properties of nickel-based single crystal superalloy by three-dimensional discrete dislocation dynamics. *Model. Simul. Mater. Sci. Eng.* **2014**, *22*, 085009. [CrossRef]

36. Hussein, A.; Rao, S.; Uchic, M.; Parthasarathay, T.; El-Awady, J. The strength and dislocation microstructure evolution in superalloy microcrystals. *J. Mech. Phys Solids* **2017**, *99*, 146–162. [CrossRef]

37. Hafez Haghighat, S.; Eggeler, G.; Raabe, D. Effect of climb on dislocation mechanisms and creep rates in γ'-strengthened Ni base superalloy single crystals: A discrete dislocation dynamics study. *Acta Mater.* **2013**, *61*, 3709–3723. [CrossRef]

38. Liu, B.; Raabe, D.; Roters, F.; Arsenlis, A. Interfacial dislocation motion and interactions in single-crystal superalloys. *Acta Mater.* **2014**, *79*, 216–233. [CrossRef]

39. Gao, S.; Fivel, M.; Ma, A.; Hartmaier, A. 3D discrete dislocation dynamics study of creep behavior in Ni-base single crystal superalloys by a combined dislocation climb and vacancy diffusion model. *J. Mech. Phys. Solids* **2017**, *102*, 209–223. [CrossRef]

40. Gu, Y.; El-Awady, J. Quantifying the effect of hydrogen on dislocation dynamics: A three-dimensional discrete dislocation dynamics framework. *J. Mech. Phys. Solids* **2018**, *112*, 491–507. [CrossRef]

41. Yu, H.; Cocks, A.; Tarleton, E. Discrete dislocation plasticity HELPs understand hydrogen effects in bcc materials. *J. Mech. Phys. Solids* **2019**, *123*, 41–60. [CrossRef]

42. Demtröder, K.; Eggeler, G.; Schreuer, J. Influence of microstructure on macroscopic elastic properties and thermal expansion of nickel-base superalloys ERBO/1 and LEK94. *Mater. Sci. Eng. Technol.* **2015**, *46*, 563–576. [CrossRef]

43. Warren, P.; Cerezo, A.; Smith, G. An atom probe study of the distribution of rhenium in a nickel-based superalloy. *Mater. Sci. Eng. A* **1998**, *250*, 88–92. [CrossRef]

44. Parsa, A.; Wollgramm, P.; Buck, H.; Somsen, C.; Kostka, A.; Povstugar, I.; Choi, P.; Raabe, D.; Dlouhy, A.; Müller, J.; et al. Advanced Scale Bridging Microstructure Analysis of Single Crystal Ni-Base Superalloys. *Adv. Eng. Mater.* **2014**, *17*, 216–230. [CrossRef]

45. Mottura, A.; Miller, M.; Reed, R. *Atom Probe Tomography Analysis of Possible Rhenium Clustering in Nickel-Based Superalloys*; Superalloys, The Minerals, Metals and Materials Society: Warrendale, PA, USA, 2008; pp. 891–900.

46. Mottura, A.; Warnken, N.; Miller, M.; Finnis, M.; Reed, R. Atom probe tomography analysis of the distribution of rhenium in nickel alloys. *Acta Mater.* **2010**, *58*, 931–942. [CrossRef]

47. Ge, B.; Luo, Y.; Li, J.; Zhu, J. Distribution of rhenium in a single crystal nickel-based superalloy. *Scr. Mater.* **2010**, *63*, 969–972. [CrossRef]

48. Gao, S.; Fivel, M.; Ma, A.; Hartmaier, A. Influence of misfit stresses on dislocation glide in single crystal superalloys: A three-dimensional discrete dislocation dynamics study. *J. Mech. Phys. Solids* **2015**, *76*, 276–290. [CrossRef]

49. Bakó, B.; Clouet, E.; Dupuy, L.; Blétry, M. Dislocation dynamics simulations with climb: kinetics of dislocation loop coarsening controlled by bulk diffusion. *Philos. Mag.* **2011**, *91*, 3173–3191. [CrossRef]

50. Ayas, C.; Deshpande, V.; Geers, M. Tensile response of passivated films with climb-assisted dislocation glide. *J. Mech. Phys. Solids* **2012**, *60*, 1626–1643. [CrossRef]

51. Danas, K.; Deshpande, V. Plane-strain discrete dislocation plasticity with climb-assisted glide motion of dislocations. *Model. Simul. Mater. Sci. Eng.* **2013**, *21*, 045008. [CrossRef]
52. Geers, M.; Cottura, M.; Appolaire, B.; Busso, E.; Forest, S.; Villani, A. Coupled glide-climb diffusion-enhanced crystal plasticity. *J. Mech. Phys. Solids* **2014**, *70*, 136–153. [CrossRef]
53. Verdier, M.; Fivel, M.; Groma, I. Mesoscopic scale simulation of dislocation dynamics in fcc metals: principles and applications. *Model. Simul. Mater. Sci. Eng.* **1998**, *6*, 755–770. [CrossRef]
54. Fivel, M.; Canova, G. Developing rigorous boundary conditions to simulations of discrete dislocation dynamics. *Model. Simul. Mater. Sci. Eng.* **1999**, *7*, 753–768. [CrossRef]
55. Shin, C. 3D Discrete Dislocation Dynamics Applied to Dislocation-Precipitate Interactions; Ph.D. Thesis, INP Grenoble and Séoul National University, Seoul, Korea, 2004.
56. Li, J. Stress field of a dislocation segment. *Philos. Mag.* **1964**, *10*, 1097–1098. [CrossRef]
57. Fivel, M. Discrete Dislocation Dynamics: Principles and Recent Applications. In *Multiscale Modeling of Heterogenous Materials*; John Wiley & Sons, Ltd.: Hoboken, NJ, USA, 2010; Chapter 2, pp. 17–36.
58. Blöchl, P. Projector augmented-wave method. *Phys. Rev. B* **1994**, *50*, 17953. [CrossRef]
59. Kresse, G.; Joubert, D. From ultrasoft pseudopotentials to the projector augmented-wave method. *Phys. Rev. B* **1999**, *59*, 1758. [CrossRef]
60. Kresse, G.; Hafner, J. Ab initio molecular dynamics for liquid metals. *Phys. Rev. B* **1993**, *47*, 558. [CrossRef]
61. Kresse, G.; Furthmüller, J. Efficiency of ab-initio total energy calculations for metals and semiconductors using a plane-wave basis set. *Comput. Mater. Sci.* **1996**, *6*, 15. [CrossRef]
62. Kresse, G.; Furthmüller, J. Efficient iterative schemes for *ab initio* total-energy calculations using a plane-wave basis set. *Phys. Rev. B* **1996**, *54*, 11169. [CrossRef]
63. Perdew, J.; Burke, K.; Ernzerhof, M. Generalized Gradient Approximation Made Simple. *Phys. Rev. Lett.* **1996**, *77*, 3865. [CrossRef]
64. Monkhorst, H.; Pack, J. Special points for Brillouin-zone integrations. *Phys. Rev. B* **1976**, *13*, 5188. [CrossRef]
65. Birch, F. Finite Elastic Strain of Cubic Crystals. *Phys. Rev.* **1947**, *71*, 809. [CrossRef]
66. Francis, G.; Payne, M. Finite basis set corrections to total energy pseudopotential calculations. *J. Phys. Condens. Matter* **1990**, *2*, 4395–4404. [CrossRef]
67. Rao, S.; Uchic, M.; Shade, P.; Woodward, C.; Parthasarathy, T.; Dimiduk, D. Critical percolation stresses of random Frank-Read sources in micrometer-sized crystals of superalloys. *Model. Simul. Mater. Sci. Eng.* **2012**, *20*, 065001. [CrossRef]
68. Harris, K.; Erickson, G.; Sikkenga, S.; Brentnall, W.; Aurrecoeche, J.; Kubarych, K. *Development of the Rhenium Containing Superalloys CMSX-4 and CM186 LC for Single Crystal Blade and Directionally Solidified Vane Applications in Advanced Turbine Engines*; Superalloys, The Minerals, Metals and Materials Society: Warrendale, PA, USA, 1992; pp. 297–306.
69. Nabarro, F. The theory of solution hardening. *Philos. Mag.* **1977**, *35*, 613–622. [CrossRef]
70. Butt, M. Review solid-solution hardening. *J. Mater. Sci.* **1993**, *28*, 2557–2576. [CrossRef]
71. Mishima, Y.; Ochiai, S.; Hamao, N.; Yodogawa, M.; Suzuki, T. Solid Solution Hardening of Nickel—Role of Transition Metal and B-subgroup Solutes. *Trans. Jpn. Inst. Metals* **1986**, *27*, 656–664. [CrossRef]
72. Grabowski, M.; Rogal, J.; Drautz, R. Kinetic Monte Carlo simulations of vacancy diffusion in nondilute Ni-X (X = Re, W, Ta) alloys. *Phys. Rev. Mater.* **2018**, *2*, 123403. [CrossRef]

metals **MDPI**

Article

The Influence of Niobium Additions on Creep Resistance of Fe-27 at. % Al Alloys

Ferdinand Dobeš [1], Petr Dymáček [1,2,*] and Martin Friák [1,2]

[1]   Institute of Physics of Materials, Academy of Sciences of the Czech Republic, Žižkova 22, CZ-61662 Brno, Czech Republic
[2]   CEITEC IPM, Institute of Physics of Materials, Academy of Sciences of the Czech Republic, Žižkova 22, CZ-61662 Brno, Czech Republic
*   Correspondence: pdymacek@ipm.cz; Tel.: +420-532-290-411

Received: 3 June 2019; Accepted: 27 June 2019; Published: 30 June 2019

Abstract: Results of creep tests of two Fe-27 at. % Al-based alloys with additions of 2.7 and 4.8 at. % of niobium conducted in the temperature range from 650 °C to 900 °C in the authors' laboratory are presented. The purpose of the study is to supplement previous work on Fe-Al-Nb alloys to obtain a more complete overview of creep properties from the dilute alloy with 1% of Nb up to the eutectic alloy with 10% of niobium. At higher temperatures and lower stresses, the creep resistance of the 10% niobium alloy is better than that of the lower niobium alloys. On the other hand, the eutectic alloy loses its preference at lower temperatures and higher deformation rates. This phenomenon is similar to that reported by Yildirim et al. for Fe-50 at. % Al-based alloys and is probably associated with an increased stress sensitivity of the eutectic alloy.

Keywords: iron aluminides; creep; stress exponent; activation energy

---

1. Introduction

High-temperature creep resistance and tensile strength of Fe-Al-based alloys can be improved by additions of niobium because it contributes to both solid solution hardening and precipitation strengthening [1–5]. The effect of niobium additions on mechanical properties was to some extent studied for both ferritic alloys with Al concentrations less than 19% (all compositions are given in at. % throughout this paper) [6–10] and alloys based on ordered compounds of Fe_3Al [8,9,11–17]. Recently, the influence of niobium additions to FeAl (Fe-(50−x)% Al-x% Nb, x = 1, 3, 5, 7, 9) [18], i.e., up to the eutectic through between FeAl and C14 Laves phase on room-temperature mechanical properties were studied. It was shown that the eutectic alloy with 9% Nb exhibited the lowest compressive strength among the heat-treated alloys. A similar systematic investigation regarding creep properties of the alloys of Fe_3Al-type was missing. To bridge this gap, it was decided to study the compressive creep of Fe-27% Al with 3 and 5% of niobium as a complement to previous studies of the alloys with 1% [19] and 10% Nb [17].

2. Materials and Methods

The alloys were prepared using vacuum induction melting and casting under Argon. The casts (dimensions 40 × 30 × 80 to 90 mm³,) were rolled at 1200 °C to 13 mm in several steps, with a 15% thickness reduction in each pass. The composition of the investigated material was determined by wet chemical analysis and is given in Table 1 and marked in a part of the ternary Fe-Al-Nb diagram in Figure 1 [20]. The mean concentration of the technological impurities coming from the metals used for the preparation of the alloys was: 0.1% Cr, 0.01% B, 0.1% Mn, and 0.06% C. The results of the microstructure investigations are described in the previous paper [21]. The grains of the FA3Nb alloy are large and have an irregular shape; their size is 200 to 500 µm. The precipitates are λ_1 Laves phase (C14) $(Fe, Al)_2Nb$. These appear as smaller irregular shapes up to 10 µm and as needle-like up to 50 µm long. They are homogeneously distributed both along the grain boundaries and inside the grains. The grains of the FA5Nb alloy are coarse with dimensions in the order of hundreds of micrometers. The Laves phase forms a eutectic with the Fe_3Al matrix and comprises about 0.4 volume fractions of the eutectic.

Table 1. Chemical composition of the studied alloys (at. %).

Alloy	Fe	Al	Nb
FA3Nb	71.5	25.8	2.7
FA5Nb	67.9	27.3	4.8

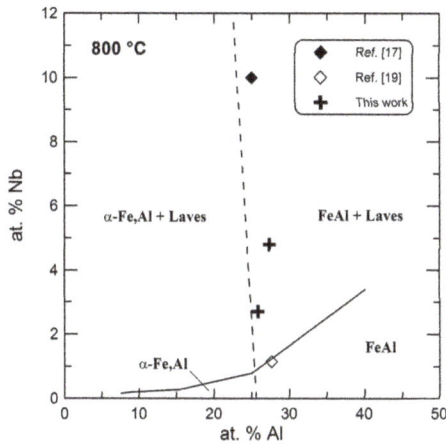

Figure 1. Detail of the isothermal section of the Fe-Al-Nb system at 800 °C according to Palm [20].

Creep tests were performed in uniaxial compression mode on prismatic samples with a height (gauge length) of 8 mm and a square base of cross-section 4×4 mm^2. The samples were prepared by traveling wire electro-discharge machining and fine grinding of the contact surfaces. The tests were performed on a dead-weight creep machine that was constructed in-house in a protective atmosphere of dry purified argon. The load was applied in a perpendicular direction to the bases. The samples were subjected to stepwise loading, where the magnitude of the load was changed after a steady creep rate was established for a given load. A substantial amount of strain occurred before a constant creep rate was registered after each loading step. The terminal values of the true stress and the creep rate, i.e., the true compressive strain rate, were evaluated for each step:

$$\sigma_i = \frac{F_i l}{S_0 l_0} \tag{1}$$

$$\varepsilon = -\ln(l/l_0) \tag{2}$$

where F_i is the force applied in the i-th step, S_0 is the initial cross-section area, and l and l_0 are the instantaneous specimen height and the initial specimen height, respectively. The creep rate was found by linear fitting of the creep strain vs. the time curve in the linear part of the respective step. Figure 2 shows several examples that document how the creep strain evolves with time after both stress decrement and increment. Final deformation of the specimens was as a rule less than 0.1. The longest test duration was approximately 727 h; the total test duration was more than 3300 h.

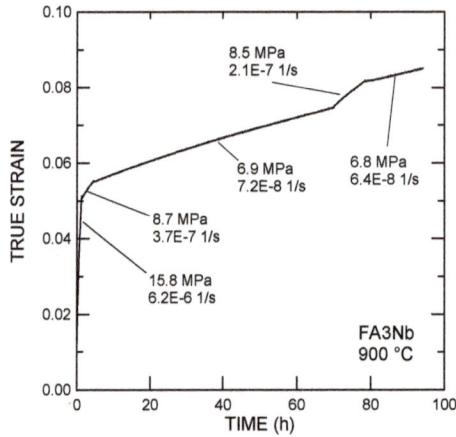

Figure 2. Example of a creep curve of the FA3Nb alloy at 900 °C showing five different segments with corresponding applied stresses and creep rates.

3. Results

The dependence of the creep rate $\dot{\varepsilon}$ on the applied stress σ at the various temperatures is shown in Figures 3 and 4, on a double logarithmic scale. The data were analyzed using the following relation:

$$\dot{\varepsilon} = A\sigma^n \tag{3}$$

where A is a parameter dependent on the temperature and, and n is the stress exponent. Its values for both investigated alloys are summarized in Figure 5. The exponent n is, albeit slightly, dependent on the temperature. It should be noted that the same temperature dependence of n is evident for the alloy Fe-Al-Nb studied by Milenkovic and Palm [17]. A detailed analysis of the exponent n is possible provided a single mechanism is rate-controlling over the experimental range of stress and temperature. The effect of the mechanism change can be taken into account through an explicit dependence of the exponent n on stress [22]. In the studied alloys, the presence of Laves phase particles introduces threshold stress responsible for an apparent increase of the exponent n at lower stresses and consequently at higher temperatures [23].

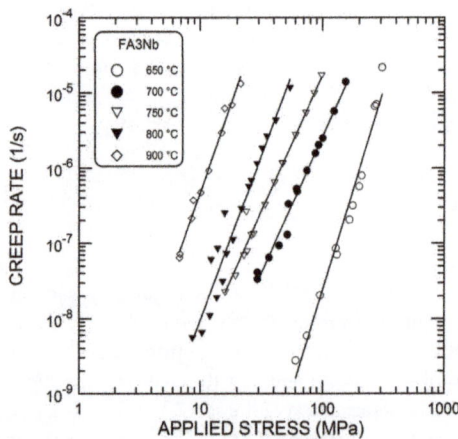

Figure 3. Dependence of the creep rate of the FA3Nb alloy on applied stress.

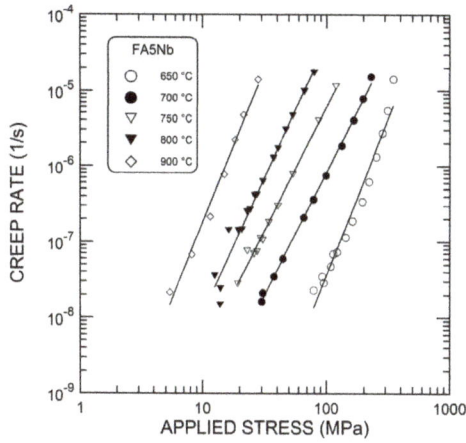

Figure 4. Dependence of the creep rate of the FA5Nb alloy on applied stress.

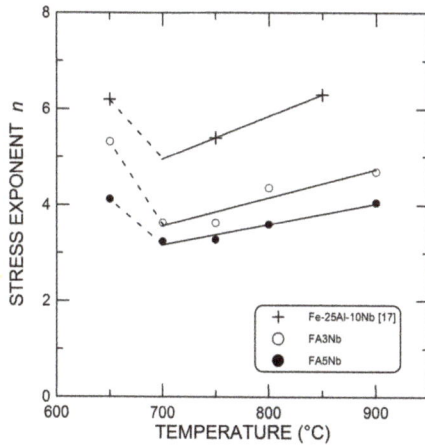

Figure 5. Dependence of stress exponent *n* on temperature.

To allow an approximate description of this behavior, the studied range of temperatures is formally divided into two areas: below and above 700 °C. Taking into account the ternary diagram, the proposed division results from a difference in the lattice structure: there would then be a DO_3 lattice at 650 °C, and a B2 lattice at 700 °C and above. Although the addition of niobium increases the temperature of the phase transition, the question is whether the niobium solubility is large enough to shift the transition to temperatures above 650 °C.

Relations between steady-state creep rate, and reciprocal temperature as derived from the $\dot{\varepsilon}$ vs. σ relations are shown in Figures 6 and 7. As in the above case of the stress exponent, we can also use the division into two temperature regions here. The apparent creep activation energy, defined as:

$$Q = -\left[\frac{\partial \ln \dot{\varepsilon}}{\partial(1/RT)}\right]_{\sigma},$$ (4)

where *T* is the absolute temperature, and *R* is the universal gas constant, can be determined separately for both regions, although this determination should be taken with great caution at temperatures below 700 °C. The estimated activation energies for the high-temperature region are given in Figure 8.

The activation energy is close to 335 kJ/mol reported by McKamey et al. [11] for the Fe-28Al-1Nb alloy. It is also comparable to the activation enthalpy of the diffusion of the niobium in Fe(Al) as estimated by Morris et al. [24], i.e., 330 kJ/mole.

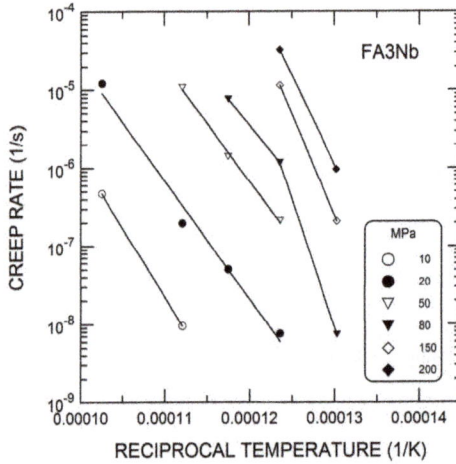

Figure 6. Dependence of minimum creep rate on reciprocal temperature for the alloy FA3Nb.

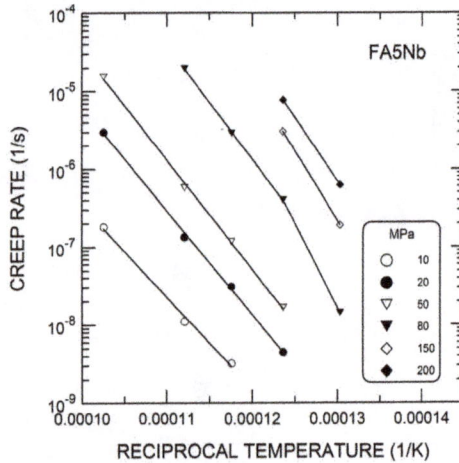

Figure 7. Dependence of minimum creep rate on reciprocal temperature for the alloy FA5Nb.

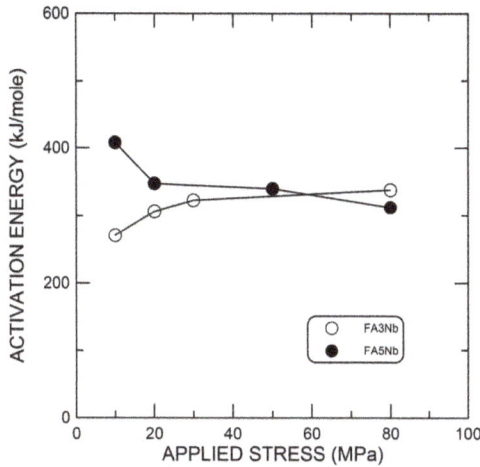

Figure 8. Dependence of activation energy Q on applied stress.

4. Discussion

Figure 9 gives the selected creep data for the present FA3Nb alloy as well as data for the alloy with 1% of niobium [19], acquired also by compressive testing in the same laboratory, for tests over a range of identical temperatures. The figure confirms that the alloy with the lower niobium content is weaker than the alloy tested in the present study, at least at high creep rates. In the alloy with 1% of Nb, probably all the niobium is dissolved in the matrix, and the formation of the Laves phase is not expected.

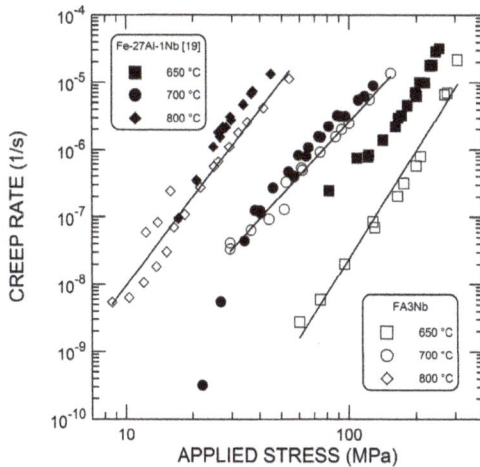

Figure 9. Comparison of the creep rate variation with the applied stress for the present alloy FA3Nb and Fe-Al-based alloy containing 1% of Nb [19].

For comparison with a previously published study on the Fe-25% Al alloy that contained the amount of niobium close to the eutectic composition [17], Figure 10 shows the creep rate vs. the stress data obtained at three temperatures. It has to be emphasized that the data in [17] were obtained by the identical testing technique, i.e., by the stepwise compressive testing. The lines for the present alloys at a temperature of 850 °C were obtained by interpolation. Two different features follow from the comparison of the data:

(i)   At higher temperatures and lower stresses, the creep resistance of the 10% niobium alloy is higher than that of the lower niobium alloys. At a temperature of 850 °C, the creep resistance of the alloy with 10% of niobium is the best over the entire stress range studied.

(ii) This effect is reversed at lower temperatures and higher stresses. At a temperature of 650 °C and stresses higher than 200 MPa, the 5% niobium alloy has the highest creep resistance.

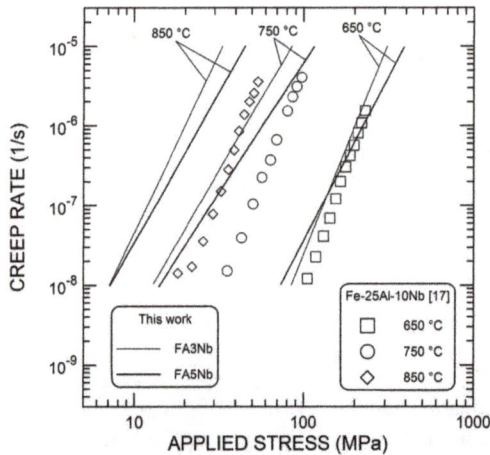

Figure 10. Comparison of the creep rate variation with the applied stress for the present alloys and Fe-Al-based eutectic alloy containing 10% Nb addition [17].

In this regard, our results fully agree with the findings of Yildirim et al. [18], whose tests were conducted at room temperature and strain rate 10^{-4} s^{-1}. They observed a similar weakening of the heat-treated eutectic alloy and attributed lower compressive strength and fracture strain of this alloy to the absence of softer Fe-Al-based primary dendrites. From the point of view of the present results, the effect can be envisaged as follows: the eutectic alloy with 10% of niobium is weaker at high strain rates, but shows a steeper dependence of creep rate on stress, i.e., higher stress exponent n, such that the alloy maintains its strength better at slow strain rates.

5. Conclusions

The creep rates of the two Fe-27 at. % Al-based alloys with additions of niobium was studied in the temperature range from 650 °C to 900 °C. The stress exponent and the apparent activation energy were estimated. The following conclusions regarding the creep resistance of the alloy could be drawn:

- The studied range of temperatures could be divided into two areas: below and above 700 °C. This division follows the existence of different crystal lattices at lower and higher temperatures.
- The stress exponent n decreases with increasing temperature at temperatures up to 700 °C and increases at higher temperatures.
- The apparent activation energy of the creep is close to 335 kJ/mol and is comparable to the activation enthalpy of niobium diffusion in Fe(Al).
- At higher temperatures and lower stresses, the creep resistance of both alloys is worse than that of a 10% niobium eutectic alloy.
- At a temperature of 650 °C and stresses higher than 200 MPa, the 5% niobium alloy has the best creep resistance.

Author Contributions: F.D. performed the creep tests and wrote the paper, P.D. performed the creep tests, and M.F. conceived the research.

Funding: This research was funded by the Czech Science Foundation, grant number 17-22139S.

Acknowledgments: The authors wish to thank P. Kratochvíl (Charles University, Prague, Czech Republic) and V. Vodičková (Technical University, Liberec, Czech Republic) for the supply of the experimental alloys.

Conflicts of Interest: The authors declare no conflict of interest.

References

1. Sauthoff, G. *Intermetallics*; VCH Verlagsgesellschaft: Weinheim, Germany, 1995; pp. 84–89.
2. Stoloff, N.S. Iron aluminides: Present status and future prospects. *Mater. Sci. Eng. A* **1998**, *258*, 1–14. [CrossRef]
3. Liu, C.T.; George, E.P.; Maziasz, P.J.; Schneibel, J.H. Recent advances in B2 iron aluminide alloys: Deformation, fracture and alloy design. *Mater. Sci. Eng. A* **1998**, *258*, 84–98. [CrossRef]
4. Palm, M. Concepts derived from phase diagram studies for the strengthening of Fe–Al-based alloys. *Intermetallics* **2005**, *13*, 1286–1295. [CrossRef]
5. Morris, D.G.; Muñoz-Morris, M.A. Development of creep-resistant iron aluminides. *Mater. Sci. Eng. A* **2007**, *462*, 45–52. [CrossRef]
6. Baligidad, R.G. Effect of niobium on microstructure and mechanical properties of hot-rolled Fe-8.5 wt% Al-0.1 wt% C alloy. *J. Mater. Sci.* **2004**, *39*, 5599–5602. [CrossRef]
7. Baligidad, R.G. Effect of niobium on microstructure and mechanical properties of high carbon Fe–10.5 wt.% Al alloys. *Mater. Sci. Eng. A* **2004**, *368*, 131–138. [CrossRef]
8. Morris, D.G.; Muñoz-Morris, M.A.; Requejo, L.M.; Baudin, C. Strengthening at high temperatures by precipitates in Fe-Al-Nb alloys. *Intermetallics* **2006**, *14*, 1204–1207. [CrossRef]
9. Morris, D.G.; Requejo, L.M.; Muñoz-Morris, M.A. Age hardening in some Fe-Al-Nb alloys. *Scripta Mater.* **2006**, *54*, 393–397. [CrossRef]
10. Morris, D.G.; Muñoz-Morris, M.A. Room and high temperature deformation behaviour of a forged Fe–15Al–5Nb alloy with a reinforcing dispersion of equiaxed Laves phase particles. *Mater. Sci. Eng. A* **2012**, *552*, 134–144. [CrossRef]
11. McKamey, C.G.; Maziasz, P.J.; Jones, J.W. Effect of addition of molybdenum or niobium on creep-rupture properties of Fe_3Al. *J. Mater. Res.* **1992**, *7*, 2089–2106. [CrossRef]
12. McKamey, C.G.; Maziasz, P.J.; Goodwin, G.M.; Zacharia, T. Effects of alloying additions on the microstructures, mechanical properties and weldability of Fe_3Al-based alloys. *Mater. Sci. Eng. A* **1994**, *174*, 59–70. [CrossRef]
13. Zhang, Z.H.; Sun, Y.S.; Guo, J. Effect of niobium addition on the mechanical properties of Fe_3Al-based alloys. *Scripta Metall. Mater.* **1995**, *33*, 2013–2017.
14. Yu, Y.Q.; Sun, Y.S. Improvement of creep resistance of Fe_3Al based alloys with tungsten and niobium additions. *J. Mater. Sci. Lett.* **2001**, *20*, 1221–1223.
15. Morris, D.G.; Muñoz-Morris, M.A.; Baudin, C. The high-temperature strength of some Fe_3Al alloys. *Acta Mater.* **2004**, *52*, 2827–2836. [CrossRef]
16. Falat, L.; Schneider, A.; Sauthoff, G.; Frommeyer, G. Mechanical properties of Fe–Al–M–C (M = Ti, V, Nb, Ta) alloys with strengthening carbides and Laves phase. *Intermetallics* **2005**, *13*, 1256–1262. [CrossRef]
17. Milenkovic, S.; Palm, M. Microstructure and mechanical properties of directionally solidified Fe–Al–Nb eutectic. *Intermetallics* **2008**, *16*, 1212–1218. [CrossRef]
18. Yildirim, M.; Vedat Akdeniz, M.; Mekhrabov, A.O. Microstructural evolution and room-temperature mechanical properties of as-cast and heat-treated $Fe_{50}Al_{50-n}Nb_n$ alloys ($n = 1, 3, 5, 7$, and 9 at%). *Mater. Sci. Eng. A* **2016**, *664*, 17–25. [CrossRef]
19. Dobeš, F.; Kratochvíl, P.; Pešička, J.; Vodičková, V. Microstructure and creep behavior of Fe-27Al-1Nb alloys with added carbon. *Metal. Mater. Trans. A* **2015**, *46*, 1580–1587. [CrossRef]
20. Palm, M. Phase equilibria in the Fe corner of the Fe–Al–Nb system between 800 and 1150 °C. *J. Alloys Compd.* **2009**, *475*, 173–177. [CrossRef]
21. Kratochvíl, P.; Švec, M.; Král, R.; Veselý, J.; Lukáč, P.; Vlasák, T. The effect of Nb addition on the microstructure and the high-temperature strength of Fe_3Al aluminide. *Metal. Mater. Trans. A* **2018**, *49*, 1598–1603. [CrossRef]

22. Bonora, N.; Esposito, L. Mechanism based creep model incorporating damage. *J. Eng. Mater. Technol.* **2010**, *132*, 021013. [CrossRef]
23. Esposito, L.; Bonora, N.; De Vita, G. Creep modelling of 316H stainless steel over a wide range of stress. *Procedia Structural Integrity* **2016**, *2*, 927–933. [CrossRef]
24. Morris, D.G.; Muñoz-Morris, M.A.; Requejo, L.M. New iron–aluminium alloy with thermally stable coherent intermetallic nanoprecipitates for enhanced high-temperature creep strength. *Acta Mater.* **2006**, *54*, 2335–2341. [CrossRef]

MDPI

St. Alban-Anlage 66

4052 Basel

Switzerland

Tel. +41 61 683 77 34

Fax +41 61 302 89 18

www.mdpi.com

Metals Editorial Office

E-mail: metals@mdpi.com

www.mdpi.com/journal/metals

www.ingramcontent.com/pod-product-compliance
Lightning Source LLC
Chambersburg PA
CBHW051847210326
41597CB00033B/5803